Professional Practice in Earth Sciences

Series Editor

James W. LaMoreaux, Tuscaloosa, AL, USA

Books in Springer's Professional Practice in Earth Sciences Series present state-of-the-art guidelines to be applied in multiple disciplines of the earth system sciences. The series portfolio contains practical training guidebooks and supporting material for academic courses, laboratory manuals, work procedures and protocols for environmental sciences and engineering. Items published in the series are directed at researchers, students, and anyone interested in the practical application of science. Books in the series cover the applied components of selected fields in the earth sciences and enable practitioners to better plan, optimize and interpret their results. The series is subdivided into the different fields of applied earth system sciences: Laboratory Manuals and work procedures, Environmental methods and protocols and training guidebooks.

Yifan Zeng · Zhenzhong Pang · Qiang Wu ·
Hui Qing Lian · Xin Du

Roof Water Disaster in Coal Mining in Ecologically Fragile Mining Areas

Formation Mechanism and Prevention and Control Measures

 Springer

Yifan Zeng
National Engineering Research Center
of Coal Mine Water Hazard Controlling
China University of Mining and Technology
Beijing, China

Zhenzhong Pang
National Engineering Research Center
of Coal Mine Water Hazard Controlling
China University of Mining and Technology
Beijing, China

Qiang Wu
National Engineering Research Center
of Coal Mine Water Hazard Controlling
China University of Mining and Technology
Beijing, China

Hui Qing Lian
North China Institute of Science
and Technology
Langfang, China

Xin Du
National Engineering Research Center
of Coal Mine Water Hazard Controlling
China University of Mining and Technology
Beijing, China

ISSN 2364-0073   ISSN 2364-0081 (electronic)
Professional Practice in Earth Sciences
ISBN 978-3-031-33142-8   ISBN 978-3-031-33140-4 (eBook)
https://doi.org/10.1007/978-3-031-33140-4

This Springer imprint is published by the registered company Springer Nature Switzerland AG
The registered company address is: Gewerbestrasse 11, 6330 Cham, Switzerland

Preface

As an important energy material in China, coal has made great contributions to China's economic development. With the depletion of coal resources in the eastern mining areas, the focus of coal development in China has gradually shifted to western mining areas such as Shanxi, Shaanxi, Inner Mongolia, Ningxia, Gansu, and Xinjiang. At present, Ordos coalfield, as an important coal accumulation area and coal production base in the Yellow River basin, mainly includes Shandong, Northern Shaanxi, Ningdong, and Huanglong coal bases. The Middle Jurassic coal resources in the Ordos Basin account for 31.9% of the country's total, which is of great significance to China's economic and social development.

The Jurassic coal resources in the Ordos Basin are characterized by large reserves, shallow burial, and relatively simple geological and hydrological conditions, which lays a foundation for the construction of a safe, efficient, green, and intelligent modern large-scale mine group in the mining area. At present, the coal mines are all located in arid and semiarid areas, and the landforms of the mining area mainly include the loess plateau in northern Shaanxi, Maowusu Desert in the Ordos Basin, loess gully in northern Shanxi and other typical landforms. The average precipitation in the mining area is only 1/6–1/3 of the evaporation, which leads to water shortage, serious soil erosion, land desertification, and other problems in the mining area. The upper aquifer of the Jurassic coal seam in Ordos Basin mainly includes bedrock pore-fissure aquifer, burnt rock pore-fissure phreatic aquifer, and Quaternary loose layer pore aquifer. The aquicludes mainly include Neogene Baode Formation laterite layer and mudstone in bedrock section. The mining process of the mines mainly adopts the high-intensity mining mode of long wall and large mining height, such as comprehensive mechanized top coal caving method. Although this mining mode has greatly improved the mining efficiency, it has an incalculable impact on the overlying aquifer structure of the coal seam roof and the surface ecological environment system. At present, these impacts are mainly reflected in coal mining subsidence, house collapse, dead surface vegetation, desertification, and other phenomena in the mining area. These phenomena seriously restrict the coordinated mining between "coal resources and ecological environment" under the condition of high-intensity mining, and even seriously restrict the implementation of the national energy strategy and the construction of ten million tons of coal mines in the west. In particular, water inrush and sand bursting accidents in the mining process often cause heavy casualties and property losses. However, the occurrence of water inrush and sand break accident in the process of mining is the result of a multifactor comprehensive disaster, such as the overburden combination characteristics of the coal seam roof, the mining thickness and mining technology of the coal seam, the characteristics of the aquifer and sand layer, the characteristics of the water resisting layer, and other factors. However, the bearing characteristics of the surface ecological environment and the characteristics of the mine stratum structure have been determined before operations of the mine, and the water inrush and sand bursting disasters induced by the mine mainly depend on whether the aquifers can meet the requirements of safe mining under the coupling effect of strong earth stress and dynamic water pressure. Therefore, in order to ensure that there are no ecological environment problems and mining safety problems in the process of coal mining in ecologically vulnerable areas, it is necessary to strengthen the research on the response characteristics of overburden and surface ecological damage patterns to the mining intensity during the mining process and carry out the research on the mechanism of water and sand inrush and the water control mining mode. The research not only enriches the formation mechanism of roof water hazards in ecologically fragile mining areas but also provides theoretical guidance for green, safe, and efficient mining in ecologically fragile mining areas.

Chinese scientific researchers have made many scientific achievements in the study of water disaster prevention and control in the central and eastern mining areas, but limited by the changes in the hydrogeological conditions and mining patterns of the western mining areas, and the prediction of the development height of the water-conducting fracture zone and the water prevention and control technology adopted are mainly derived from the experience gained in the central and eastern mining areas. Therefore, on the basis of current scientific research achievements, it is urgent for ecologically fragile mining areas to conduct systematic research on the occurrence mechanism of mine water disaster, the prediction method of water hazard zoning, and the prevention and control measures. On the basis of systematic analysis of the geology and hydrogeology of the Yushenfu mining area in the Ordos Basin, the book studies the mechanism of water and sand inrush from overburden rock failure under strong mining conditions in the Yushenfu mining area and the related content of water control mining mode by triaxial mechanical test, theoretical analysis, physical simulation test of similar materials, fluid–solid coupling numerical simulation test, and in situ field test. At the same time, the water inrush model under the bedrock fissure aquifer, the water inrush model of bedrock fissure+loose pore aquifer and the water inrush model under the water body proposed in this book and the "coal water" dual resource coordinated mining model proposed in the ecologically fragile mining area of Yushenfu have achieved good field application results. This book describes the achievements of years of scientific research and engineering practice in the prevention and control of coal seam roof water disasters by the National Research Center for Coal Mine Water Disaster Prevention and Control Engineering of China

University of Mining and Technology (Beijing). Yifan Zeng was responsible for the general idea, basic framework, and technical content of research on the formation mechanism and prevention and control measures of coal seam roof water disasters in ecologically fragile mining areas. Of the eleven chapters in the book, other authors made respective contributions.

Beijing, China   Yifan Zeng
Beijing, China   Zhenzhong Pang
Beijing, China   Qiang Wu
Langfang, China   Hui Qing Lian
Beijing, China   Xin Du
March 2023

Acknowledgments

The research work included in this book is supported by the National Key R&D Program of China (2021YFC2902004), the National Natural Science Foundation of China (42072284, 42027801), the Major Science and Technology Projects of Inner Mongolia Autonomous Region (2020ZD0020-4, 2020ZD0021). Academician Qiang Wu provided guidance in the process of planning and compiling this book. National Engineering Research Center of Coal Mine Water Hazard Controlling, Key Laboratory of Mine Water Hazard Controlling, National Mine Safety Administration, Hebei Key Laboratory of Mountain Geological Environment, Shaanxi Coal Caojiatan Co., Ltd., Inner Mongolia Research Institute, China University of Mining and Technology (Beijing), China Coal Energy Group Co., Ltd., North China University of Science and Technology, Anhui Wanbei Coal Power Group Co., Ltd., provided assistance and are much appreciated! I also would like to thank the co-authors for their contributions.

As the knowledge grows in mine water control and management, there are rooms to expand the technical contents of the book. We sincerely invite readers to criticize and correct and put forward valuable suggestions, so as to promote the safe, efficient, and green development of ecologically fragile mining areas in western China.

We would like to thank the staff at Springer Nature, for their help and support.

Yifan Zeng

Contents

Chapter 1
Introduction

1.1 Research Purpose and Significance

As an important energy material in China, coal has made great contributions to China's economic development. However, with the gradual westward shift of the focus of coal mining in China, the problems of coal mining and ecological environment faced by mines have become increasingly prominent. According to the data of raw coal production of Shaanxi Province, a big coal producing province in western China in the past decade, its coal production has been increasing year by year, and the ratio of coal production to national coal production has been stable at approximately 16.5% since 2018 (Fig. 1.1). However, as the main coal production base of Shaanxi, the Yushenfu mining area has made great contributions to the coal production capacity of Shaanxi Province and even the whole country. The Yushenfu mining area is located in an arid and semi-arid area, with scarce water resources and extremely fragile ecological surface environment. The coal seam roof of the mining area is mainly composed of sandstone, soil layer and sand layer. The aquiclude in this area is mainly composed of impermeable bedrock and clay. Currently, the mining process of the mine mainly adopts the long-wall high-strength mining mode, such as comprehensive mechanized top coal caving. This method has a large disturbance and damage range to the overburden of coal seam. The first sand barrier of bedrock is broken and becomes unstable under the influence of the intensive mining, resulting in water conduction fractures with different damage degrees. Meanwhile, the most important second soil impermeable barrier to realize water control mining in the Yushenfu mining area is also affected with various degrees of damage. Therefore, the current high-intensity mining in the Yushenfu mining area not only causes damage to the overlying rock containing water-barrier structure of coal seam roof, but also causes immeasurable impact on the groundwater system and ecological environment system. Some of the adverse impacts include surface subsidence, collapse of buildings, withered vegetation on the surface, and increased desertification range. Even major water inrushes and sand outburst accidents occurred in some mining areas

(Fig. 1.2). The occurrence of water inrush and sand outburst accident in the process of mining results from multiple factors, such as the characteristics of the overburden combination of the roof of coal seam, the thickness of coal seam mining and mining technology, the characteristics of aquifer and sand layer, and the characteristics of the waterproof layer. However, under the condition that the bearing characteristics of the surface ecological environment of the mine and the characteristics of the stratum structure of the mine have been determined, the water–sand inrush accident induced by the mine mainly depends on whether the aquifers can meet the requirements of safe mining under the coupling effect of strong earth stress and dynamic water pressure.

There is no doubt that the above-mentioned ecological environment problems and mine safety mining problems have become the "obstacles" to the coordinated development between coal resource development and ecological environment protection in the mines. The main reason for their occurrence is that the mines have not formulated an appropriate mining mode before mining. After the overburden of coal seam roof is disturbed by high-intensity mining, the overburden appears periodic breaking subsidence, which eventually leads to a large range of surface subsidence. At the same time, mining-induced stress leads to the expansion of rock fractures and changes the permeability of rock mass, so that the rock with weak permeability increases the water conductivity, thereby increasing the mine water inflow. After the bedrock of the immediate coal seam roof is broken, whether the water–sand inrush accident occurs mainly depends on the water-proof performance of the key stratum of laterite

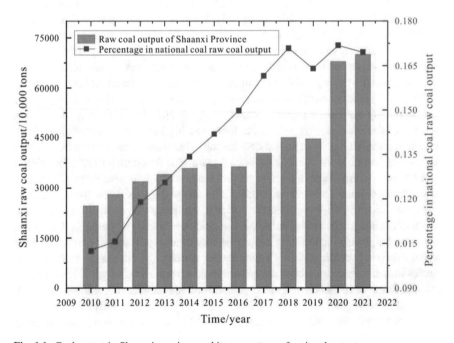

Fig. 1.1 Coal output in Shaanxi province and its percentage of national output

(a) House collapse   (b) Underground water-sand inrush accident

Fig. 1.2 Disasters caused by coal mining

in the Baode group under the influence of strong mining pressure and hydrodynamic pressure. Therefore, under the premise of determining the carrying capacity of the ecological environment, in order to ensure that the above-mentioned ecological environment problems and mine safety mining problems do not occur in the process of coal seam mining in the ecologically fragile areas, it is necessary to strengthen the research on the response characteristics of the overburden and surface ecological damage patterns to the mining intensity during the mining process, as well as the mechanism of water–sand inrush in the working face and the water control mining mode when the bedrock is completely broken. The development of this research work can not only enrich the formation mechanism of roof water disaster in the Yushenfu mining area, but also provide theoretical guidance for green, safe and efficient mining in ecologically fragile mining areas.

1.2 Research Status

1.2.1 Research Progress of Rock Mass Damage and Permeability Under the Influence of Mining

Various discontinuities such as micro-cracks, pores, joints and fissures are produced in the underground rock mass under the long-term action of geological structure and rock movement, which creates certain space conditions for the migration of groundwater, and further produces different magnitudes of pore-water pressure. The pore-water pressure acts on the rock mass, thereby changing the stress distribution of the rock mass, and ultimately causing the expansion and penetration of pores, cracks, and other defects, and ultimately causing changes in the permeability of the rock. This process of interaction and inter-dependence between solid and liquid is called fluid–solid coupling. In recent years, with the increase of underground geotechnical engineering quantity such as mine exploitation and tunnel construction, the actual environment and problems encountered have become more and more complex, so the fluid–solid coupling problem caused by groundwater and mining damage is more and more serious, and the number of accidents is also increasing. For example, the underground tunnel construction has encountered karst water problems caused by karst caves, roof and floor water inrushes caused by high-intensity mining. With the increase in the above-mentioned macroscopic fluid–solid coupling accidents, the research on the mechanical properties, seepage characteristics and failure mechanism of intact rock and defective rock mass under microscopic conditions have become a hot topic. Relevant experts studied the evolution patterns of permeability characteristics of rocks under different lithology, different paths, different initial fractures, different confining pressures, different grain sizes and other conditions through indoor triaxial servo apparatus, acoustic emission and other experimental equipment.

Zhu [105] obtained that the change of rock permeability during the test is closely related to its failure mode by conducting seepage tests under different confining pressure settings. Wang [66] studied the relationship between strain and rock permeability in the process of rock deformation and failure through laboratory tests, and found that the response of circumferential strain to permeability change is faster than that of axial strain. Sun [52] obtained from the permeability test that under the low confining pressure, the permeability of limestone can still be stabilized at a higher level after instability and failure, while the permeability of mudstone will decrease significantly after failure. Guo [14] found that the relationship between temperature and rock permeability is related to the initial permeability of rock through laboratory tests, For rocks with low or high permeability, the temperature permeability relationship is negatively nonlinear. In super permeable rocks, there is a positive nonlinear correlation between temperature and permeability. On this basis, a theoretical model is proposed. Ji [20] studied the change rule of permeability of coal measures rocks through triaxial permeability test, and obtained that the change of rock permeability is mainly divided into five stages: primary fracture compaction and closure—fracture generation and development—fracture expansion and connection—macro fracture generation—fracture re-closure. On the basis of triaxial stress rock movement and deformation model, Tian [60] conducted permeability experiments of different lithological combinations, analyzed the law of permeability coefficient changing with confining pressure and water pressure difference, and deduced the calculation formula of permeability coefficient of different lithological rocks on both sides of single fracture rock mass.

Sunwen [54] conducted experimental research on shale and found that the permeability change of shale is highly correlated with the change of its axial stress, and its permeability has obvious stage characteristics at each stage of rock deformation and failure. Yang [81] studied the change of permeability characteristics during rock deformation and failure through energy dissipation theory and further studied the change process of seepage path through numerical simulation. Yang [76] analyzed the law of permeability change under the condition of changing the initial confining pressure for fractured limestone, and studied the change of rock permeability under the conditions of confining pressure loading and unloading and complex conditions such as osmotic pressure change. Lv [39] revealed the control mechanism of effective stress on the change of permeability of different surrounding rock combinations through the sensitivity test between permeability and effective stress of different surrounding rock combinations in coal seam roof.

Based on the above understanding of the deformation and failure process of rock and the law of permeability change, relevant experts and scholars carried out mechanical tests to improve and revise the applicability of relevant theories. Based on the fluid solid coupling theory and rock failure process, Yang [78] obtained the relationship between the size, seepage direction and randomness through numerical simulation research, and finally compared the sensitivity between the strength of rock and the seepage scale effect. Sun [49] established a rock seepage-damage-stress coupling model based on the characteristics of roadway failure and deformation and its influencing factors, combined with the relevant theories of elastoplastic mechanics, fluid mechanics and rock damage. Zheng [101] independently established a set of theoretical models of the relationship between porosity, permeability and effective stress of low-permeability rocks based on the concepts of the two Hooke models, which were verified by relevant reference data. Wang [63] proposed a conceptual model of rock permeability evolution and the formation mechanism of micro fracture development according to the relevant basic theories of micromechanics and showed that the model can accurately predict the osmotic variation of brittle rocks and describe its experimental phenomena through field tests and numerical simulation results. Hu [16] proposed the calculation method of permeability coefficient through indoor tests and microscopic theoretical analysis. Xie [71] added the damage value based on deformation to establish a new permeability evolution model.

In terms of numerical simulation, Yang [79] team based on damage mechanics, Biot classical seepage mechanics and fluid mechanics, analyzed the parameters of the seepage coupling equation of the continuum model, obtained the equation of the stress seepage deformation relationship in the rock fracture process, developed the seepage stress coupling analysis system for the rock fracture process, and conducted numerical simulation research of triaxial tests under different confining pressures, water pressure differences, and other conditions, The simulation results are compared with the indoor test results to further enhance the understanding of the rock deformation and failure mechanism under the fluid solid coupling condition. Zhang [93] studied the micro process of rock failure and damage based on fluid solid coupling theory and COMSOL numerical simulation software, and found that there was an "indirect coupling" effect between seepage and stress. Zhang [94] conducted an experimental study on the damage characteristics of rock fractures using the fluid solid coupling analysis system based on the finite element, and obtained the influence of crack propagation on the seepage field of rock and the functional relationship between rock damage and seepage.

1.2.2 Research Status of Evolution Characteristics of Overburden Fissure Channel

Mining induced fractures in the overlying rock are mainly formed by the redistribution of the stress field of the upper rock stratum caused by the goaf formed by coal mining. According to the formation reason and distribution position, the overlying rock fractures can be divided into mining failure fractures and stress redistribution fractures. Mining failure fractures, distributed above and around the goaf, are mainly formed by the loss of support of the rock stratum above the goaf after coal mining and the downward collapse under the action of gravity. The cracks formed by the redistribution of stress are mainly the stress field cracks formed due to the leading stress concentration phenomenon at the goaf boundary after the coal seam mining, resulting in the compression and other damage of the rock stratum under the action of the concentrated stress. As the most important research medium in underground engineering, rock mass has different physical and mechanical properties. After the completion of coal seam mining, the stress field in the overburden begins to redistribute to stability due to the loss of coal seam support. However, due to the nature of rock beam, the density and number of fractures above the goaf are much more than those on both sides. In terms of the distribution law and formation characteristics of rock fractures, Bai, Palchik, Karmis and Hasenfus all believed that there were three different distribution zones in the overburden. At present, a large number of experts have studied the fracture distribution law of mining overburden and the distribution characteristics of water flowing fracture zone by means of theoretical research, numerical simulation, physical simulation and field measurement.

Liu [32] put forward the theory of "three horizontal zones" and "three vertical zones" through physical similarity simulation test and field measurement. The results show that the overlying formation can be roughly divided into support influence zone, separation zone and recompaction zone along the strike, and can be divided into collapse zone, fracture zone and bending subsidence zone in the longitudinal direction. At the same time, through a large number of field data and simulation experiment results, the empirical calculation formula of water conducting fracture zone is summarized.

Qian [45] conducted further detailed research on the distribution law of mining overlying rock fractures by using physical simulation tests, image feature analysis and induction, numerical simulation and other means. The research results show that the development law of mining overlying rock fractures is mainly divided into two stages, and at the same time, it finally forms the distribution characteristics of "O" rings. Chen [3] introduced the fracture failure criterion of soft rock stratum and the spatial constraint condition of expansive rock stratum through the classification standard of soft and hard rock, and then established the fracture model of soft and hard rock stratum, and proposed the calculation method of the failure height of overlying rock stratum based on the fracture model.

Xu [74] studied the influence of the movement of the main key layer on the development characteristics of the final water conducting fracture zone through physical simulation model test, image analysis summary, numerical simulation and engineering exploration, and proposed a new method to predict the development height of the water conducting fracture zone through the location and damage characteristics of the key overburden layer. According to the geological conditions of Yili extra-thick coal seam, Liu [36] analyzed the structure and mechanical characteristics of overlying strata, and studied the failure characteristics of overlying strata, the development law of water-conducting fractures and the coupling relationship between key stratum structure and water-resisting layer through numerical simulation. Huang [18] put forward the concept and calculation formula of rock stratum fracture connectivity through physical similar simulation test research and mechanical test analysis and explained the rock stratum fracture connectivity under different stiffness and thickness conditions.

Based on the observations of roof breaking water gushing and water gushing reduction caused by coal mining, Wang [64] used the empirical formula of surface deformation and the concept model of multi-porosity media to evaluate the permeability distribution of the fracture zone. The research results improved the understanding of the influence of overburden stress reconstruction on permeability changes after mining. Ning [44] put forward the mechanical model and activation conditions of the secondary "activation" of the fractured overburden based on the research on the mechanical mechanism of the overburden fracture in the close coal seam mining and the development law of the fracture zone and revised them according to the field measurement.

Zhong [102] used the coupled numerical simulation of discrete element PFC3D and finite element GID to study the water and sand migration law in the water bearing loose layer overlying the coal seam mining. The results show that the opening and dip angle of the fracture determine the type of water–sand inrush in the process of water–sand inrush, and have a greater impact on the water velocity in the overburden and the stability practice of the osmotic press, among which the size of the fracture angle has the largest impact on the drag force between water and sand. Liang [30] studied the influencing factors of water–sand inrush accident when the collapse zone ripples to the loose sand layer through numerical simulation. The results show that the broken rocks in the collapse zone have a certain inhibition on the sand break speed, and the flow rate of the water sand mixture in the loose layer is roughly spindle shaped. Li [27] conducted physical simulation experiments with the independently developed water–sand inrush test equipment, and revealed the mechanism of water–sand inrush and roof cutting disasters under the condition of high-intensity mining in thin bedrock, thick loose layers. The test results show that increasing the initial support force of the support and accelerating the advancing speed of the working face can effectively reduce the probability of water–sand inrush and roof cutting disasters. Tan [57] established the beam theory and the influence of coal seam spacing on the basis of the fracture criterion of overlying strata for the failure evolution law of overlying strata in multi-layer mining.

According to the summary of the field measurements on the failure form and fracture development of the overburden, Lai [23] comprehensively used total station monitoring, high-definition borehole television monitoring, 3DEC numerical simulation software, SPSS professional statistical analysis software to study the overburden movement law, fracture channel evolution process and fracture zone distribution characteristics, and concluded that the overall failure form of overburden presents a "quasi-hyperbolic" shape. Fan [11] used theoretical analysis, numerical simulation, field measurement and other methods to analyze the law of rock failure. At the same time, they used RFPA software to simulate the development process of water conducting fracture zone in the process of coal mining and verified it through field measurement. In addition, they analyzed the seepage characteristics and plastic characteristics of the Quaternary clay layer through laboratory tests. Wang [67] analyzed the movement mechanism of overburden during mining under bedrock ditch and sand ditch, and concluded that the direction of advance, ditch slope angle, ditch erosion coefficient and mining height have important effects on slope stability. Zhang Guoqi, Liu Shiqi, Gao Yu, Yue Qian and Li Jingjing measured the overburden failure characteristics and the development height of the water conducting fracture zone under different engineering geological structures and mining conditions through the consumption of surface drilling fluid and the borehole television observation method, and studied the development rules of the water conducting fracture zone.

1.2.3 Research Status of Influence of Mining Intensity on Aquifer and Aquifuge

In coal mining industry, there is little research worldwide on the impact of mining strength on aquifers (aquiclude). Some scientific researchers mainly study the impact of water content or loading intensity on the stability of coal pillars during mining. Li [29] studied the development of fractures of different overburden types under high mining strength. The research results show that the collapse mining ratio of high-strength mining of hard roof is between 9.39 and 9.62, and the fracture mining ratio is more than 17.80, which is nearly twice that of soft overburden. Zhu [104] showed that the stress and strain change significantly under the condition of different water content of coal, and emphasized that the critical point of stress and strain changes with the change of water content. Wang [62] studied the influence conditions of the fully mechanized coal panel on the stability of the soaked coal pillar in the high-strength mining strength. The research shows that there is a great relationship between the loading rate and the water content on the brittle or ductile failure of coal near the peak strength. Tarasov [56] found that the higher the water content of coal and rock mass, the higher its resistivity and the stronger its stability.

In the study of mining strength, the uniaxial tensile and compressive properties of rock mass with different water content are studied to obtain its geotechnical mechanical properties. Yu [83] used the water injection pressurization simulation experimental device to study the uniaxial test of coal and rock samples under natural and pressurized conditions. The results show that the mechanical properties of coal rock samples under natural conditions are relatively good, and the failure time is relatively long. According to the geological and hydrological characteristics of the mining area. Zhao [100] studied the water inflow pattern of roof aquifer caused by coal mining disturbance, and combined with the mining coal seam depth, put forward three types of aquifer water inflow: water-conducting fissure caused by mining distur- bance, distribution of aquiclude and water inflow characteristics. According to the geological data of five mining areas in the east of Ningxia, Zhai [86] analyzed the distribution characteristics and water filling conditions of the aquifer through data analysis and numerical simulation. Xu [73] studied the failure of coal samples in the loading process under different water content through triaxial experiments; The results show that the higher the water content of coal sample is, the stronger the plasticity is.

In the study of the mechanical relationship between the aquifer and the aquifuge in the coal seam, the main focus is on the study of the mining process in which the water abundance and other properties of the aquifer at the bottom of the coal seam are influenced by the pressure at the top of the coal seam. Fan [12] elaborated the key factors and mechanisms of overburden aquifer protection for groundwater in the process of underground mining through three examples. It is proposed that controlling the deformation and damage of the key impermeable layer and ensuring the appropriate interlayer thickness can effectively protect the underground aquifer and surface water. Xu [75] investigated the deformation characteristics, crack distri- bution law and water resistance change of viscous aquifer under loose aquifer by physical simulation method in view of the water flow problem of seepage aquifer. In application of numerical simulations to predicting the flow behavior of loose aquifer under different conditions, Tian [59] showed that the roof pressure in the mining area has a certain relationship with the water resisting floor, which transmits the pressure through the coal pillar, thus causing the floor to bulge and damage, further inducing the water inrush from the coal seam floor. Xu [72] studied the deforma- tion and fracture distribution characteristics after mining at different depths through physical simulation tests in view of the influence mechanism of the subsidence of loose aquifer caused by the movement of overburden in deep mining. By analyzing the deformation of overburden and water flow data, they analyzed the impact of mining at different depths on the loose aquifer and determined the limit value of due deformation. Shi [48] collected aquifer data from 61 sample points of Hengyuan coal mine from 1988 to 2019, and found that the water quality of the aquifer basically did not change much in the early stage of mining, but in the late stage of mining, with the increase of mining intensity, the water quality of the aquifer changed significantly. Based on the SCWID mechanism, Wang [68] studied that during the mining process under UCA, when it is close to the aquifer, the load impact caused by the aquifer will cause damage to the overburden. In order to solve the contradiction between high intensity mining of coal mines and groundwater protection, Zhang [88] conducted aquifer protection mining technology tests, and monitored overburden damage, loose aquifer water level and surface deformation and damage using borehole observation and ground measurement. The results show that aquifer protection mining technology can be successfully applied by changing a few mining parameters such as mining height or advance speed.

In studying changes in the aquifer properties in response to mining, the relationship between the mining depth and the hydraulic properties of the aquifer or the traceability analysis of the mine water are mainly studied by means of multivariate statistics such as cluster analysis, comparison of simple water quality types and similar material simulation experiments. Based on Peigou Coal Mine, Yang [77] established the source of water source after mining by Fisher discrimination and BP neural network discrimination, and found the development of fractures between aquifer and water source. Sun [50] used similar material simulation experiments to study the fracture development and the control mechanism of seepage channels or paths in the key aquifers under the fluid solid coupling. Zhang [92] through the study of 26 groups of aquifers, using factor analysis and Fisher discriminant analysis, obtained that the factors that are easy to produce water inrush in loose aquifers during mining generally include the thickness of bottom aquifer, water pressure, height of caving area, thickness of bedrock, etc. Wu [70] took Huangling No. 2 Coal Mine as the background, used system cluster analysis and grey relational analysis to trace the source of the water body above the goaf, and the relevant model obtained was verified in Yukou Coal Mine. Chi [7] studied the change characteristics of water resources in the aquifer after mining disturbance for different mining heights, different aquifer locations and multiple numerical models, and analyzed the relationship between groundwater pressure changes and mining heights and aquifers locations.

1.2.4 Status of Research on the Prediction Method and Prevention Technology of Mine Water Inrush

Progresses have been made in recent years on prediction method and prevention technology of mine water inrush. The research object of mine water inrush is mainly on Ordovician limestone karst water inrush in coal seam floor. However, the research on prediction and prevention method of water inrush in coal seam roof lags behind. Back in the nineteenth century, European countries have carried out relevant research, including Doris-Gorno theory, span equation and suspension beam hypothesis. The earlier studies laid a firm foundation for the subsequent theories related to mine water inrush.

In recent years, mathematical inference has occupied a larger proportion of the research such as multivariate statistical analysis and the water inrush coefficient method. Zhang [96] established a prediction model for roof water inrush in shallow coal seam mining by combining AHP-GRA methods. Combined with the actual project, they determined the water inrush points on the roof of the coal seam and predicted the corresponding water inrush risk. The prediction results are consistent with the actual measurement results. In view of the water inrush from the roof of longwall mining, Liu [37] proposed a time-varying risk coefficient to characterize the mechanical criterion of water inrush from the roof of longwall mining through numerical simulation and physical test data, combined with the theory of stress intensity factor (SIF). Lv [38] proposed the "multi-type four double method" evaluation and prediction technology, referred to as MTFD method, on the premise that water abundance and conductivity are the main criteria for judging the occurrence of water inrush from the coal seam roof. Bi [2] used factor analysis (FA) and radial basis function (RBF) neural network to establish a prediction model of mining height of the coal mining belt based on five factors, including mining depth, coal seam dip angle, mining height, overburden compressive strength and working face length. The predictions were tested and evaluated using new sample data for average absolute error, root mean square error and average relative error. Feng [13] took the coal field in north China as the research object, and predicted the distribution range of roof water inrush area through Fisher discriminant analysis, GIS-AHP multivariate information fusion method and UDEC fluid–solid coupling simulation. Liu [35] established a prediction model of water inrush from coal seam roof through computer program simulation, using stochastic finite element theory and seepage mechanics principle, and compared it with the actual situation of Okhobrak Coal Mine to verify the accuracy of the model. Cheng [6] established a geological model for roof water hazard evaluation in multi-seam mining according to the deformation and failure characteristics of the overlying rock in coal seam mining. Taking eight factors as control factors, they determined the weight of the factors by AHP-EM method and established a prediction and evaluation model for roof water inrush risk in multi-seam mining. Zhou [103] conducted coupling analysis on multi-source geological information based on the analytic hierarchy process (AHP), and divided the roof aquifer, hydrogeological conditions, and structural characteristics into seven independent evaluation indicators as the main control factors for evaluating the water yield of K-2 limestone, and finally built a normalized aquifer water yield coupling evaluation model. In order to improve the accuracy of the prediction of water inrush from the combined coal mine floor, Du [9] used the AHP to establish the vulnerability index method to analyze the risk of water inrush from the combined coal mine floor.

In addition to the theoretical analysis, it is also a hot research topic to predict the water inrush in mines by using the field measurements. Meng [42] found that different filling materials and structures have different effects on mine water inrush by studying the seepage properties in the tunnel filling karst structure and came up with the optimal solution to solve the sudden water inflow in mines. Zhao [99] established a mature framework for rapid response to water inrush through the study of a series of water inrush cases and proposed a visual water inrush solution. Ji [19] systematically analyzed the mechanism of water inrush in the mine by theoretical analysis, numerical simulation, and borehole measurement, and determined the best time and location of separation layer drainage through the study of key aquifers, and finally applied it to the site. According to the research on the level of water inrush prevention in mines at the present stage, Zhang [97] proposed the "ground + underground" linkage pre-control method of drainage (interception) to cope with the prediction and prevention of water inrush in mines.

Although the above methods are not applicable to all situations, the prediction and control methods have been verified in practice and have good operability. In addition, many models have been established in combination with the development of the times to effectively predict the water inrush phenomenon in mines. Fang can quickly identify mine water inrush and its water source by establishing convolutional neural network system and combining AF algorithm. Li [27] invented the near-field dynamics method and system for the simulation of sudden water surge disaster caused by rock mass failure in tunnels, which can accurately calculate the real-time parameter changes of mines and predict the risk of water inrush through computer simulation. Zhang [90] can effectively predict the mine water inrush and identify the water inrush source by establishing the multivariate matrix model for identifying the water inrush source, whereas the requirements for water source samples are not strict. Shao [46] established a BP neural network, which can establish a good nonlinear model between the water inflow and inrush of the mine and the influencing factors through computer calculation and simulation to predict the water inflow and inrush of the mine. Most of these models are closely combined with computers, and most of them have realized semi-automatic detection and prediction of water inrush in mines, which is now a mainstream research direction in the world.

Research on water inrush from coal seam roof evolves over time. In the early stage of research, most of the studies were measured in the field, and simple mathematical analysis methods were used to predict and prevent occurrence of water inrushes. In recent years, with the rapid development of computer technology, the research on water inrush from the roof of coal seam has gradually become more mature. Scientific researchers use numerical simulation tests and computer programs to predict the water inrush from the roof of coal seam, and design the most suitable mining plan. Meanwhile, scientific researchers also use computer software to conduct real-time monitoring and early warning of mine disasters, and provide real-time solutions to emergency rescue plans for water disasters.

1.2.5 Research Status of Response Characteristics of Soil Damage to Mining Intensity

Liu [34] studied the mechanical properties and permeability of the clay aquifuge through triaxial seepage tests and studied the fracture development of the overlying bedrock and aquifuge through physical simulation. The results showed that the height of the water-conducting fracture zone would be significantly reduced by the inhibition of weathered rock and clay layer. Ma [40] used red mudstone as the surface soil water-resisting material, and obtained the composition of red mudstone by XRD diffraction experiment. The three indexes, i.e., initial infiltration rate, stable infiltration rate and average infiltration rate, were obtained by laboratory tests. The permeability law and permeability coefficient of each group were also obtained. The results show that the red mudstone has strong anti-seepage ability, and the smaller the particle size, the better the anti-seepage performance. Zhao [98] obtained through similar simulation tests that the key to bridging laterite mining cracks is montmorillonite. The higher the content of montmorillonite, the stronger the healing ability of the cracks. Sun [53] analyzed the rock fracture process by using RFPA (2D) numerical simulation software, and studied the fracture evolution and seepage response mechanism of key aquifers (KAS) under different structural characteristics, different rock thickness and filling ratio. The simulation results show that soft rock plays a key role in the crack repair of KAS with different structural characteristics. Miao [43] proposed that once the key stratum is damaged, the height of the water-conducting fracture zone will be much higher.

Zhang [95] has done extensive research on the failure patterns of the soil layer, and believes that the appropriate instability of the coal pillar can effectively change the failure form of the key stratum, prevent the water-conducting fracture zone from penetrating the soil layer, and provide a judgment basis through coupling evaluation on whether the soil layer can maintain a relatively complete water barrier. Yi [82] calculated the tensile limit of clay, expounded the reason why the mining induced water flowing fracture zone ends at the bottom of the soil layer, and again verified that the soil layer still has certain water resistance after mining. Huang [17] analyzed the overburden failure process and groundwater seepage changes during the mining of shallow coal seams through physical simulation tests. The test results showed that the deformation of the clay aquifers was closely related to the movement of the underlying bedrock layer, and the water conducting fractures in the fracture zone would affect the stability of the aquifers.

Cheng [5] sampled soil samples from different parts of the overlying formation in the mining area. The statistical analysis suggests that significant organic carbon content of the soil on the subsidence slope was lost due to mining cracks, and the organic carbon loss in the middle of the subsidence slope was the earliest and the fastest. At the same time, mining will also cause a series of geological disasters. Jiao [22] classified loess landslides caused by mining according to their characteristics. In order to analyze the change of pore-water pressure in the water resisting soil layer, Yang [77] simplified the water resisting layer to establish a viscoelastic dynamic model for analyzing the pressure distribution. He established a fluid–solid coupled model using FLAC3D and analyzed the characteristics of pressure change and overburden change under different mining intensities, and then determined the space–time characteristics of water inrush.

By triaxial tests, Wang [61] found that after the laterite was loaded and then unloaded, its surface was not apparently cracked, but its permeability was greatly improved. Deck [8] takes the mining damage of the soil layer into account in the surface rock movement in the mining subsidence area. Through data monitoring, it is believed that the deformation of the soil layer itself offsets some of the deformation above, and interacts with the above. The deformation of its own structure is transmitted partially to the upper part, and finally manifested by surface deformation. Bai [1] further investigated the effect of soil damage on the seismic performance of buildings above on this basis. Huang [18] established a downward crack model to explain the phenomenon of soil layer integration, and believed that downward cracks were conducive to protecting aquifers. Suo [55] conducted similar material simulation on the failure of the overburden of frozen soil above Niangmute mine and determined the upper limit of mining. Yu [84] studied the mining damage of the collapsible loess layer with a thickness of more than 60 m, analyzed the damage to the soil derived from the cracks in the loess layer caused by mining, and divided it into four grades. Liang proposed a method to predict surface deformation for the loess layer overlying Tongchuan mining area. Tang [58] divided the thick loess layer into two, referred it as two media of rock stratum and soil layer, and then simulated it. Li [25] also predicted the surface movement better from the mathematical point of view by using the probability integration method. Yang [80] studied the deformation and failure mechanism of overlying loess in the mining area through field observation and numerical simulation. Zhang [87] comprehensively analyzed the damage of coal mining to the surrounding ecological environment and gave corresponding prevention and control measures. Chen [4] described in more detail the periodic stepped cracks formed in the overlying soil layer of Zhaojiazhai Coal Mine after mining, and then concluded that the discontinuous deformation was caused by stress change with the help of numerical simulation software. Jia [21] used polygon evaluation method to evaluate the mining damage of the Yushenfu mining area in a simple and convenient way. Li [24] evaluated the mining damage of Daliuta mining area with the gray correlation method, and gave the mining optimization scheme to reduce the mining damage and suggestions for later repair.

1.2.6   "Coal-Water" Dual Resource Coordinated Mining Theory and Technology

Although the contradiction between coal mining and water resources protection has always existed, it has not attracted enough attention for a long time, resulting in a number of mining accidents. After coal mining entered the western mining area, the fragile ecological environment put forward more strict requirements for the mining process. For this reason, Academician Wu [62] proposed the use of coal-water dual resource mining technology and method to solve the contradiction of "coal resources-water resources-ecological environment protection." This technology is based on comprehensive consideration of the relationship among the three, and proposes a reasonable mining method to achieve coordinated mining among the three and achieve the goal of win–win. Shen [47] applied the proposed mining mode of "coal-water" dual resource mining under the threat of roof water disaster to Xingyuan Mine and Jinjie Mine, achieving green and safe mining in the two mines. Huang [17] proposed to conserve the impermeable soil layer by controlling the subsidence speed of the impermeable soil layer, which was verified by bench-scale physical simulation tests. Fan [10] believes that it is particularly important to alleviate and even solve such contradictions, and puts forward the viewpoint of "water conservation and coal mining". The main means is to reduce the height of the water-conducting fracture zone so that it does not develop into the overlying aquifer and avoids the formation of a water diversion channel between the aquifer and the coal mining face. It gives a more specific guidance scheme for water conservation and coal mining in the Jurassic coal field in Northern Shaanxi. Zhang [89] proposed a method to evaluate the feasibility of water retaining coal mining from the perspective of the overall overburden, and selected typical working faces for verification. Wang [69] completed the regional division of coal mining methods in Northern Shaanxi and proposed that coal mining measures with limited mining height and water conservation should be taken in ecologically fragile areas. Zhang [91] proposed to protect regional water resources by filling and reforming the key stratum of water barrier. Liang [31] discussed the importance of aquifuge stability for water retaining mining by establishing an index system for evaluating aquifuge stability, and analyzed the factors affecting aquifuge stability through mathematical analysis. Wang [65] used numerical simulation software to study the relationship between the height of the water-conducting fracture zone and the upper aquifer under different mining heights, and analyzed the reasonable mining height and mining mode in the study area. Wang [62] believes that the mining height and the water-conducting fracture zone are not linearly correlated, but in a stepped manner, which provides a certain basis for raising the upper limit of mining height on the basis of water conservation. Sun [50] proposed the backfilling technology to realize the safe mining of the coal seam under the water bearing body and analyzed and studied the relevant filling materials. Li [26] compared the estimated development height of the water-conducting fracture zone under the two mining methods through empirical formulas from the perspective of water conservation and coal mining, and finally decided to reduce the height of layered mining in Zhundong Dajing mining area. Li [28] discussed and analyzed the influence of grouting on the stability of aquifuge through F-RFPA2D numerical simulation software and applied the research results to Dongjiahe Mine. Yu [85] proposed to solve the roof subsidence and protect the mine water resources by using the roadway side filling method in coal mining and carried out a successful test in Yuyang Coal Mine, reducing the decline of groundwater level and realizing water-saving coal mining.

Ordos Basin is an important coal-rich basin in China. Many scholars have done extensive research on water-preserved coal mining in this area. Han [15] realized the importance of the water resources of Salawusu formation in the northern mining area of Shenmu to the local area in the last century and proposed that backfilling mining can be used to reduce the height of the induced water-conducting fracture zone. Backfilling mining is also a commonly used and effective means of water retention, and many researchers have achieved desirable results by using the strip filling method supplemented by numerical simulations. Ma [41] proposed a coal mining process with wall layout and continuous mining and filling, which solved the problem of difficult water-preserved coal mining under the condition of thin aquifuge. Sun [51] used the function of sealing and aquifuge of clay layer above coal seam in Shendong mining area, and then used strip filling to stabilize the structure of the water resisting layer, which greatly increased the advancing distance of the working face and made the mining process more stable. Zhang [97] proposed that the water resources in the upper aquifer should be drained for protection and utilization before coal mining.

For the side mining and discharge that cannot be discharged in advance, the water resources should be protected to the maximum extent. Liu [33] summarized a set of methods for the water retention mining of shallow buried coal seams on sand bases, believing that it should be mined quickly, and timely protection, and then certain treatments should be carried out for local water seepage areas. At the same time, with the continuous improvement of the emphasis on the theory of coal-water dual resource mining, the detection and evaluation methods of aquifers have also been improved to a certain extent, and the technical means of electric exploration of the damage of coal-seam roof aquifers have also matured. Miao [43] has divided the types of ecological disturbance caused by coal mining in the Yushenfu mining area. New technologies are being proposed to coal mining, such as cloud computing and big data analysis, making the geological conditions in the mining process transparent, and monitoring the ecological environment in real time.

1.3 Main Contents of This Book

1.3.1 Main Research Contents

This book takes the Yushenfu mining area, located in an arid and semi-arid area and has fragile ecological environment, as the research background. The mine has been challenged with the water–sand inrush accident caused by the use of high-intensity mining technology and surface ecological environment problems. The authors used a variety of research methods, such as indoor triaxial rock mass seepage meter, fluid solid coupling numerical simulation, physical simulation of similar materials, theoretical analysis and field measurement to determine the most appropriate mining method that can overcome these challenges.

The specific research contents are as follows:

(I)   **Regional geological survey of study area**: The general situation of production geology in the Yushenfu mining area is deeply analyzed. The structural characteristics of coal and rock are summarized, and the types and physical and mechanical properties of overlying rock mass in the study area are determined. The storage characteristics and hydraulic connection characteristics of groundwater are identified, summarized and analyzed. The water-containing and water-isolating performance is comprehensively evaluated.

(II) **Study on deformation, damage and seepage characteristics of rock mass under the influence of mining stress**: The strength characteristics and seepage characteristics of sandstone overlying the coal seam under different mining stress paths are revealed by using triaxial compression seepage meter, and the laws of rock damage and seepage change under different confining pressure and water pressure are discussed and analyzed.

(III) **Simulation and analysis of the impact of mining intensity on the law of ecological damage of overlying rock and surface**: This book mainly introduces the method of surveying the height development characteristics of the water flowing fracture zone after mining in the high-intensity mining face with the field measurement method, and through collecting and summarizing the development height of the water flowing fracture zone after high-intensity mining in the Yushenfu mining area, the calculation equation of its development height is fitted and revised. At the same time, the bedrock thickness is defined and divided according to the relationship between the bedrock sedimentary thickness and the fracture development height after mining.

(IV) **Impact of high-intensity mining on water resisting soil layer**: According to the similarity theory of physical simulation and the fluid–solid coupling numerical simulation theory, this book has carried out the research on the impact of high-intensity mining on the mining of N2 water resisting soil layer. Firstly, by using bench-scale physical simulation test, the fracture evolution and rock displacement field evolution of overburden and N2 water resisting soil layer under high strength mining conditions are analyzed. Secondly, the fluid–solid coupling numerical simulation is used to analyze the seepage evolution pattern of overlying rock mass and aquifer under different mining conditions. Finally, the degree of surface ecological damage under high-intensity mining is analyzed through field measurements.

(V) **Disaster mechanism of the water–sand inrush in overburden rock under mining and the criteria for risk assessment and discrimination**: According to the research results of research contents (I), (II), (III) and (IV), the aquifer sand layer, the water yield and recharge characteristics of the aquifer, the thickness of effective clay layer, the thickness ratio of bedrock to unconsolidated layer, and mining fissures are systematically summarized and selected as the key influencing factors for water and sand inrush. The catastrophic criteria for the key water resisting soil layer for water and sand inrush are deduced and calculated based on the theory of material mechanics. Based on the AHP theory, geological statistics, and spatial statistical analysis are used to determine the weight of each key influencing factor, and a quantitative evaluation model of water and sand inrush from mining induced fractures based on AHP is constructed. Finally, the water inrush mode under bedrock fissure aquifer is constructed. Also proposed are bedrock fissure + loose pore aquifer and water inrush mode under water body, the corresponding zonal evaluation method and technical system of roof water inrush and sand break risk.

(VI) **Theoretical method and engineering application of coordinated mining of "coal-water" dual resources**: Based on the coordinated mining theory of "coal-water" dual resources, the coexistence characteristics of coal-water and the ecological environmental problems caused by high-intensity mining in the Yushenfu mining area are collected and statistically analyzed, and the water-controlled mining mode of ecologically fragile mining area is established.

1.3.2 Technical Route

The main research content of this book is to take the mechanism of water–sand inrush and water control mining mode of high-strength mining overburden full-breakage type in thick coal seams in the Yushenfu mining area as the main research line. The research starts from the analysis of geological and hydrogeological character-istics and mechanical characteristics of overburden soil in the study area. Firstly, the damage and permeability characteristics of rock under different mining stress paths are analyzed through triaxial mechanical tests. Secondly, combined with the field measurement results, the physical simulation test of similar materials and the numerical simulation of fluid–solid coupling are used to analyze the law of ecological damage of mining strength to the overlying rock and surface. Thirdly, the risk evalua-tion model and evaluation method of strong mining induced overburden full breakage and lead to mine water and sand breakout are proposed. Finally, the mechanism of water–sand inrush and water control mining mode of strong mining overburden full breakage in the Yushenfu mining area are obtained, and the application analysis is carried out. Figure 1.3 shows technical approaches.

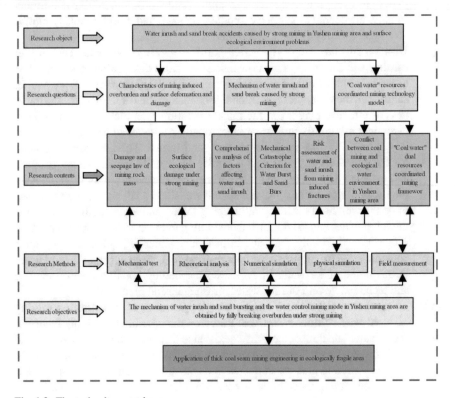

Fig. 1.3 The technology roadmap

References

1. Bai C (2020) Influence and analysis of coal mining considering soil-structure interaction on seismic performance of RC frame structure model. Liaoning University of Engineering and Technology, Fuxin
2. Bi Y, Wu J, Zhai X et al (2022) A prediction model for the height of the water-conducting fractured zone in the roof of coal mines based on factor analysis and RBF neural network. Arab J Geosci 15(3):1–15
3. Chen L, Fan S, Zhao C et al (2019) Calculation method of overburden damage height based on fracture mechanics analysis of soft and hard rock layers. Geofluids 2019:1–15
4. Chen J, East L, Huang C (2013) Research on the mechanism of surface damage caused by mining under extremely thick loose layer. Chin J Geol Hazards Prevention 24(01):51–55
5. Cheng JX, Nie XJ, Liu CH (2014) Spatial change of soil organic carbon in coal mining subsidence area. J Coal 39(12):2495–2500
6. Cheng X, Qiao W, Li G et al (2021) Risk assessment of roof water disaster due to multi-seam mining at Wulunshan Coal Mine in China. Arab J Geosci 14(12):1–15
7. Chi M, Zhang D, Liu H et al (2019) Simulation analysis of water resource damage feature and development degree of mining-induced fracture at ecologically fragile mining area. Environ Earth Sci 78(3):1–15
8. Deck O, Ai Heib M, Homand F (2003) Taking the soil-structure interaction into account in assessing the loading of a structure in a mining subsidence area. Eng Struct 25(04):435–448
9. Du W, Jiang Y, Ma Z et al (2017) Assessment of water inrush and factor sensitivity analysis in an amalgamated coal mine in China. Arab J Geosci 10(21):471
10. Fan LM (2005) On water conservation and coal mining. Coal Field Geol Explor 05:53–56
11. Fan H, Wang L, Lu Y et al (2020) Height of water-conducting fractured zone in a coal seam overlain by thin bedrock and thick clay layer: a case study from the Sanyuan coal mine in North China. Environ Earth Sci 79(6):125
12. Fan G, Zhang D (2015) Mechanisms of aquifer protection in underground coal mining. Mine Water Environ 34(1):95–104
13. Feng XQ (2016) Research on prediction and prevention technology of roof water inrush from mining under loose confined aquifer in concealed coal fields in North China. Hefei University of Technology, Hefei
14. Guo X, Zou GF, Wang YH et al (2017) Investigation of the temperature effect on rock permeability sensitivity. J Petrol Sci Eng 156(03):616–622
15. Han SQ, Fan LM, Yang BG (1992) Analysis of several hydrogeological and engineering geological problems in the development of Jurassic coalfields in northern Shaanxi. China Coalfield Geol 01:53–56
16. Hu DW, Zhu QZ, Zhou H et al (2008) Study on anisotropic damage and permeability evolution of brittle rocks. J Rock Mech Eng 27(09):1822–1827
17. Huang QX (2007) Experimental research of overburden movement and subsurface water seeping in shallow seam mining. J Univ Sci Technol Beijing 14(06):483–489
18. Huang BX, Liu CY, Xu JL (2010) Study on the penetrability of mining overburden fractures. J China Univ Min Technol 39(01):45–49
19. Ji Y, Cao H, Zhao B (2021) Mechanism and control of water inrush from separated roof layers in the Jurassic coalfields. Mine Water Environ 40(2):357–365
20. Ji XK, Guo JB, Xing TJ et al (2015) Characteristics of stress strain and strain permeability of coal measures sedimentary rocks. Coal Field Geol Explor 43(03):66–71
21. Jia Z, Song SJ, Zhao XG (2012) Quick assessment of mining subsidence damage in Yushenfu mining area based on geometric polygon method. Coal Mine Safety 43(07):200–202
22. Jiao S, Long JH, Yu HL (2017) Characteristics and classification of landslides in Shanxi coal mine area. Coal Geol Explor 45(03):101–106
23. Lai XP, Zhang XD, Shan PF et al (2021) The development law of water-conducting cracks in overlying strata during mining of thick loose layer and three soft coal seams. J Rock Mech Eng 40(09):1739–1750

24. Li YP, Bian ZF, Lei SG et al (2017) Evaluation of land damage caused by mining of super large working face in western aeolian sand area. Coal Sci Technol 45(04):188–194
25. Li DH, Chen XG, Li DS (2002) Prediction of surface movement and analysis of rock movement parameters in mining under thick loose layer. Mine Pressure Roof Manage 19(01):90–92
26. Li GS, Zeng Q, Yang J et al (2019) Research on the development and utilization of water-conserving coal mining in Zhundong Dajing Mining Area. China Mining 28(07):142–147
27. Li SC, Gao CL, Zhou ZQ et al (2021) Near-field dynamic method and system for simulation of tunnel rock mass failure and sudden water burst disaster. Shandong Province: CN111368405B, 2021-12-28
28. Li A, Liu Y, Sun, Mou L et al (2018) Numerical analysis and case study on the mitigation of mining damage to the floor of no. 5 coal seam of Taiyuan Group by grouting. J S Afr Inst Min Metall 118(5):461–470
29. Li JH, Wang DH, Li L et al (2021) Comparative study on the development characteristics of high-strength mining fractures of different overburden types. Coal Sci Technol 49(10):9–15
30. Liang YK, Sui WH, Zhu T, Zhang XJ (2017) Study on discrete element numerical simulation of broken rock mass in the collapse zone of Halagou Coal Mine. J Coal 42(02):470–476
31. Liang SS, Zhang DS, Fan GW et al (2021) Aquiclude stability evaluation and significance analysis of influencing factors of close-distance coal seams: a case study of the Yili No. 4 Coal Mine in Xinjiang, China. Geofluids 2021(3):3518271
32. Liu TQ (1995) Mining influence and control engineering of mine rock mass and its application. J Coal 20(1):1–5
33. Liu YD (2008) Water-conserving mining technology and classification of its application conditions in basic shallow coal seam. China University of Mining and Technology, Xuzhou
34. Liu Z, Fan Z, Zhang Y (2019) Fracture characteristics of overlying bedrock and clay aquiclude subjected to shallow coal seam mining. Mine Water Environ 38(1):136–147
35. Liu WQ, Tang J, Lin HX (2012) Stochastic finite element program design of seepage and prediction of water inrush from coal seam roof. J Liaoning Univ Eng Technol (Nat Sci Edn) 31(03):310–314
36. Liu HL, Wang HF, Zhang DS et al (2014) Study on the development law of water flowing fractured zone in extremely thick coal seam mining at oasis mining area. Legislation Technol Pract Mine Land Reclam 28(01):321
37. Liu M, Qiao W, Zhao S et al (2022) Stress intensity factor-based prediction method and influential factors of roof water inrush under longwall mining. Arab J Geosci 15(3)
38. Lv YG, Li HJ, Xia YJ et al (2019) Research on prediction and evaluation of water inrush from coal seam roof based on multi-type four-double method. Coal Sci Technol 47(09):219–228
39. Lv R, Xue J, Zhang Z et al (2022) Experimental study on permeability and stress sensitivity of different lithological surrounding rock combinations. Front Earth Sci 9:762106
40. Ma L, Liu C, Bi Y et al (2021) Experimental study on impermeability law of aquiclude reconstructed by mudstone of external dump in arid zone. Adv Civil Eng 2021(4):1–14
41. Ma LQ, Wang SK, Yu YH et al (2021) The technology and practice of continuous wall mining and water filling coal mining. J Min Saf Eng 38(05):902–910
42. Meng YY, Jing HW, Yin Q et al (2020) Experimental study on seepage characteristics and water inrush of filled karst structure in tunnel. Arab J Geosci 13(9):450
43. Miao XX, Cui XM, Wang J et al (2011) The height of fractured water-conducting zone in undermined rock strata. Eng Geol 120(04):32–39
44. Ning JG, Wang J, Tan YL et al (2020) Mechanical mechanism of overlying strata breaking and development of fractured zone during close-distance coal seam group mining. Int J Min Sci Technol 30(2):207–215
45. Qian MG, Xu JL (1998) Study on the characteristics of "O" ring in the distribution of mining fractures in overburden. J Coal 23(5):466–469
46. Shao AJ, Meng QX, Wang SW et al (2014) Prediction of mine inrush water based on BP neural network method. Adv Mater Res 3326:989–994
47. Shen JJ (2017) Mining mode and application of "coal water" dual-resource mine under the threat of roof water disaster. China University of Mining and Technology (Beijing), Beijing

48. Shi YL, Wu JW, Zhai XR et al (2021) Study on the chemical characteristics of the eighth aquifer of Hengyuan Coal Mine in northern Anhui Province under the influence of mining. J Anhui Univ Technol (Nat Sci Edn) 41(01):31–40

49. Sun QH, Ma FS, Zhao HJ et al (2019) Deformation and failure analysis of roadway surrounding rock considering the weakening of mechanical parameters under the coupled effect of seepage damage stress. J Eng Geol 27(05):955–965

50. Sun Q, Meng GH, Sun K et al (2020) Physical simulation experiment on prevention and control of water inrush disaster by backfilling mining under aquifer. Environ Earth Sci 79(18):429

51. Sun J, Wang LG, Zhao GM (2018) Stability criteria of the overlying aquifer filled with strips of Shendong special water-retaining mining coal seam. J China Univ Min Technol 47(05):957–968

52. Sun XZ, Zhang L, Liu KM et al (2012) Experimental study on permeability characteristics of low confining pressure sedimentary rocks. Coal Eng 06:94–96

53. Sun Q, Chen Y, Huang J et al (2021) Numerical simulation and field measurement analysis of fracture evolution and seepage response of key aquiclude strata in backfill mining. Geofluids 2021

54. Sunwen JB, Zuo YJ, Wu ZH et al (2017) Experimental study on shale seepage damage evolution characteristics. China Mining 26(03):142–145

55. Suo YL, Fan QQ (2014) Experimental study on the upper limit of mining in mines covered by permafrost. Coal Technol 33(05):149–151

56. Taasov BG, Ivanov VV, Dyrdin VV (1976) A mathematical model of the electrical resistivity of a deformed rock. Sov Min Sci 11(4):323–327

57. Tan YW, LY, et al. (2017) In Situ Investigations on Failure Evolution of Overlying Strata Induced by Mining Multiple Coal Seams[J]. Geotechnical testing journal, 40(2):20160090

58. Tang FQ (2005) Prediction method of surface movement in northwest thick loess mining area. J Xi'an Univ Sci Technol 25(03):317–320

59. Tian F (2018) Application of surface microseismic fracturing monitoring technology in coalbed methane development. China Coal Geol 30(08):75–78

60. Tian Z, Zhang W, Dai C et al (2019) Permeability model analysis of combined rock mass with different lithology. Arab J Geosci 12(24):755

61. Wang QQ (2017) Research on the evolution mechanism of mining damage disaster of N2 laterite under the northwest gully cushion. China University of Mining and Technology, Xuzhou

62. Wang ZW (2019) Study on the stability of submerged coal pillar under the influence of high intensity mining in fully mechanized mining face. China University of Mining and Technology, Xuzhou

63. Wang HL, Chu WJ, He M (2012) Anisotropic permeability evolution model of rock in the process of deformation and failure. J Hydrodyn 24(01):25–31

64. Wang WX, Sui WH, Faybishenko B et al (2016) Permeability variations within mining-induced fractured rock mass and its influence on groundwater inrush. Environ Earth Sci 75(4):326

65. Wang Y, Xia YC, Du RJ (2014) Discussion on the maximum mining height of water-retaining coal mining in a mine field in northern Shaanxi. J Min Saf Eng 31(04):558–563

66. Wang HL, Xu WY, Yang SQ (2006) Experimental study on permeability evolution during rock deformation and failure. Geotech Mech 10:1703–1708

67. Wang XF, Zhang DS, Zhang CG et al (2013) Mechanism of Mining-induced slope movement for gullies overlying shallow coal seams. J Mt Sci 10(3):388–397

68. Wang XZ, Zhu WB, Xu JL et al (2021) Mechanism of overlying strata structure instability during mining below unconsolidated confined aquifer and disaster prevention. Appl Sci 11(4):1778

69. Wang SM, Fan LM, Ma XD (2010) Coal development and ecological water level protection in ecologically fragile areas. In: Proceedings of the 2010 national mining science and technology summit forum, pp 232–236

70. Wu QH (2021) Study on the chemical characteristics and water source discrimination of the ponding water in the goaf of the multi-aquifer mine. Xi'an University of Science and Technology, Xi'an

71. Xie XH, Zheng YR, Zhang MF (2009) Study on the relationship between rock deformation and permeability change. J Rock Mech Eng 28(S1):2657–2661

72. Xu S, Zhang Y, Hong S et al (2008) Physical simulation of strata failure and its impact on overlying unconsolidated aquifer at various mining depths. Water 10(5):650

73. Xu Y (2016) Experimental study on the effect of water content on the mechanical properties of formed coal samples. North China University of Science and Technology, Langfang

74. Xu JL, Wang X, Liu W et al (2009) The influence of the location of the main key layer of the overburden on the height of the water-conducting fracture zone. J Rock Mech Eng 28(02):380–385

75. Xu S, Zhang Y, Shi H et al (2020) Impacts of aquitard properties on an overlying unconsolidated aquifer in a mining area of the loess plateau: case study of the Changcun Colliery, Shanxi. Mine Water Environ 39(1):121–134

76. Yang BT (2019) Research on loess damage and its environmental effects under mining. China University of Mining and Technology, Xuzhou

77. Yang T, Li J, Wan L et al (2021) A simulation study on the spatial-temporal characteristics of pore water pressure and roof water inrush in an aquiclude. Shock Vib 2021(2):1–8

78. Yang T, Liu HY, Tang CA (2017) Scale effect in macroscopic permeability of jointed rock mass using a coupled stress–damage–flow method. Eng Geol 228:121–136

79. Yang TH (2001) Study on the permeability property of rock fracture process and its coupling with stress. Northeast University, Shenyang

80. Yang X R, Jiang Konan, Zhang F R, et al. Study on seepage characteristics of fractured limestone under the coupling of seepage pressure and stress. J Coal 44(S1):101–109

81. Yang T, Tang C, Liang ZZ (2018) Permeability evolution and energy dissipation in damage and fracture of jointed rock mass. J Underground Space Eng 14(03):622–628

82. Yi SH, Zhu W, Liu DM (2019) Study on overburden failure law under the condition of thin bedrock and thick loose layer. Coal Eng 51(11):86–91

83. Yu YB, Jiang YJ, Cheng WM et al (2015) Experimental study on permeability characteristics of high-pressure water injection in coal seams. J Min Saf Eng 32(01):144–149

84. Yu XY, Li BB, Li RB et al (2008) Analysis of mining damage degree of huge thick collapsible loess in the west. J China Univ Min Technol 01:43–47

85. Yu YH, Ma LQ (2019) Application of roadway backfill mining in water-conservation coal mining: a case study in Northern Shaanxi, China. Sustainability 11(13):3719

86. Zhai W, Li W, Huang Y et al (2020) A case study of the water abundance evaluation of roof aquifer based on the development height of water-conducting fracture zone. Energies 13

87. Zhang HX (2015) Assessment and protection of coal mining damage. Coal Mining 20(03):1–2

88. Zhang D, Fan G, Ma L et al (2011) Aquifer protection during longwall mining of shallow coal seams: a case study in the Shendong Coalfield of China. Int J Coal Geol 86(2–3):190–196

89. Zhang DS, Fan GW, Zhang SZ, Ma LQ, Wang XF (2022) Connotation, algorithm and application of equivalent water barrier thickness of overburden rock in water conservation mining. J Coal 47(01):0128–0136

90. Zhang LY, Qu ZQ, Chen Y et al (2018) Numerical simulation study on the influence of fracture structure on the fluid-solid coupling failure process of rock. Sci Technol Innov Guide 15(25):76–83

91. Zhang JX, Sun Q, Li M et al (2019) The mining-induced seepage effect and reconstruction of key aquiclude strata during backfill mining. Mine Water Environ 38(3):590–601

92. Zhang WQ, Wang ZY, Zhu XX et al (2020) A Risk Assessment of a Water-Sand Inrush during Coal Mining under a Loose Aquifer Based on a Factor Analysis and the Fisher Model. J Hydrol Eng 25(8):04020033

93. Zhang YJ, Xu T (2013) Hydro-mechanical coupled analysis of the variable permeability coefficient of fractured rock mass. Appl Mech Mater 477–478:531–534

94. Zhang J, Yang T (2018) Study of a roof water inrush prediction model in shallow seam mining based on an analytic hierarchy process using a grey relational analysis method. Arab J Geosci 11(7):153

95. Zhang J, Yang T, Suo YL et al (2017) Research on the prediction of roof water inrush disaster based on the instability model of water-resisting soil layer. J Coal 42(10):2718–2724

96. Zhang J, Yao DX, Su Y (2018) Multivariate matrix model for source identification of inrush water: a case study from Renlou and Tongting coal mine in northern Anhui province, China. IOP Conf Ser Earth Environ Sci 113(1):012212

97. Zhang PS, Zhu HC, Wu YH et al (2021) Research progress in the mechanism of water inrush from coal seams and key technologies for prevention and control. J Eng Geol 29(04):1057–1070

98. Zhao C, Jin D, Wang Q et al (2021) Water inflow characteristics of coal seam mining aquifer in Yushen mining area, China. Arab J Geosci 14(4)

99. Zhao YW, Wu Q, Chen T et al (2020) Location and flux discrimination of water inrush using its spreading process in underground coal mine. Saf Sci 124(C):104566

100. Zhao BC, Yang X, Guo YX, Sun H, Zhai D (2021) The effect of montmorillonite content on N2. Research on the influence law of crack closure in laterite layer. Min Res Dev 41(08):88–93

101. Zheng J, Zheng L, Liu HH et al (2015) Relationships between permeability, porosity and effective stress for low-permeability sedimentary rock. Int J Rock Mech Min Sci 78(05):304–318

102. Zhong JC, Zhou HW, Zhao YF, Liu YQ, Yi HY, Xue DJ (2017) Coupled numerical study of water inrush and sand burst two-phase flow in shallow coal seam mining. Eng Mech 34(12):229–238

103. Zhou K (2021) Water richness zoning and evaluation of the coal seam roof aquifer based on AHP and multisource geological information fusion. Geofluids

104. Zhu CJ, Li CM, Yin ZQ (2017) Experimental study on mechanical characteristics of soft coal under different water content and clay content. J Rock Mech Eng 36(S1):3258–3265

105. Zhu ZD, Zhang Ai J, Xu WY (2002) Experimental study on seepage characteristics of brittle rock during full stress-strain process. Geotechn Mech 05:555–558

Chapter 2
Overview of the Study Area

2.1 Physical Profile

2.1.1 Location, Scope and Traffic

Yushen Mining area is under the jurisdiction of Yuyang District and Shenmu County of Yulin City and is located in the north of Yulin City, Shaanxi Province. The northwest of Yushen mining area is bounded by Shaanxi-Mongolia border; the south is bounded by the north border of Yuheng mining area; and the northeast is bounded by the west border of Shenfu mining area. The Yushen mining area is irregularly polygonal in shape, with a maximum length of 97 km from east to west and a maximum width of 95 km from north to south, covering an area of 5265 km^2. The Yushen mining area consists of four planning districts (Fig. 2.1).

2.1.2 Topography and Geomorphology

Yushenfu Mine is located in the southeast of Ordos Plateau, southeast edge of Maowusu Desert and northern part of Loess Plateau in northern Shaanxi Province. The overall topography of the area is high in the northwest and low in the southeast. The highest elevation of 1385 m is in Leijialiang in the north of the mine area. The lowest elevation of 950 m is in the valley of Tuwei River in the east of the area. The maximum relative height difference is 435 m. The existing geomorphic features of this area are shaped under the joint action of regional new tectonic movement and exogenic forces. According to the morphological genesis of each geomorphic unit, the Yushenfu mining area is divided into three geomorphic types: desert beach area, loess hilly and gully area, and river valley area, as shown in Fig. 2.2.

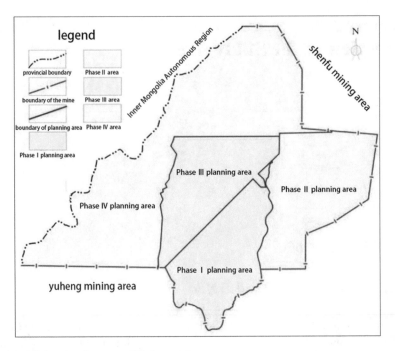

Fig. 2.1 Mining planning map of Yushen mining area

2.1.2.1   Desert Beach Area

Desert beach area is distributed in the middle and west of Yushen mining area, accounting for approximately 5/6 of the total mining area. This area is a part of Maowusu Desert, consisting of undulating dune sand land, semi closed or closed relatively low-lying beach land, which are interdependent and intermittently distributed.

(1) **Desert area**: It is mainly distributed in the areas outside the larger beaches, composed of fixed and semi-fixed dunes and a small number of mobile dunes. The overall distribution of barchan dunes, sand ridges, sand beams and depressions hallow concave in a chain to the north-east (Fig. 2.3). The terrain in this area is undulating with a relative elevation difference of 1–15 m. The windward slope is relatively slow and the leeward slope is steep. The vegetation in this area is mainly drought-tolerant plants, such as *Artemisia sphaerocephala* Krasch, *Caragana korshinskii*, *Amorpha fruticosa*, *Astragalus adsurgens* and *Salix psammophila*. The vegetation coverage ranges from 10 to 40%.

This area is controlled and affected by the ups and downs of the ancient terrain. The thickness of the accumulated sand layer varies greatly, showing different hydrogeological units. Some units consist of watery zones, and the groundwater table is shallow.

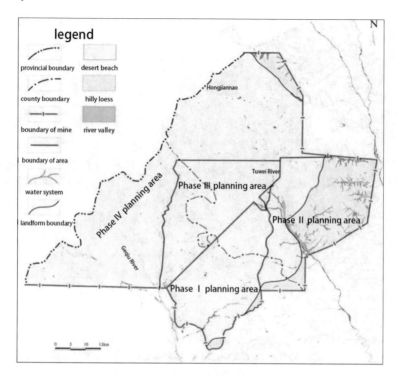

Fig. 2.2 Topography of the study area

Fig. 2.3 Desert landform

Fig. 2.4 Beach landform

(2) **Beach area**: It is distributed in the low-lying areas of the Yushenfu mining area, surrounded by sand dunes, mainly in Baxiacaidang, Shibantai and Yuanjiagedu of Erlintu. The area of the upper reaches of the Bulang River through Mahe to the southwest side of Hekou Reservoir in Mengjiawan, the area of Zhongji Nanalin Caidang, Jinji Beach and Dabaodang area. The distribution area varies in size, generally 1–2 km in width, a few km to a dozen km in length, with poor regularity in spreading direction and irregular shape, with predominating stripes (see Fig. 2.4).

The area is relatively flat, mostly crop planting areas, with grass beaches and poplar, willow, and other mixed forests in the middle. The diving level in the area is shallow and the burial depth is less than 10 m. The vegetation coverage is between 50 and 85%.

2.1.2.2   Loess Hilly Area

It is mainly distributed in the eastern margin of the Yushenfu mining area. Such a landform accounts for approximately 1/6 of the mining area. The area is dominated by earth beam and loess hills, with alternate distribution of earth beam and loess hills and development of dendritic gullies. The sources and the upstream sections are mostly 'V' shaped. The downstream sections and the main gullies and valleys are mostly 'U' shaped, whereas the bank slopes are gentle above and steep below. The overall topography of the area is fragmented. Figure 2.5 shows the general terrain. Other characteristics include scarce vegetation, serious soil erosion, and collapse, landslide and other geological disasters due to heavy rain or continuous rain.

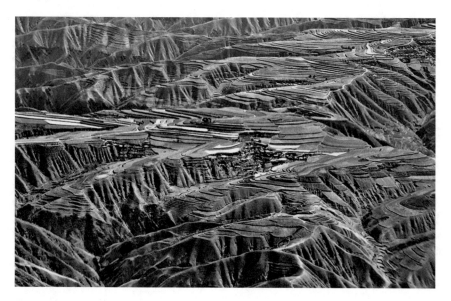

Fig. 2.5 Tidal land geomorphology

The loess (Q_{3+2}), laterite (N_2) cover and part of the bedrock in this area are exposed by gully cutting. The groundwater storage conditions are poor. They are mainly distributed in river valleys and their tributaries, such as the right banks of the Ulan Mulun River and Kuye River, on both sides of the Tuwei River and the Erdao River, which is a tributary of the Yuxi River.

2.1.2.3 River Valley Area

It is mainly distributed in the Tuwei River, Yuxi River and its tributaries. The upper reaches of the loess river valley are mainly seasonal gully flows. The valley is 'V' shaped. The terraces and beaches are not well developed, and the gully beds are narrow. From the middle reaches to the lower reaches of the loess valley, the valley gradually broadens, mostly in the shape of 'U'. The first-order terraces are well developed, and the second-order terraces are sporadically exposed with different types of floodplain micro-topography. The valley floor is 300–700 m wide, the widest is 1000 m, the narrowest is 100–200 m, the maximum 900 m. The valley slope in the loess valley area is steep, and the riverbed in the middle and lower reaches is mostly perennial water, with large seasonal differences. The deposits at the bottom of the valley are mostly composed of alluvial and diluvial, slope sand and a small amount of gravel. The thickness gradually increases from upstream to downstream with an average thickness of 3 m.

Located in the desert beach valley area, the valley is generally wide and shallow, partially controlled by terrain and artificial embankment constraints. Valley section is shallow 'U' type with the width ranging from 50 to 1000 m. The low floodplain is more developed, while the continuity is poor, and first terrace is not obvious. The riverbed slope in this area is reduced. The riverbed and beach are mostly flat, and the alluvial layer consists of mostly fine sand.

2.1.3 Hydrographic Net

2.1.3.1 River

The surface water system of the Yushenfu mining area can be divided into the Kuye River and Tuwei River, the first-level tributaries of the Yellow River, and Yuxi River, the second-level tributaries of the Yellow River, from east to west. Although the main channels of Kuye River and Yuxi River do not pass through the Yushenfu mining area, their tributaries are distributed in the Yushenfu mining area to varying degrees (Fig. 2.6).

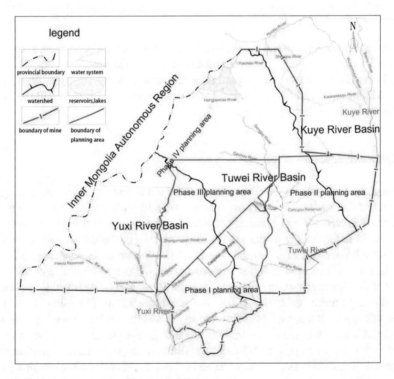

Fig. 2.6 Surface water system of Yushenfu Mining Area

(1) Kuye River

Kuye River originates in Yijinhuoluo county, Inner Mongolia, and is one of the largest rivers in northern Shaanxi. The surface water system in the area of Zhongji and exploration area of the southern Zhongji in the north of the Yushenfu Mining area is mainly in the Kuye River basin. The first-level tributaries Kaokaowusu Ditch, second-level tributaries Huojitu Ditch and Zhugai Ditch originate in the north of the Yushenfu mining area.

Kaokaowusu Ditch: As a first-order tributary of Kuye River, it originates from the vicinity of Chaohaishili in the area and flows from west to east at Chenjiagou bifurcations into Kuye River. The total length of the river is 41.9 km. The width of the basin is approximately 6.2 km, and the drainage area is 259.5 km^2. The slope of the river channel is 7.9‰. The length of the river in the area is approximately 10 km. The river slope is 3.4‰, and the width of the floodplain and terraces ranges from 200 to 400 m. According to the measurements at Qiaojiata Station, the average annual discharge is 0.2277 m^3/s with the maximum of 0.5171 m^3/s and the minimum of 0.0685 m^3/s. The discharge observed in Kaokaowusu Ditch in 2012 was 0.3081 m^3/s.

Huojitu Ditch: It is the secondary tributary of the Kuye River. It originates from Yijinhuoluo Banner of Inner Mongolia and is the boundary river between Yijinhuoluo Banner of Inner Mongolia and Shenmu County of Shaanxi Province, with a drainage area of 309.2 km^2. The area belongs to the loess hilly-gully region. The main channel is 35.9 km long with an average slope of 8.6‰. It is a typical seasonal river with a shallow and wide riverbed, with an average width of 150 m. The headwaters of Shenshu Ditch, Gongge Ditch and Mima Ditch, which are the branches of Huojitu Ditch, are in the north of Yushenfu mining area.

In the east of Yushenfu mining area, the tributaries of Kuye River, such as Huichang Ditch and Yin Ditch, are also developed, and the water flows into Kuye River from west to east.

(2) Tuwei River

The Tuwei River originates from Gongpo Haizi north of Yaozhen Township, Shenmu County, and is called the Tuwei River after the confluence of Gongpo Ditch and Gechou Ditch at Wuji Beach. It flows southeast through Yaozhen, Gaojiapu and other places, and flows into the Yellow River at the southernmost estuary in the Shenmu County. The river is 140 km long, and the drainage area is 3294 km^2. The river slope is 3.87‰. The river valley in the area is open. The riverbed is wide and shallow. There are roaming beaches and terraces developed. The interannual variation of the flow in the Tuwei River is very small. The G value of runoff variation coefficient of Gaojiapu station is 0.14. The measured maximum annual average flow is 14.2 m^3/s (1967), and the minimum annual average flow is 11.6 m^3/s (1969). The ratio between maximum and minimum is 1.2, which is much smaller than 3.8 of Kuye River in the north and 3.5 of Jialu River in the south. From the perspective of seasonal variation of runoff, the summer and autumn runoff accounts for 58% of the annual discharge at Gaojiapu Station, and there are generally two flood seasons: spring flood (March to April) accounting for 16.5% of the annual discharge and summer flood (July to August) accounting for 26.0%. Because the upper reaches are located in the desert area, the precipitation recharge the river in the form of groundwater, so the seasonal variation of river runoff is also not significant.

The Tuwei Wei River flows from the northern part of the Phase IV Planning Area of Yushenfu Mining Area through the northeast corner of the Phase III Planning Area and obliquely through the middle of the Phase II Planning Area, flowing through the mining area with a length of approximately 50 km. The main tributaries in the area are the Gechou Ditch, the Gongpo Ditch and the Hongliu River.

Gechou Ditch: The river is 17.5 km long. The drainage area is 344.80 km^2. The slope of the river is 4.84‰, and the measured discharge of the ditch is 0.8470 m^3/s.

Gongpo Ditch: Originated from Gongpohaizi in the north, the river length is 23.4 km. The drainage area is 323 km^2, and the river slope is 3.85‰.

Hongliu River: It originates from the west of Wangjiamao Village, Mahuangliang Township, and flows through Mahuangliang and Daheta Township from south to northeast into the Tuwei River in Gaojiapu, Shenmu County. The total length is 22 km with 18 km in the district. The drainage area is 120 km^2. The perennial flow is 1.1 m^3/s (Xiangshui station). Its larger tributaries are the Dongqingshui River and so on.

(3)   Yuxi River

Yuxi River originates from Daotuhaizi, north of Yulin. Wudao River, Sidao River, Sandao River, Erdao River, Toudao River, Geqiu River and Bai River in the Yushenfu mining area are all its tributaries, which converge to form Yuxi River at the southern boundary of the Yushenfu mining area. The river is 155 km long, and the drainage area is 5537 km^2. With a river slope of 3.07‰, it flows through Yulin in the southeast and into Wuding River at Yuhebao. The river flows through the sandy area. The river is wide and shallow. The floodplain and terraces are developed. The amount of sediment content is small, and the water volume is stable. The annual average runoff is 7.44 m^3/s. The monthly average maximum discharge is 14.1 m^3/s, and the monthly average minimum discharge is 2.38 m^3/s. The annual flow modulus is 2.38 m^3/s km^2.

The Wudao River is the eastern tributary of the upper source of the Yuxi River. It originates from the Hezhang Spring in the west of Daotu Haizi, Xiaohaotu Township, Yuyang District, flowing south through Mengjiawan township. In the northwest end of Niujialiang township and east of Wanghuagedu village it joins the Geqiu river to form the Yuxi river. It is 32 km long, and its perennial flow rate is 2.09 m^3/s. The drainage area is 524 km^2.

The Sidao River originates from Dongbanchengtan, Mengjiawan Township, and flows into Wudaohe in the south of Dahaizewan. It is 6 km long. The annual flow is 0.3 m^3/s, and the drainage area is 205 km^2.

The Sandao River originates from the north of the Damiantu Village, east of Mengjiawan Township, Yuyang District. It runs through Mengjiawan Township to the southwest, and enters Yuxi River in the south of Wanghuagedu Village, Niujialiang Township. The river is 15 km long with a perennial flow of 0.2 m^3/s and a drainage area of 130 km^2.

The Erdao River originates from Majiahuochang, Jinjitan Township, Yuyang District, passes through Jinjitan Township to the southwest, and enters Yuxi River in the west of Lijiahuochang Village, Niujialiang Township. The river is 18 km in length. The perennial flow is 0.13 m³/s, and the drainage area is 15 km².

The Toudao River originates from the north of Yinshanjie Village, Mahuangliang Township, Yuyang District. It flows from northeast to southwest through Mahuangliang and Niujialiang Townships, and enters Yuxi River in the west of Toudaoheze Village, with a total length of 30 km and a perennial flow of 0.3 m³/s. The area of the watershed is 262 km².

The Geqiu River is a tributary of the upper source of the Yuxi River. It originates from the Halajie Spring in Geqiuhe Village, Mengjiawan Township, Yuyang District. It flows southeastward through Mengjiawan Township and meets the Wudo River in the east of Wanghuafudu Village at the northwest end of Niujialiang Township as the Yuxi River. The river is 20.5 km long, with a perennial flow of 2.2 m³/s and a drainage area of 128.6 km².

The Baihe River, a western tributary of the upper source of the Yuxi River, originates from the Laohai Temple of Urat in Wuxin Banner, Inner Mongolia. It flows southeastward into Hekou Reservoir in Yuyang District, passes through the Chaheze Township and enters the Yuxi River in the south of Miaozui Village, Niujialiang Township. The total length of the river is 54 km, and the perennial flow is 1.3 m³/s. The drainage area is 809 km². The river is seasonal in the Inner Mongolia region, whereas the flow in the study area becomes perennial.

The basic characteristics of the rivers in the Yushenfu mining area are described below. Although the surface runoff is mainly recharged by rainwater, the western part of the area has a wide area of wind-blown sand. There are also heavy rainstorms, and abundant phreatic water in the sandy area. Groundwater recharge accounts for a large proportion, in the range from 30 to 80% of the total runoff. The flow of each river is not only affected by the interannual variation of precipitation (usually 3–6 times), but also greatly affected by seasonal changes, with the least flow in winter and the largest in summer. The rivers are of an inconspicuous bimodal type. Generally, the flow increases in March and April each year due to the melting of ice and snow. The flow in spring flood (March–April) accounts for approximately 15.5% of the average annual flow. Floods often occur in July and August due to concentrated precipitation, and the summer flood (July–August) flow accounts for approximately 19.3–40.9% of the annual average flow (Fig. 2.7).

In addition, the Yeji River, which is a larger tributary of the Tuwei River, is an inland ditch, which originates in the desert area and flows through the desert beach area. The flow path is short and the water volume is small. It is a seasonal gully flow.

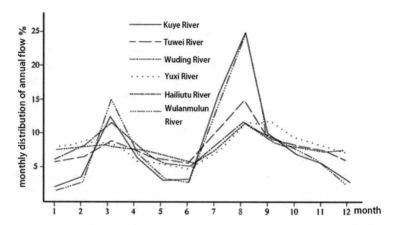

Fig. 2.7 Multi-year average flow characteristics of major rivers

2.1.3.2   Lakes and Reservoirs

There are many lakes and reservoirs in the Yushenfu mining area. The man-made reservoirs in the area are Hekou Reservoir, Zhongyingpan Reservoir, Caitugou Reservoir and Lijialiang Reservoir.

(1)   Hongjiannao

Hongjiannao is located at the junction of Shaanxi Province and Inner Mongolia Autonomous Region, at the northwest edge of the Yushenfu mining area. It is the largest inland freshwater lake in Shaanxi Province and the largest desert freshwater lake in China. The lake is approximately triangular in shape. The length of shoreline of the lake is 43.7 km. The average water depth is 8.2 m with the maximum of 10.5 m. The elevation of the lake is 1100 m. The water level of Hongjiannao decreased at an annual rate of 20–30 cm, and the water area of Hongjiannao lake decreased from 67 km^2 in 1996 to 25.5 km^2 in December 2015. The pH value was approximately 7.2 in the 1960s and 1980s. Approximately the pH increased to 8.2 in the 1990s, and now it is 9.6. The increase of the pH value led to the gradual deterioration of water quality. The main reason is that the supply river is intercepted upstream. Yingpan River and Naogaitu River account for more than 60–70% of the water recharging into Hongjiannao lake. However, Inner Mongolia built reservoirs in the upper reaches of Yingpan River and Manggaitu River. More than 67% of Hongjiannao water recharge, as measured in 2004 and 2009, was cut off from these two rivers. Nowadays, Hongjiannao does not have these two water supply arteries, and most of other smaller recharge rivers have also been cut off.

The surface water in this area is greatly affected by the seasons. The general rule is that the end of winter (March) and rainy season (July–September) are the wet seasons, and the time at the end of winter and at the beginning of spring is the dry season.

(2) Hekou Reservoir

The Hekou Reservoir is located at the junction of the western part of Yulin City, Shaanxi Province and Wushen Banner, Inner Mongolia Autonomous Region. It is located in Hekou Village, Chaheze Township, on the tributary of Yuxi River of Wuding River System, approximately 60 km from Yulin City. Built in 1959, it controls a watershed area of 1400 km^2. It is a medium-sized reservoir with benefits such as flood control, irrigation, and aquatic products. The hydro-junction consists of three parts: a flood-control dam, a water tunnel and a spillway. The dam is 12.7 m high and 1000 m long. The total storage capacity is 94 million m^3, of which the effective storage capacity is 39 million m^3. The normal water level is 1223.73 m above mean sea level, and the average annual runoff is 56.5 million m^3. Reservoir irrigation area of 500 mu. It serves to protect the Yulin city and 3 townships of more than 50,000 people along the river, and 34,000 mu of arable land from flooding. The annual water intake for agricultural irrigation is 1 million m^3.

(3) Zhongyingpan Reservoir

Zhongyingpan Reservoir is located in Zhongyingpan Village, Mengjiawan Township, Yuyang District, on the Wudao River, a tributary of Yuxi River in Wuding River system. It is 45 km north of Yulin City and managed by the Yudong Canal Management Office. The reservoir was completed in 1972, with a drainage area of 606.7 km^2 and a total storage capacity of 19 million m^3. It is an annual regulation medium-sized reservoir with comprehensive utilization of flood control, irrigation, fishery, power generation and sand control.

The constant flow of Zhongyingpan Reservoir is 0.8 m^3/s. The total runoff is 36.72 million m^3, and the average annual output of water from the reservoir is 29 million m^3. The reservoir not only ensures the irrigation water of 10,000 mu of farmland in Yudong Canal, but also supplies 3–8 million m^3 of water to the downstream annually. It also ensures the flood control safety of 20,000 mu of farmland along the river from the reservoir to the 40 km embankment of Hongshixia and the safety of Hongshixia Reservoir.

(4) Caitugou Reservoir

Caitugou Reservoir is located on the main stream of the middle reaches of Tuwei River in Caitugou Village, Jinjie Town, Shenmu County, 40 km away from Shenmu urban area and 70 km from Yulin City. The reservoir has a total design capacity of 72.81 million m^3, with a drainage area of 1339 km^2. It has a maximum inundation area of 5 km^2.

The reservoir is a medium-sized reservoir with industrial water supply, agricultural irrigation and domestic water. It is the key water source project of energy and chemical base construction planning in northern Shaanxi Province. The reservoir was completed in November 2008. The original design of the reservoir had a water supply capacity of 150,000 tons/day. Currently, the water is mainly supplied to Jinjie Industrial Park. In the future, the reservoir will be the water supply to the Clearwater Industrial Park. Because of the consideration of downstream 10,000 mu farmland's irrigation and ecological water demand, the water supply is adjusted to 130,000 m^3/ day with 20,000 m^3/day releasing to the river to conserve the ecological environment. Topographically, the reservoir is located in the southern edge of Maowusu Desert. Most of the watershed is wind-sand grassland, with a small number of loess gullies distributed around the reservoir. The two banks are sandy slopes, and the reservoir basin is a river bottom landform.

(5) Lijialiang Reservoir

Lijialiang Reservoir is located in Caojialiang Village, Mengjiawan, 40 km north of Yulin City. It is located in the lower reaches of the Geqiu River, the first-class tributary of the right bank of the Yuxi River in the Wuding River Basin. The normal flow of the Geqiu River channel is 1.68 m^3/s. The average annual runoff is 52.57 million m^3, and the control basin area above the reservoir dam site is 848 km^2. The reservoir is a typical desert reservoir. It is a medium-sized water conservancy project that focuses on irrigation and takes into account comprehensive benefits such as flood control and agriculture. The hydro-junction consists of the dam project, the diversion cave project, the diversion canal and the road project.

The dam is 25 m high, with a total design capacity of 23.4 million m^3. It is responsible for the task of water supply for large industrial projects such as Jinma Industrial Zone in Yuyang District and Yankuang 600,000 tons of coal to methanol facility. It is one of the important water source projects in Yulin Energy and Chemical Industry Base. The annual water supply is 28 million m^3, irrigating 100,000 mu of farmland. The flood control standard is once every fifty years. At present, the reservoir migration and other projects were completed. The reservoir water has reached 1167 m above mean sea level in elevation. It can supply 1.7 million m^3 of water to Shaanxi Galaxy Power Plant every day. Yanzhou 600,000 tons of methanol project, Yulin Economic Development Zone water supply pipelines have been paved to Lijialiang Reservoir, water supply prospects are very broad.

2.1.4 Meteorology

The Yushenfu mining area is located in the middle temperate zone of the mid-latitude area, which belongs to the arid and semi-arid continental monsoon climate. In winter, it is controlled by the dry and cold denatured polar continental air masses, forming the climate characteristics of low temperature, cold and scarce precipitation. Influenced by the tropical maritime air mass with high temperature and humidity, the precipitation increases in summer. There is the activity of polar cold air from time to time. When it meets with the warm and humid southeast air current in the Pacific Ocean, heavy rain and hail weather may occur. It is prone to cold wave and strong

wind in spring, and the turbidity drops obviously in autumn. It is a typical continental monsoon climate. The climate varies from north to south, so does the terrain.
In general the temperature is warm in the south and cool in the north, high in the
east and low in the west. There is more precipitation in the south and less precipitation in the north, and more precipitation in the east and less precipitation in the
west. The difference between east and west exceeds the difference between north
and south. In addition, the Yellow River and the Wuding River Valley, due to the
concave terrain, are prone to heat accumulation but difficult to disperse, forming
a relatively high temperature climate condition. Lastly, the temperature difference
in the northwest is larger than that in the southeast, both in annual range and the
diurnal range. The annual average precipitation is 316–513 mm, which is the lowest
in Shaanxi Province. The precipitation is mostly concentrated in July, August and
September, accounting for approximately two-thirds of the annual precipitation. The
main natural disasters are drought and low temperature. The spring drought is very
serious, which is known as "nine droughts in ten years". Other disasters, such as
frost, hail, gale and sandstorm, also occur from time to time (Fig. 2.8).

The variation pattern of precipitation in this area is decreasing from southeast
to northwest, making it the area with the lowest precipitation in Shaanxi Province;
Contrary to the distribution of precipitation, evaporation increases from southeast to
northwest.

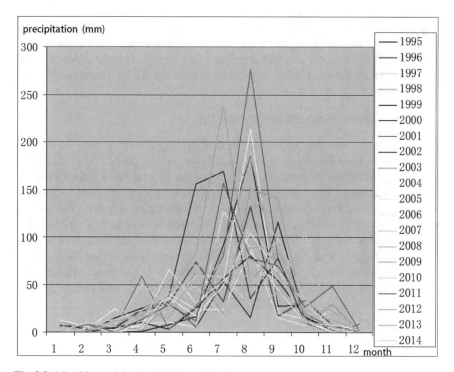

Fig. 2.8 Monthly precipitations in Yuyang District

The main meteorological parameters are as follows:

- Extreme highest temperature: 38.9 °C (1996.6)
- Extreme lowest temperature: −29.0 °C (2003.1)
- Annual average temperature: 8.6 °C (1961–2013)
- Annual average precipitation: 403.5 mm (1995–2014)
- Dry year precipitation: 108.6 mm (1965)
- Wet year precipitation: 819.1 mm (1967)
- Maximum daily rainfall: 141.1 mm (1971.7.21)
- Annual average evaporation: 1712.0 mm (1961–2013)
- Annual average relative humidity: 56% (1961–2013)
- Extreme maximum wind speed: 19.0 m/s (1970.7)
- Maximum frozen soil depth: 146 cm (1968).

2.2 Geological Overview

2.2.1 Stratum

Surface of the Yushenfu mining area is mostly covered by the Quaternary and Neogene strata. There are scattered bedrock outcrops in gullies. According to the borehole and previous geological data, the strata in the area are as follows from old to new: Upper Triassic Yongping Formation (T_3y), Lower Jurassic Fuxian Formation (J_1f), Middle Jurassic Yan'an Formation (J_2y), Zhiluo Formation (J_2z), Anding Formation (J_2a), Lower Cretaceous Luohe Formation ($K_1 1$), Neogene (N), and Quaternary (Q).

1. **Upper Triassic Yongping Formation (T_3y)**

This group is the sedimentary base of the coal-bearing rock system of the Jurassic coalfield in northern Shaanxi. The formation is spread throughout the region. Due to the Indo-China tectonic movement, this stratum was once uplifted and subjected to denudation, resulting in an undulating top surface. Bedrock is exposed in the gully near Dabai Fort in the northeastern part of the Phase II planning area. The stratum was not penetrated in the previous and current supplementary exploration boreholes. According to the regional information, its thickness ranges from 80 to 200 m.

The lithology is a set of gray-green thick-layered, medium and fine-grained feldspar quartz sandstone. It contains a large amount of mica and chlorite with scattered quartz gravel, gray-green argillaceous inclusions and pyrite nodules.

The sorting and roundness are medium. Well developed are large-scale plate-shaped cross-bedding, trough-shaped cross-bedding, wedge-shaped cross-bedding, and block-shaped bedding and wavy bedding.

2. **Lower Jurassic Fuxian Formation (J$_1$f)**

This group is distributed in Xiaohaotu and Xiaobaodang areas in the middle of the third-phase planning area of the Yushenfu mining area. The formation is not exposed. According to the regional data, its thickness ranges from 0 to 147.86 m. The stratum is deposited on the undulating Yongping Formation and is in discomfortable contact with the underlying Yongping Formation. This group of strata consists of fluvial and lacustrine facies deposits. According to the sedimentary characteristics, the lithologic assemblage roughly divides the Fuxian Formation into the following two sub-cycles.

Lower sub-cycle: The lower lithology is mainly coarse-grained quartz sandstone, gravel-bearing coarse-grained quartz sandstone, with quartz fine conglomerate, followed by medium-grained and fine-grained feldspar quartz sandstone. The conglomerate is developed at the bottom of the local section. The composition of the conglomerate is composed of vein quartz, flint, siliceous rock, etc. The conglomerate varies from a few mm to 150 mm, with medium grinding and poor sorting. The upper lithology is greenish-gray, brownish-gray, purple-hybrid siltstone, sandy mudstone, and locally blackish-gray, dark gray sandy mudstone.

Upper sub-cycle: The lower and central part is a giant thickly laminated grayish-white coarse-grained feldspathic quartz sandstone and gravelly coarse-grained sandstone, interspersed with medium-grained and fine-grained feldspathic quartz sandstone. The top is gray-green and purple siltstone and sandy mudstone in blocky bedding, with local black mudstone and coal line at the top.

2. **Middle Jurassic Yan'an Formation (J$_2$y)**

The Yan'an Formation of the Jurassic Middle System is the coal-bearing strata in the Yushenfu mining area. The strata are overclad sedimentary with the Lower Fufu County Formation, in parallel unintegrated contact with the upper Triassic system, and also in parallel unintegrated contact with the overlying Zhiluo Formation. Scouring by the Zhiluo River or erosion by the Cenozoic denudation, the strata are exposed to varying degrees in the valleys along the east of the Xigou and Tuwei River in the second-phase planning area. In the valley near Dabaibao in the northeast, the stratum was completely eroded. The stratum is fully preserved in the Phase III and Phase IV planning areas. Thirty-one boreholes were drilled to the Yan'an Formation in the supplementary survey. According to the previous geological data, the thickness of the Yan'an Formation varies from 0 to 329.69 m. The thickness of deposition is restricted by the pattern of the basement and influenced by the denudation of the Cenozoic boundary, showing the characteristics of being thinner in the east and thicker in the west, and being thinner in the north and thicker in the south.

The formation is a set of terrigenous clastic deposits. Its lithology is mainly grayish white to light grayish white coarse, medium and fine-grained feldspar quartz sandstone, lithic feldspar sandstone and calcareous sandstone, followed by gray to grayish black siltstone, sandy mudstone, mudstone and coal seam. Local sections with a small amount of carbonaceous mudstone are sandwiched with lenticular marl and pyrite nodules. Following are its macroscopic characteristics:

(1) Sandstones are mostly fine and medium-grained sandstones. There are a few coarse-grained sandstones and conglomerate sandstones mostly concentrated in the upper and lower part of the coal seam.
(2) Fine clastic rocks are dominated by siltstones. Sandy mudstones and mudstones are secondary and mostly concentrated in the top and bottom of the coal seam.
(3) Sandstones are relatively less in the middle, while siltstones and mudstones are more abundant.
(4) There are few clay rocks, which appear in thin layers. There are marl lens bodies. Sandstone is mostly calcareous cementation, which reflects the weak alkaline water deposition environment.
(5) The section contains more than 20 coal seams, of which 1–12 layers can be mined. The coal in the middle and upper part of the section is better and thicker, whereas the coal seams in the middle and lower part are thinner.

4. Middle Jurassic Zhiluo Formation (J_2z)

The formation is distributed in the Phase III and IV planning area, the western part of the Phase I planning area and the northwestern part of the Phase II planning area. The rest of the area is stripped away by the Cenozoic denudation. The strata are sporadically exposed in the Kaokaowusugou and Miaojiagou in the northeast of the Phase IV planning area and the Xigoupan area in the east of the Phase III planning area. According to this supplementary survey and previous data, the thickness of the strata of this group ranges from 0 to 237 m. The overall trend is thicker in the west and gradually thinned to the east by the denudation of the Cenozoic boundary until it is exhausted (Fig. 2.9). It is in parallel unconformity contact with the underlying coal strata. The stratigraphy of this group is a set of deposits of fluvial system under semi-arid conditions, which can be roughly divided into upper and lower sections according to lithology:

   Lower section: The upper part is grayish green and bluish gray agglomerate mudstone and siltstone interspersed with fine-grained feldspathic sandstone. The lower part is gray-white medium and coarse-grained feldspathic sandstone lithic feldspar sandstone with greenish gray mudstone and siltstone. The bottom part is composed of thickly laminated coarse-grained feldspathic sandstone. In some parts, it is conglomerate sandstone or conglomerate. In the section, it is equivalent to the "Qili Township Sandstone" in that certain region, with giant tabular cross-bedding and massive bedding. Scouring at the bottom is obvious.

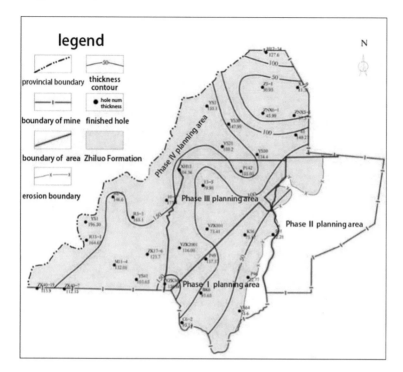

Fig. 2.9 Distribution range and thickness contour map of Zhiluo Formation

Upper section: This section consists of a set of greyish green, blueish grey silty mudstone and siltstone, mixed with greyish green, grayish white, dark purple mica-rich fine-grained feldspar sandstone and lithic feldspar sandstone. Mudstone and siltstone are mostly massive bedding.

5. **Middle Jurassic Anding Formation (J_2a)**

It is distributed in the western part of the Phase I planning area and most of the Phase III and Phase IV planning areas. The rest of the area is missing from the stratum by the Cenozoic denudation. It is sporadically exposed in Shila Ditch, Hejia Ditch and Miaojia Ditch in the northeast of the Phase IV planning area, and in the area of Baodaoshili in the northeast of the Phase III planning area. According to this supplementary survey and previous data, the thickness of this group varies from 0 to 297.87 m. The overall trend is that it is thicker in the north and west, and gradually becomes thinner to the east by the Cenozoic denudation until it is exhausted (Fig. 2.10).

The formation is composed of a set of fluvial system sediment under semi-arid conditions. The lithology is dominated by purplish red, purple-hybrid, dark purple clumpy mudstone and siltstone in the upper part, and light red, purplish gray giant thick laminated coarse-grained, medium-grained, fine-grained feldspar quartz sandstone and lithic feldspar sandstone in the middle and bottom. The sandstone is poorly

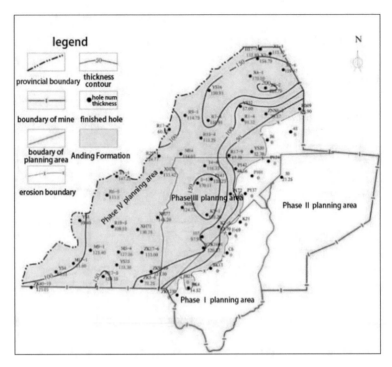

Fig. 2.10   Distribution range and thickness contour map of Anding Formation

sorted, angular in grinding, looser in cementation, with porphyritic structure and tubercular prominence. It is in integrated contact with the underlying Zhiluo Group strata.

6.   The Lower Cretaceous Luohe Formation (K_1l)

The stratigraphy is continuously distributed in the western part of the Phase III planning area and the western part of the Phase IV planning area in the Yushenfu mine, and is missing in the rest of the area. It is sporadically exposed in the area of Shenshu Ditch, Gongge Ditch, Mima Ditch and Hejiagetai in the north of the Phase IV planning area, Houliang in the central part and Yejialiang in the northwest, as well as in the area of Niudinghao in the north of the Phase III planning area. According to this supplementary survey and previous data, the thickness of the strata of this group ranges from 0 to 336.87 m. The overall trend is thicker in the west and gradually thinned to the east by Cenozoic denudation until it is exhausted (Fig. 2.11). The thickest place is located at the junction of the southwestern Yushenfu mining area and Yuheng mining area, reaching 336.87 m. It is in parallel unconformity contact with the underlying Anding formation.

The lithology is a set of purple-red, brown-red giant thick stratified and medium grained, coarse grained quartz feldspar sandstone. It is well sorted and poorly

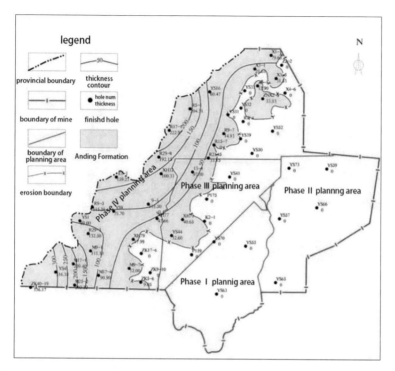

Fig. 2.11 Contour map of distribution range and thickness of Anding group

rounded, showing sub-angular to sub-circular shape, loose cementation, poor consolidation and large cross-bedding.

7. Neogene Pliocene Baode Formation (N_2b)

The strata are mostly distributed in sheets and bands. They are exposed in the upper reaches of valleys. In the mining area, along the upper reaches of valleys such as Tuwei River, Gongbo Ditch, Gechou Ditch, Geqiu River, Yudongqu, Bai River and Sidao River, the thickness changes greatly due to the erosion in the early stage of Quaternary sedimentation. The thickness ranges from 0 to 129.02 m. The regularity is not obvious (Fig. 2.12). It is in unconformity contact with the underlying strata.

The lithology is mainly light red, brownish red clay and sub-clay, containing irregular calcareous nodules, and some of them are layered. Occasionally, there is a thin layer of gravel at the bottom. The gravel is mostly composed of quartz sandstone, conglomerate, etc., which is hard and dense with calcareous cementation.

8. Quaternary Middle Pleistocene Lishi Formation (Q_2l)

Most of the formation is flaky and widely distributed. The loess ridge landform is mostly exposed area. The thickness changes greatly, ranging from 0 to 154 m, with an average of 23 m (Fig. 2.13). The formation is in unconformity contact with the

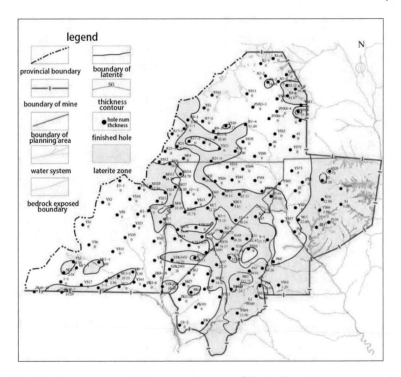

Fig. 2.12 Distribution range and thickness contour map of Baode Formation

underlying strata. The lithology is mainly grayish yellow, light brownish yellow powder clay and powder soil, interspersed with multiple layers of ancient soil and 2–5 m thick powder sand layer. The lower part contains dispersed calcareous nodules of different sizes with vertical joints. The underlying strata are in unconformable contact.

In the Yushenfu mining area, the total thickness of the soil layer ranges from 0 to 174.71 m. The overall change trend is disorderly. It is thicker in the southeast of the Phase I planning area and the southwest of the Phase II planning area, and the rest of the area is thinner.

9. **Quaternary Upper Pleistocene Salawusu Formation (Q_3s)**

This formation is mainly distributed in the desert beach areas. It is found in the beach areas such as Baxiacaidang, Shibantai, Xihulusu, Nalincaidang and Qianer-lintu. The formation is thick in the central part of the beach and gradually thinned to the surrounding areas. The thickness varied greatly, ranging from 0 to 166.00 m. It is in unconformity contact with the underlying layer.

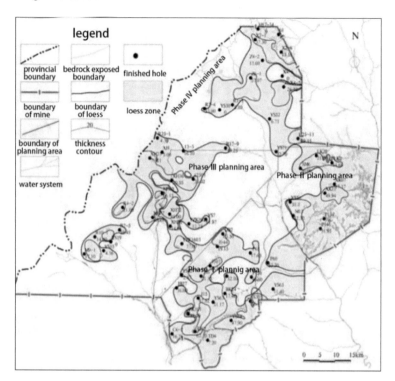

Fig. 2.13 Distribution range and thickness contour map of Lishi Formation

The lithology is mainly composed of grayish yellow, grayish brown and grayish black silt, fine sand and medium sand, with sub-sandy soil, sub-clay and peat layers.

10. Quaternary Holocene Eluvium (Q_4^{eol}) and Alluvium (Q_4^{al})

(1) Eluvium (Q_4^{eol}): It is widely distributed in Yushenfu Mine, in the form of fixed dunes, semi-fixed dunes and mobile dunes overlying other strata. The lithology is mainly light yellow, brownish yellow fine sand and powder sand, with uniform texture, medium sorting and poor rounding. The thickness ranges from 0 to 146.15 m. It is in unconformity contact with the underlying strata.

(2) Alluvium (Q_4^{al}): It is mainly distributed in the valleys of rivers in the area. The lithology is mainly grayish yellow and grayish brown fine sand, powder sand, sub-sand and sub-clay, containing a small amount of humus. Most of the bottom part contains gravel layer, with poor sorting and poor rounding. According to the geological filling data, the thickness varies from 0 to 27.30 m. The formation is in unconformity contact with the underlying strata.

Within the Yushenfu mining area, the total thickness of the sand layer ranges from 0 to 172.10 m. The overall trend is thicker in the west and gradually thinned to the east. The southeastern part of the Phase I planning area and the eastern part of the

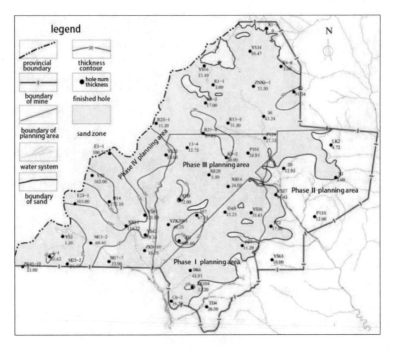

Fig. 2.14 Contour map of distribution range and thickness of sand layer

Phase II planning area are characterized with loess hilly terrains, with a small area
of localized sheet sand coverage (Fig. 2.14).

2.2.2   Coal Seam

There are more than 20 coal beds (including coal lines) in Yan'an Formation in the
area. Fifteen numbered coal beds have commercial values. From top to bottom, they
are numbered 1–1 upper, 1^{-1}, 1^{-2} upper, 1^{-2} (1^{-2} lower), 2^{-2} upper, 2^{-2} (2^{-2} lower),
3^{-1}, 4^{-1}, 4^{-2}, 4^{-3}, 4^{-4}, 5^{-2} (5^{-2} upper), 5^{-2} lower, 5^{-3} upper, 5^{-3} lower. The main coal
seam can be mined in 5 layers: 1–2, 2–2, 3–1, 4–2, 5–2. Nine layers are secondary
recoverable coal seams: 1^{-1} upper, 1^{-1}, 1^{-2} upper, 2^{-2} upper, 4^{-1}, 4^{-3}, 4^{-4}, 5^{-2} lower,
5^{-3} upper, 5^{-3} lower. The 4^{-4} coal seam and some other unnumbered coal seams are
thin coal lines, or only sporadically recoverable points, and are not contiguous. These
coal seams are unminable coal seams (Fig. 2.15).

   The five main mineable coal seams are in different blocks. The distribution of the
first coal seam in each planning area is shown in Table 2.1.

Fig. 2.15 Strata division and coal seam numbering of Yan'an Formation

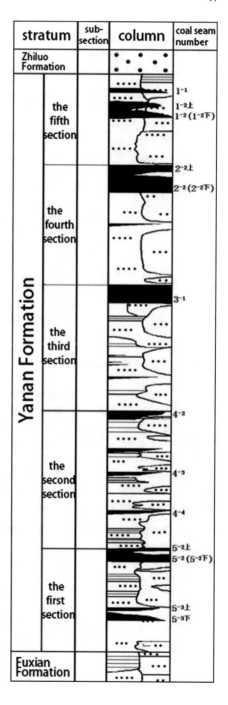

Table 2.1 List of distribution of the first mined coal seam in Yushenfu Mining Area

The first coal seam	Distribution range			
	Phase I planning area	Phase II planning area	Phase 3 planning area	Phase 4 planning area
1^{-2} coal			Distributed in the northwestern part	Distributed in the northeastern, central, southwestern parts
2^{-2} coal	The whole district	The whole area was basically denuded	Distributed in the eastern part	Distributed in the northeastern, central, southwestern parts
3^{-1} coal		Distributed on the western boundary and northwest corner	Located in the northeast corner	Distributed in the northeast corner
4^{-2} coal		Distributed in the north-central and west-central part		
5^{-2} coal		Distributed in the central and northeastern parts		

2.2.2.1 1^{-2} Coal Seam

This coal seam deposits in the upper part of the fifth section of the Yan'an Formation and is one of the first coal seams to be mined in Yushenfu Mine. Due to the erosion of the Zhiluo Formation, the coal seam was eroded in the Phase I and Phase II planning areas, and only remained in the northwest of the Phase III planning area and the northeast, middle and southwest of the Phase IV planning area (Fig. 2.16).

The thickness of the 1^{-2} coal seam ranges from 0.80 to 9.72 m, which is a thin to thick coal seam. The burial depth varies from 214.53 to 673.55 m. The overall change trend of coal seam thickness is as follows:

- In the northeast of the Phase IV planning area, the thickness gradually thickens from east to west.
- The thickness of coal seam gradually thickens from northeast to southwest in the northwest of the Phase III planning area and the middle and southwest of Phase IV planning area.

The coal seam is generally a single coal seam without gangue. Occasionally, it contains 1^{-2} layers of gangue. The structure is relatively simple. The coal types are mainly long-flame coal 41 (CY41) and nonstick coal 31 (BN31). The ash and sulfur contents change moderately.

The coal seam is thin to thick coal seam in the distribution area, while most of the coal seam thickness is medium thick and thick. The thickness changes greatly, but the regularity is obvious. The coal seam is stable.

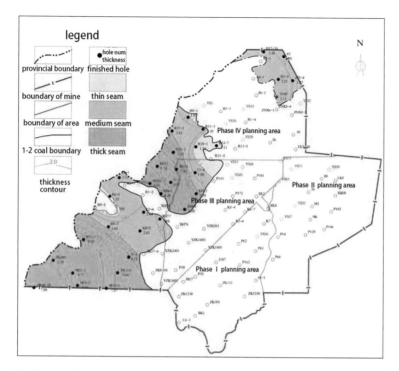

Fig. 2.16 Schematic diagram of the distribution and thickness contour of the first mining 1^{-2} coal seam

2.2.2.2 2^{-2} Coal Seam

The coal seam is located at the top of the fourth section of the Yan'an Formation and is distributed in the Phase I planning area, the eastern part of the Phase III planning area, and the northwestern part of the Phase IV planning area. Due to the overlying strata denudation, this coal seam is missing in the Phase II planning area and the northeastern part of the Phase IV planning area in the eastern part of the mine area (Fig. 2.17).

The thickness of the 2^{-2} coal seam ranges from 0.27 to 12.99 m, which is a thin to thick coal seam. The burial depth ranges from 40.68 to 603.72 m. The overall trend of the coal seam thickness is as follows:

- In the Phase I planning area, the thickness gradually thickens from west to east.
- In the eastern part of the Phase III planning area, the thickness gradually thickens from northeast and southwest to the east-central part.
- In the northeast and central part of the Phase IV planning area, the thickness gradually thickens from southeast to northwest. The coal seam is not mineable or

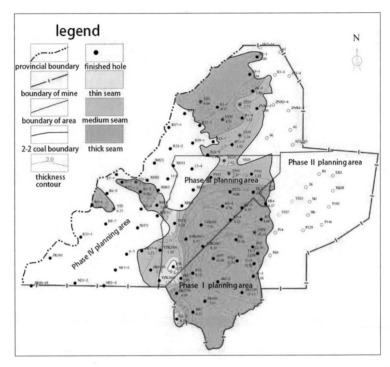

Fig. 2.17 Schematic diagram of the distribution and thickness contour of the first mining 2^{-2} coal seam

stripped by the overlying strata outside the boundary of the east of the distribution area.

Generally, the coal seam does not contain gangue. Occasionally, it contain one or two layers of gangue, and the structure is simple. The coal types are mainly long-flame coal 41 (CY41) and nonstick coal 31 (BN31). The ash and sulfur contents vary moderately.

The coal seam thickness varies greatly, but the regularity is obvious. The coal seam is stable.

2.2.2.3 3^{-1} Coal Seam

The coal seam is located at the top of the third section of the Yan'an Formation, mainly at the western boundary and northwest corner of the Phase II planning area, the northeast corner of the Phase III planning area, and the northeast corner of the Phase IV planning area. The coal seam was stripped away in the central and eastern part of the Phase II planning area in the eastern part of the mine area due to the stripping of the overlying strata (Fig. 2.18).

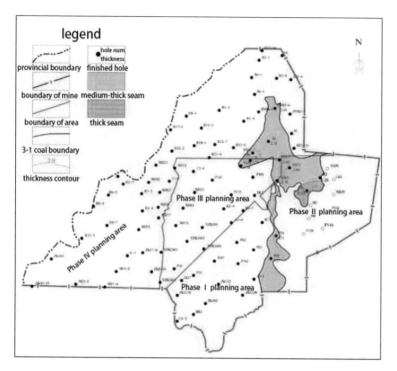

Fig. 2.18 Schematic diagram of the distribution and thickness contour of the first mining 3–1 coal seam

The thickness of 3^{-1} coal seam ranges from 1.72 to 4.32 m, which is medium to thick coal seam with medium thick coal seam dominating. The burial depth varies from 65.27 to 267.53 m. The overall trend of coal seam thickness is thick in the east and thin in the west. The coal seam is not mineable or stripped by the overlying strata to the east of the distribution area boundary.

The coal seam is generally a single coal seam without gangue, and a few of areas contain one layer of gangue. The structure is simple. Coal types are mainly long-flame coal 41 (CY41) and nonstick coal 31 (BN31), and the ash and sulfur contents vary moderately.

The coal seam in the distribution area is characterized with small thickness variation and obvious regularity. The coal seam is stable.

2.2.2.4 4^{-2} Coal Seam

The coal seam is located in the middle and upper part of the second section of the Yan'an Group, mainly in the north-central and west-central part of the Phase II planning area. The coal seam was stripped out in the southeast of the Phase II planning area in the eastern part of the mine due to overlying strata denudation (Fig. 2.19).

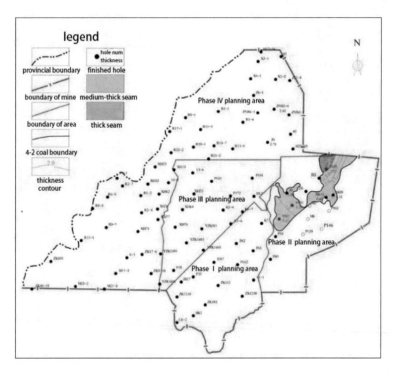

Fig. 2.19 Schematic diagram of the distribution and thickness contour of the first mining 4^{-2} coal seam

The thickness of 4^{-2} coal seam ranges from 0.62 to 3.72 m, which is a thin to thick coal seam, mainly medium-thick coal seam. The burial depth varies from 38.07 to 140.65 m. The overall variation trend of coal seam thickness is gradually thickening from south to north. The coal seam is not recoverable or eroded by the overlying strata to the east of the boundary of the distribution area.

The coal seam is generally a single coal seam without gangue and has a simple structure. The coal types are mainly long-flame coal 41 (CY41) and nonstick coal 31 (BN31), and the ash and sulfur contents vary moderately.

In the distribution area, the coal seam thickness varies greatly, but the regularity is obvious. The coal seam is stable.

2.2.2.5 5^{-2} Coal Seam

The coal seam is located at the top of the first section of the Yan'an Group, mainly in the central and northeastern part of the Phase II planning area. The coal seam was stripped out in the southeast of the Phase II planning area in the eastern part of the mine due to overlying strata stripping (Fig. 2.20).

Fig. 2.20 Schematic diagram of the distribution and thickness contour of the first mining 5^{-2} coal seam

The thickness of coal seam 5^{-2} ranges from 0.80 to 2.06 m, which is a thin to medium thick coal seam with the burial depth varying from 32.33 to 167.30 m. The overall trend of the thickness of the coal seam is gradual increase from the northeast to the middle. To the east of the distribution area boundary, the coal seam is not mineable or eroded by the overlying strata.

Most areas of the coal seam do not contain gangue, whereas some areas contain 1^{-2} layers of gangue. The structure is simple. The coal types are mainly long-flame coal 41 (CY41) and nonstick coal 31 (BN31), and the ash and sulfur contents vary moderately.

The coal seam is characterized with small thickness change and obvious regularity. The coal seam is stable.

2.3   Hydrogeological Conditions

2.3.1   Aquifer

There are four main aquifers in Yushenfu Mine. They are the Quaternary Upper Pleistocene Salawusu Formation (Q_3s) alluvial-lacustrine phreatic aquifer, weathered bedrock pore and fissure confined aquifer, unconfined and confined aquifer of sandstone in lower Cretaceous Luohe Formation (K_1l) and burnt rock cavity fissure unconfined aquifer. Their details are described as follows:

1. **Alluvial-lacustrine phreatic aquifer in Salawusu Formation (Q_3s)**

The alluvial-lacustrine phreatic aquifer is widely distributed in the area, locally exposed on the surface, generally covered by wind-deposited sand layers. The total thickness of the Salawusu Formation and the overlying wind-deposited sand varies from 0 to 166 m. The depth to groundwater is mostly within 10 m. The thickness is generally thinning from southwest to northeast. The lithology is mainly composed of powder, fine and medium sand, with sub-sandy soil, local silt or carbonaceous lens. It is a set of gray to black gray alluvial-lacustrine sedimentary layer. The middle and lower parts are characterized with partial consolidation, poor sorting, loose and high porosity.

2. **Weathered bedrock pore and fissure confined aquifer**

The weathered bedrock is located at the top of bedrock approximately 30 m range of rock layers and distributed in the whole area. The lithology is mainly fine-grained sandstone and siltstone, followed by medium-grained sandstone and mudstone thin layer. Rock fissures are well developed, but show the characteristics of gradually weakening from top to bottom. The upper 6–10 m of the rock is strongly weathered, resulting in secondary structure surface and weathered fissure network development, and a good space for groundwater storage. The water-bearing capacity is stronger than the lower competent rock section. The water yield is generally poor and moderate in local areas.

3. **Unconfined-confined water aquifer in Luohe Formation (K_1l)**

Distributed in the western part of the Phase III planning area and Phase IV planning area of Yushenfu Mine, thickness of this aquifer ranges from 0 to 336.87 m. Scattered outcrops occur in local areas. The lithology is mainly brownish red and purplish red, fine, medium, and coarse-grained feldspathic sandstone, with quartz and feldspar as the main mineral composition, thickly laminated to mega-thickly laminated, with cross bedding, muddy cementation, loose structure and local fissure development, especially in the top 30 m. The rock in the weathering zone is broken, and weathering fractures are developed. Groundwater exhibits the nature of transitioning to pressurized water. The aquifer is unconfined In its outcrop area and the shallow part with thin thickness and the shallow burial areas. In the area with large thickness and deep burial with large overburden, the hydraulic characteristics belong to confined aquifer. Water richness is weak to moderate. The areas with moderate water abundance are mainly distributed in the southwest of the Phase IV planning area and the northwest of the Phase III planning area of Yushenfu Mine. With a strip-like distribution, the thickness of the Luohe Formation is generally greater than 60 m. The areas with poor water-richness are mainly distributed in the north and southeast of the Phase IV planning area and the northwest of the Phase III planning area of Yushenfu Mine, and along the eastern and northern margins of the Luohe Formation deposit area. The overall thickness is less than 60 m.

4. **Burnt rock cavity fissure unconfined aquifer**

This fissure aquifer is mainly distributed in the Phase II planning area on both sides of each large valley showing a strip distribution, due to roof collapses caused by 2^{-2}, 3^{-1}, 4^{-2}, 5^{-2} coal spontaneous combustion and fractures/cavities developed from weathering of burnt rock zone. The degree of burnt metamorphism decreases from the spontaneous combustion coal seam upward to the downside, affecting a thickness of 30–50 m. The thickness of the aquifer varies from 1 to 30 m, with unstable distribution. Because the rock stratum is broken and permeable to water, it has sufficient water storage space and water supply. It is located at the edge of the desert beach, and partially exposed in the steep slope section of the big valley. and the groundwater is mostly drained. According to the pumping tests of three boreholes of the burnt rock aquifer section, the water richness is moderate.

2.3.2 Aquifuge

The aquifuge in the Yushenfu mine area is mainly composed of loess (Q_2l) of the Quaternary Pleistocene Lishi Formation and Sanzhima laterite (N_2b) of the Neogene Pliocene Baode Formation. The loess layer thickness of the Lishi Formation varies greatly, ranging from 1.20 to 145.56 m. It is thicker in the southeast of the Phase I planning area and the southwest of the Phase II planning area, but thinner in other areas. The lithology of the Sanzhima laterite in Baode Formation is mainly light red and brownish red clay and mild clay. The clay minerals in Baode Formation are mainly chlorite, containing a small amount of kaolinite and illite, a small amount of montmorillonite. Te crude minerals are calcite, quartz and feldspar. There are irregular calcareous nodules, and occasionally thin gravel layers at the bottom. The gravel compositions are mostly quartz sandstone and conglomerate. With calcareous cementation, they are hard and dense.

The Sanzhima laterite in the Yushenfu mining area is mainly distributed in the upper reaches of Tuwei River, Gongpogou, Geqiu River, Yudongqu River, Baihe River, Sidao River and other gullies. The thickness ranges from 0.50 to 139.50 m within the distribution range, showing a pattern of sheets and bands. Influenced by the erosion at the early stage of Quaternary deposition, the thickness of the laterite varies greatly and the variation pattern is not obvious. The laterite is in unconformity contact with the underlying strata. The impermeable clay layer composed of loess of Lishi Formation and laterite of Baode Formation in Yushenfu Mine is the main barrier in the area. Reasonable use of clay water-resisting performance is of great significance during coal mining to protect the groundwater resources in the mining area and prevent mining-induced geological disasters.

2.3.3 Groundwater Recharge, Runoff and Discharge Conditions

The unique geological conditions of Yushenfu Mine determine the generality and particularity of groundwater recharge, flow, and discharge conditions. In terms of phreatic water replenishment, the main recharge sources in the Yushenfu mining area include precipitation, lateral runoff recharge, irrigation return and a small amount of interlayer water recharge. In terms of phreatic water runoff, the direction of phreatic water runoff is mainly controlled by regional topography, while the local part is controlled by geomorphology. The overall movement is from northwest to southeast. In the aspect of groundwater discharge, it mainly discharges in the form of springs or groundwater flows, followed by vertical seepage and evaporation.

Due to the influence of the unstable layer, the confined water becomes multi-layer local confined water. There is no unified recharge area for the confined water in the mining area, and it mainly receives precipitation recharge and vertical infiltration recharge by transmitting in the interstream area. The general trend of runoff direction is from east to west. In addition to confined water transmission in local areas, the confined water mostly discharges in the valley.

1. Recharge of groundwater

The Quaternary phreatic water in the Yushenfu mining area has a wide distribution range. The terrain in the distribution area is flat. Due to the loose structure, weak capillary effect and strong water permeability, the aeolian sand layer distributed on the surface is very conducive to the infiltration recharge of atmospheric precipitation and the regression recharge of farmland irrigation water. There is no surface runoff even in the event of heavy rains.

(1) Infiltration recharge of atmospheric precipitation

The annual mean value of atmospheric precipitation in the region is 434.1 mm. The recharge characteristic is planar infiltration. The vadose zone in the aeolian sand beach area and river valley area is composed of aeolian and lacustrine materials. The lithology is mainly fine sand and silt fine sand. The structure is loose. The water permeability is strong. The thickness varies from 0 to 172.10 m. It is easy to receive precipitation infiltration recharge. In the area of loess ridges, the lithology of the lower part of the vadose zone is mainly loess with dense structure and poor water permeability. The upper part consists of the sand layers of the vadose zone, which is conducive to precipitation infiltration. According to the meteorological data of Yulin City from January 1995 to December 2014, the monthly average precipitation in July, August and September was between 16.1 and 276.4 mm. Most of rainfall occurred in the form of rainstorm with high precipitation intensity, accounting for approximately 60% of the total annual precipitation, which was the primary recharge of groundwater. The average monthly precipitation in April, May, June and October was between 0.5 and 156.1 mm, which was the secondary recharge source of groundwater. Most of the daily precipitation was less than 10 mm and had little effect on groundwater recharge. From December to March, the average monthly precipitation was only between 0 and 10.8 mm and had limited effect on groundwater recharge. Correspondingly, the groundwater level showed a downward trend from January to July. With the increase of precipitation in July, August and September, the groundwater level rose after August. These data reflect the significant recharge effect of atmospheric precipitation on groundwater. The effect is especially strong during the rainy season.

(2) Lateral runoff recharge

Yushenfu Mine is located in the downstream of the regional surface water and ground-water basins. Most of the Shaanxi-Mongolia boundary is the lateral groundwater recharge boundary with flow exchange. The recharge section length is 138.72 km. The aquifer thickness ranges from 6.78 to 162 m. The hydraulic conductivity hydraulic conductivity varies from 3.5 to 10 m/d. The hydraulic gradient is between 4 and 6‰. The recharge volume is approximately 70,936 m^3/d. It is one of the recharge sources to the Quaternary unconfined water in the area.

(3) Farmland irrigation regression replenishment

In Yushenfu Mine, a large area of farmland is distributed in desert beaches and river valleys, mostly planting corn. It is the main area of irrigation water regression recharge. The irrigation time is generally from April to September each year with concentrated irrigation from May to August accounts for 90%. The annual water consumption per mu of irrigated land is 350 m^3. In normal years, the sandy beach area irrigation occurs approximately 7 times, whereas the sand-covered loess area beach irrigation occurs approximately 8 times. In drought years, the annual irrigation times increased to 9–10 times, and the annual water consumption per mu of land was between 400 and 450 m^3. Regression coefficients of irrigation were between 0.21 and 0.30. According to the analysis and statistics of the investigation data of farmland irrigation, the regression recharge of farmland irrigation is approximately 8.18 × 10^4 m^3/d.

2. Groundwater runoff

Groundwater runoff is controlled by topography and geomorphology. The boundary of the surface water basin in the region is basically the same as the boundary of the groundwater system. The total runoff direction of the Quaternary phreatic ground-water is from northwest and northeast to the lowest base point of Yuxi River, Tuwei River and Kuye River where the corresponding water flow system is formed. In each basin, under the control of the discharge datum of the secondary water flow system, local runoff flows to the tributaries of the secondary rivers, and the corresponding water flow subsystems are formed. Because of the terrain fluctuation, the runoff velocity of groundwater is different. The local minimum discharge datum formed by the trace-to-source erosion in the steep river-fronted gully area makes the ground-water flow more strongly. In areas with flat terrain where the hydraulic head contour lines are relatively sparse, and the hydraulic gradient is relatively small, the runoff is relatively slow. Due to the influence of unstable aquifuge, the local confined water is multi-layered, and there is no unified recharge area for the local confined water. The joint fissures function as channels through which the vertical infiltration of transmit-ting and receiving the recharge of atmospheric precipitation occurs. The direction of runoff is mainly controlled by topography, from northeast to southwest. Finally, the water discharges into river valleys or is replenished for transmission.

3. Groundwater Discharge

The ways of groundwater discharge in the Yushenfu mining area mainly include: spring water discharge, groundwater seepage, evaporation and dewatering in mines. Under natural circumstances, the groundwater discharges to the river in the form of submerged flow and spring water in the relatively low-lying valley area. In the low-lying windy beach area or the section where the groundwater table is shallower, the groundwater discharges in the form of evaporation. In addition, there is a certain amount of dewatering during mining and excretion.

(1) Spring water discharge

When the groundwater runoff from the water flow system in the mining area reaches the regional or local gully discharge benchmark, it discharges into the river sin the form of springs or submerged flow, and becomes the source of many rivers, such as Bai River, a tributary of Yuxi River, the Huojitu Ditch of Kuye River, and the Kaokaowusu Ditch. The overflow along the Tuwei River formed many large springs, such as Qingshui Spring, Caitugou Spring and Hailonggou Spring. The spring flow became the main source of water in the Tuwei River.

In the eastern part of the mining area, the groundwater runoff from the bottom of the valley replenished the burnt rock area. Because the burnt rock developed the fissure cavities, it could function as a strong runoff zone. Large scale burnt rock springs were present and discharged to the surface. In the deep river valley where the burnt rock was exposed, the springs discharged in groups. However, in the Yuxi River basin and the northern part of the Tuwei River Basin where there is no distribution of burnt rocks, there are few channels for centralized confluence and discharge of groundwater. The groundwater is mainly discharged in the form of subsurface flow.

(2) Evaporation discharge

The Yushenfu mine area mainly belongs to the semi-arid climate area. The annual average evaporation is 1712 mm. In the area where the groundwater level is shallowly buried in the blown-sand region, the transpiration effect is strong. According to the results of previous studies on vegetation ecology, the extreme evaporation depth of bare sandy land in blown sand region is 1.2 m. The evapotranspiration depth of wet and mesophytic vegetation is generally less than 3.8 m, and the extreme transpiration depth of vegetation is 8.0 m. The depth to groundwater level is mostly within 2–3 m. The area of the Quaternary phreatic water distribution area is less than the evaporation limit groundwater level depth area, which is within the range of capillary rise height of soil. The evaporation intensity is high in spring, summer and autumn. The vegetation that grow on the beach has a certain transpiration effect.

(3) Artificial groundwater discharge

Dewatering in the mine pits during coal mining is one of the important ways of groundwater drainage within the Yushenfu mine area. Coal-related industries in the Tuwei River and Kuye River basin mostly use seepage drains and other forms to exploit groundwater in the valley area. Under the influence of coal mining in the local area, fractures formed in the roof of the coal mining area will communicate with the phreatic aquifers of the Quaternary and fracture aquifers of the Jurassic clastic rocks. The surrounding fissure groundwater is directed to flow and discharge to the artificially formed lower datum such as wells and goafs. The water is then collect in the drainage chamber of the coal pits and pumped out to the surface. This constitutes the main drainage method of groundwater in the area.

The residential water in the mining area is also mostly supplied from dewatering activities, such as seepage canals, large-diameter wells and spring diversions. In addition, in the areas of Daxiaobaodang, Erlintu and Baxiacaidang, groundwater is mainly extracted by tube wells for agricultural irrigation, and artificial excretion is another important way of groundwater discharge in this area.

2.4 Summary

(1) The strata in the Yushenfu mine area from old to new are: Yongping Formation (T_3y) of the Upper Triassic, Fuxian Formation (J_1f) of the Lower Jurassic, Yan'an Formation (J_2y), Zhiluo Formation (J_2z), and Anding Formation (J_2a) of the Middle Jurassic, Lower Cretaceous Luohe Formation (K_1l), and the strata in Neogene (N) and Quaternary (Q). The main coal seams are 2^{-2}, 3^{-1}, 4^{-2}, 4^{-3}, and 5^{-3}.

(2) There are four main aquifers in the Yushenfu mine area. They are the alluvial-lacustrine phreatic aquifer in the Quaternary Upper Pleistocene Salawusu Formation (Q_3s), weathered bedrock pore fracture confined aquifer, unconfined and confined sandstone aquifer in lower Cretaceous Luohe Formation (K_1l) and burnt rock hole fissure unconfined aquifer. The aquifuge is mainly composed of loess (Q_2l) of the Quaternary Pleistocene Lishi Formation and Sanzhima laterite (N_2b) of the Neogene Pliocene Baode Formation.

Chapter 3
Research on Deformation Damage and Permeability Evolution of Overlying Formations

The main coal-containing formation in the Yu Shenfu mining area is the Mid-Jurassic Yan'an Formation. The middle and upper parts of the section have good coal content and large thickness, and the lower coal seam is thin. The mining area mainly recovers 2^{-2} coal seam in the Yan'an Formation. According to the scientific research scholars analysis of the combined characteristics of the roof rock layer of 2^{-2} main coal layer in the Yushenfu mining area, the rock layers in this area are mainly divided into the following five types: sand soil base type (I), sand base type (II), soil base type (III), bedrock type (IV) and burning rock type (V) [2, 6]. The characteristics of the rock combination studied in this chapter are sand soil base type (I).i.e. The coal seam cap rock is mainly composed of sand layer, soil layer and bedrock. The spatial and temporal evolution characteristics of the rock stress field and seepage field during the mining process are analyzed.

The stress state of the coal seam and the rock cover formation is a regular complex change state, which is mainly manifested as the different influence on the mechanical strength and permeability characteristics of the coal rock formation under different dynamic load conditions. According to the combination characteristics of sandy soil type (I) rock strata, the integrity of the bedrock during the mining process can be regarded as the first barrier to prevent the occurrence of water–sand inrush accidents in the working face. The integrity of the aquiclude can be regarded as the second barrier to prevent the occurrence of water–sand inrush accidents in the working face. The mining failure time of the second barrier is after the first barrier is destroyed. Therefore, indoor triaxial mechanical tests were conducted to analyze the characteristics of the deformation damage and permeability evolution of the first barrier under the influence of different loading and mining stresses [1, 5]. Theoretical guidance is provided in this chapter for preventing water–sand inrush disaster of coal seam roof and protecting the ecologically fragile mining area.

Y. Zeng et al., *Roof Water Disaster in Coal Mining in Ecologically Fragile Mining Areas*, Professional Practice in Earth Sciences, https://doi.org/10.1007/978-3-031-33140-4_3

3.1 Analysis of Stress-Seepage Evolution Zoning Characteristics of Mining Overburden Rock

3.1.1 Characteristic Response of the Mine Water Damage to the Rock Cover Damage

In the water–sand inrush accidents of the roof of the coal seam of the mines, Qian Minggao, Xu Jialin and other scientific researchers [3, 4, 7] found that the mechanical properties of the water-insulated rock layer between the working face and the water-bearing body and the infiltration characteristics after mining play a decisive role in the water inrush and sand inrush of the working face. The mechanical strength of the water-insulating rock layer also determines the development height of the water-conducting channel of the overlying rock under the influence of mining stress. In particular, the modern mines in western China have high mechanization level and high mining intensity. The movement of roof strata shows strong mine pressure in both horizontal and vertical directions. Some mining areas' roof mining cracks even directly developed to the surface, which leads to the pressure containing water and directing water into the mine, causing mine flooding and water inrush accidents. According to the theory of key strata, the failure mode of roof overburden during coal mining is layer-by-layer upward development and failure. The development of rock mass fissures and the deformation and damage of rock mass show the changes of the whole life cycle along with this process. But, under the influence of the same mining stress, the rock properties of different sedimentary processes show different response characteristics, and the rock strata of different layers have different initiation and fracture pressures. However, the water diversion characteristics of water-conducting fissure channels in the overlying rock mainly depends on the degree of rock layer fracture damage and the change characteristics of rock layer permeability after mining. Therefore, the experimental research on the deformation damage and permeability characteristics of rock mass under the influence of dynamic stress is of great significance to analyze and prevent the occurrence of mine water disasters.

3.1.2 Analysis of Fracture Propagation and Permeability Zoning Characteristics of Mining Overburden Rock

According to the analysis of the above section, the covered rock fissure development characteristics and their permeability change have an important effect on the occurrence of mine water damage. According to the theory of "upper three zones" and the three elements of the occurrence of water gushing at the working face, the direct top and basic top of the coal seam gradually form the upper three zones, i.e., caving zone, fractured zone, and bending zone, as the working face advances. Under certain conditions, there are only two zones, i.e., caving zone and fractured zone. According

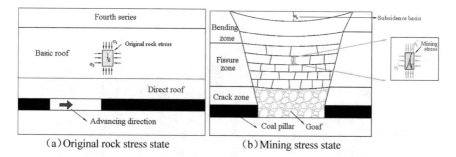

Fig. 3.1 Schematic diagram of stress evolution of overlying strata fissures during mining

to the analysis of the characteristics of the formation of the three zones, the caving zone is mainly formed by the crushing and filling of the rock mass in the immediate roof. The fractured zone is mainly formed by the rock beam through the process of periodic breaking and fracture closure, and the bending zone is mainly formed by the overall bending and sinking of the rock mass. According to the formation process of the above three zones, the rock mass in the fracture zone has good hydraulic conductivity. The hydraulic conductivity characteristics in the fractured zone mainly depend on the mining fissure closure degree and the change of the permeability of the rock mass after the mining influence. The hydraulic conductivity in the bending zone mainly depends on the cracking degree in the overall sinking process of the rock mass (Fig. 3.1). Therefore, under the premise of determining the rich water area of the working surface, the main inducing factors cause the rock fracture expansion and change the permeability of the rock mass. The rock layer with weak permeability increases the water conductivity and increases the water surge of the mine, and thus increases the risk of roof water accident in the mine.

3.2 Materials and Methods of Overburden Stress-Seepage Test

3.2.1 Collection and Preparation of Test Rock Mass

The rock samples used in this triaxial test came from Yan'an Group, Zhiluozu and Anding Group of the roof of the 122109 working face of Caojiatan Coal Mine in the phase I planning mining area of Yushenfu Mining Area. They are characterized with light gray, mud cemented and local calcium cementing, and native cracks in some sections (Fig. 3.2). The triaxial test standard rock sample was prepared according to the mechanical test requirements of Determination Method of Physical and Mechanical Properties of Coal and Rock (GB/T23561-2009). The test rock sample is made by the standard cylinder rock sample (diameter of 50 mm, height of 100 mm). The unevenness of the two ends of the cylindrical rock sample is in accordance with the

Fig. 3.2 Field rock samples

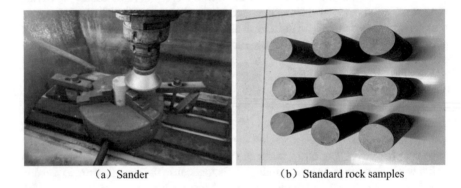

(a) Sander   (b) Standard rock samples

Fig. 3.3 Preparation of standard rock sample

standard and using the grinding machine in Fig. 3.3a to grind the production. The unevenness of the specimen is in accordance with less than 0.05 mm implementation. The standard rock samples are shown in Fig. 3.3b.

3.2.2   Test Equipment and Test Loading Path

3.2.2.1   Test Equipment and Parameters

In order to ensure the accuracy of the analogous simulation test, a new type of controllable confining pressure triaxial apparatus is adopted. The principle diagram and physical diagram of the triaxial testing machine are shown in Fig. 3.4. The triaxial apparatus is composed of an axial pressure control system, a confining pressure control system, an oil pressure loading system, a water pressure control system, a water source cooling system, and a computer data acquisition system. The axial pressure control system mainly sends load signals to the sample through the control

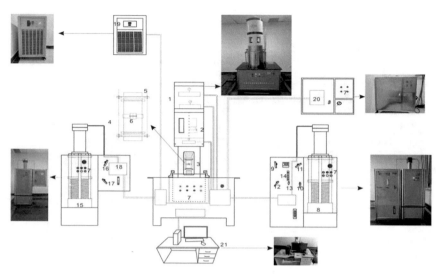

Fig. 3.4 Test system physical diagram

system. The maximum axial pressure can provide 100 kN with a measurement accuracy of 1%. The peripheral pressure control system is to inject high-pressure hydraulic oil into the surrounding pressure chamber through the oil pressure loading system and apply a flexible load to the specimen. The maximum surrounding pressure load can be applied to 100 MPa. The pressure control system mainly adopts intelligent controller and servo motor to accurately control the pressure loading. The pore water pressure control system accurately controls the pore water pressure of the specimen through a dedicated pore water loading device and a controller. The water cooling system is mainly realized by the water pressure order. All test data are automatically acquired by the computer acquisition system. The above control systems can be manually or automatically controlled by a computer. The technical specifications of the triaxial are shown in Table 3.1.

(1) axial compression warehouse; (2) oil cylinder; (3) closed silo; (4) oil pipeline; (5) axial deformer; (6) circumferential deformer; (7) manual console; (8) oil pressure loading system; (9) air valve; (10) oil inlet valve; (11) oil return valve; (12) oil supply valve; (13) air valve; (14) oil standard; (15) hydraulic loading system; (16) back-water valve; (17) water injection valve; (18) water injection tank; (19) cooling system; (20) oil storage tank.

3.2.2.2 Test Principles and Loading Path

(1) Experimental principle

According to the conventional test method of rock permeability, the current researchers mainly use the steady-state method and the transient method to determine

Table 3.1 Technical specifications of the triaxial testing device

Technical index	System and unit		
	Axial pressure control system	Surrounding pressure control system	Water pressure control system
Maximum load	1000 kN	100 MPa	100 MPa
Measurement accuracy	1%	1%	1%
Resolution ratio	0.01 kN	0.01 kPa	0.01 kPa
Loading rate	0.001–50 kN/s	0.001–2 MPa/s	0.001–2 MPa/s
Rock sample size	50 mm × 100 mm		
Deformation range radial	Radial 0–4 mm, axial 0–12.5 mm		
Deformation measurement accuracy	0.001 mm		
Deformation resolution	0.1 μm		

the rock permeability. The permeability are calculated in Eqs. 3.1 and 3.2.

$$K_i = \frac{\mu L \Delta Q_i}{A \Delta P \Delta t_i} \tag{3.1}$$

$$\Delta P_t = (\Delta P_0)e^{-\alpha t}$$

$$K_i = \frac{\alpha \mu L (C_1 C_2)}{A(C_1 + C_2)} \tag{3.2}$$

where

K_i is the average permeability of rock in Δt_i time, m²;
μ is the fluid viscosity coefficient, Pa s; Δ.
Q_i is the water flow volume through rock sample during Δt_i time, m³;
L is the specimen height, m;
A is the contact area of specimen and pressure head, m²;
ΔP is the loaded pressure difference, MPa;
Δt_i is the sampling interval, s;
ΔP_t is the upstream and downstream pressure difference, MPa;
ΔP_0 is the initial pressure difference, MPa;
t is the elapsed time, t;
α is the slope of half log pressure differential-time curve.

C_1, C_2–Capacity of upstream and downstream pressure vessels. C_1 is defined as: C_1 dv_1/dp_1, with an order of magnitude of 10–14 m^3/Pa.

The two test methods are different. The seepage pressure does not change during the steady state test while the water flow through the rock sample during the seepage time is observed. The transient test mainly measures the attenuation of the pressure head at both ends of the rock sample during the seepage time. The permeability of the rock samples was tested by the steady-state method.

In order to eliminate the influence of rock sample adoption and the test machine on the calculation of rock sample permeability, the following assumptions are made:

(1) The microcracks and pores in the rock sample used in the test are regarded as porous media; (2) The water is regarded as incompressible during the rock sample permeability test; (3) The pore water flow in the rock samples conforms to Darcy's law. The water pressure loading in this test is injected into the rock sample through the water injection of the seepage pressure head, and the outlet end is the atmospheric pressure. The principle process of rock seepage test is shown in Fig. 3.5a, and the physical diagram of rock loading is shown in Fig. 3.5b. The rock samples used in this experiment were collected from the 122109 working face of Caojiatan Coal Mine in the Yushenfu Mining Area. The overlying strata of the working face are mainly composed of medium-fine grained sandstone, siltstone, sandy mudstone and mudstone. It was difficult to extract and take standard rock samples in the process of drilling and coring. Therefore, this book mainly analyzes the strength and permeability evolution of sandstone rock mass of the 122109 working face under different mining stress paths.

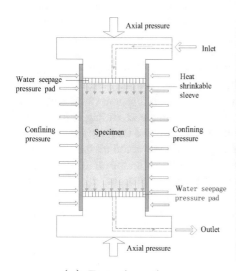

(a) Test schematic   (b) Test setup

Fig. 3.5 Schematic diagram of test principle

(2) **Test steps**

Because this experiment measures the change rules of the damage characteristics and permeability of rock strength under different triaxial stress paths, the careful degree of the preparation work directly determines the success of the test. The sealing nature of the test sample determines whether the hydraulic oil in the test chamber will be mixed with the water body, ultimately leading to the test failure. These experiment also requires that the surrounding pressure of the rock sample loading must be greater than the pore water pressure; otherwise, the high pressure water in the heat contraction tube will enter the test chamber, resulting in test failure and test machine damage, and the high pressure hydraulic oil into the heat contraction tube to mix the oil and water. Therefore, for the success of the test, the following detailed test steps are developed and should be followed:

1. Soaking specimens: According to the water content of the strata and the relevant test requirements, this test used soaking method to carry out water recovery soaking test on the specimen.
2. Specimen number: The soaked rock samples were selected to number the specimens, and the diameter and length of the specimens were measured by Vernier Caliper. The size of the rock samples in this test is 50 * 100 mm standard rock samples.
3. Assemble the specimen: In order to isolate the standard rock samples in the test chamber from high-pressure hydraulic oil, the rock samples fixed on the osmotic pressure head were wrapped with high-temperature and high-pressure heat-shrinkable pipes by hot air gun. After the sample loading was completed, the contact area between the heat shrinkable tube and the osmotic pressure head was tightened by using a tablet. After the wrapping hoop work was completed, the axial displacement meter and the circumferential displacement meter were installed on the infiltration pressure head and the rock sample, respectively.
4. Connect to the pressurized system: Put the assembled test pieces into the test chamber, and connect them to the information collection and transmission system of the water pressure, confining pressure and axial pressure control system respectively. Applied pre-tightening force to the rock sample according to the vertical stress of the rock sample in the formation. Raised the confining pressure silo and open the confining pressure oil filling control system to fill the surrounding pressure silo with oil, and close the oil filling system after the oil filling was over.
5. Apply the initial pressure: According to the horizontal stress and vertical stress of the test rock sample in the stratum, the initial confining pressure, hydrostatic pressure and axial pressure were applied to the rock sample by the pressure control system. The penetration test and the pretest of rock damage were carried out to determine whether the tightness of the rock samples in the test chamber meets the test requirements.
6. Test phase: After the above preparation work was completed, the test started according to the set test scheme. The data collection system was started during the test process to collect the data for the whole test process.

7. Analysis of the test results: Saved the test data after the test sample was damaged. Unloaded the water pressure and shaft pressure, and started the oil recovery system of the surrounding silo. After the hydraulic oil return in the confining pressure silo was completed, the confining pressure silo and the axial pressure head were lowered. After the broken rock sample test was completed, the unloading was performed according to the test plan to obtain the specimen, and the rock damage characteristics were observed and recorded.

(3) **Test and loading path**

In order to analyze the crack evolution characteristics of the roof rock strata of the coal seam under different mining stress paths and the permeability change pattern of the rock strata affected by mining, this book studies the influence of the coal seam on the safe mining of the coal seam under different mining intensity conditions and the change characteristics of the hydraulic conductivity under different mining stress paths by designing different loading tests. However, the process of coal seam excavation belongs to the unloading process with complex but regular rock mass. The patterns of roof overburden fracture damage and permeability change are different under the conditions of different coal seam mining intensity. Therefore, this triaxial mechanical test is based on the different deformation damage characteristics and permeability evolution characteristics of the overlying rock beam cracks in the process of mining with different strengths. The following triaxial loading and unloading mechanical tests with different mining stress paths are designed. The rock samples in the fracture zone of working face 122109 are used for the test, and the different loading path schemes are as follows:

(1) Conventional triaxial compression test

According to the research results of Yuyi River on the stress path in the "three zone" deformation zone, the loading path of this conventional triaxial compression test mainly refers to the mining stress loading path of the fracture zone in its research results (Fig. 3.6). In Fig. 3.6, section AB is the loading compaction stage of the rock sample, and BC section is the crack stage of the rock sample. Section CD is the stress unloading stage of the ultimate bearing strength, which can be divided into passive unloading and active unloading. The passive unloading occurs when the rock reaches the ultimate strength, and the stress drops rapidly. The active unloading occurs when the stress drop of the rock is unloaded to point D according to the set rate. Section DE represents the cyclic loading and unloading.

(2) Pre-peak constant axial pressure unloading confining pressure

The main implementation process of the loading and unloading path of the pre-peak constant axial pressure unloading confining pressure is shown in Fig. 3.7. There are three loading steps. The first loading phase of section A'B' is the loading phase to the original rock stress level, which is similar to the initial loading phase of the conventional triaxial test. The second loading stage is the B'C' section. The strength of the position C at the general termination time of the loading process is 80% of

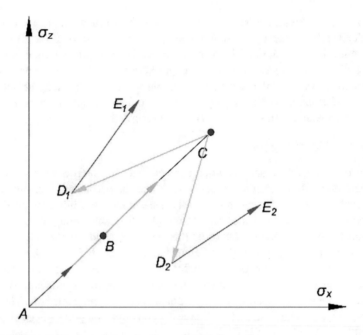

Fig. 3.6 Schematic diagram of stress loading and unloading path

the uniaxial strength. It should be noted that the loading termination time can be designed according to the specific test scheme. The final stage is the C′D′ stage, which is mainly based on the designed unloading confining pressure scheme for constant rate unloading. The unloading rate can be designed according to the actual coal seam mining strength.

(3) Pre-peak unloading axial pressure unloading confining pressure

The biggest difference between this loading path and the above-mentioned pre-peak constant axial pressure unloading confining pressure is that when loading to the unloading point C, the loading path is to simultaneously unload the axial pressure and confining pressure. The axial pressure unloading link in the loading path shows that the field rock beam loses the lower rock beam support function, and the unloading confining shows that the rock beam fault loses the lateral support of the rock beam. The loading process (Fig. 3.8) is also divided into three steps. The first and second steps are the same as the first two steps in the above-mentioned constant axial pressure unloading confining pressure.

The third step is the C″D″ section, which is to unload the axial pressure and the confining pressure at the same time. The specific unloading rate can be designed according to the actual mining conditions.

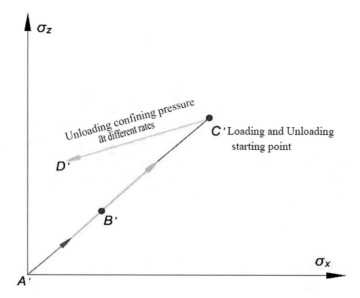

Fig. 3.7 Schematic diagram of confining pressure path of constant axial pressure unloading at the pre-peak under triaxiale compression

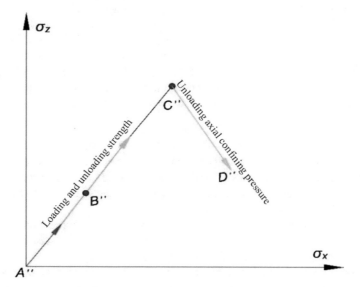

Fig. 3.8 Schematic diagram of axial pressure and confining pressure path at the pre-peak under triaxiale compression unloading

3.3 Analysis of Sandstone Deformation Damage and Permeability Characteristics Under the Influence of Mining Stress Path

This section mainly studies on sandstone and analyzes the sandstone stress-seepage coupling test results under different surrounding pressure and osmotic pressure conditions by using the above conventional triaxial compression test loading path. According to the horizontal stress distribution characteristics of 122109 working face and the water pressure data, the confining pressure applied in the conventional triaxial compression test is 6 MPa, 8 MPa, and 10 MPa, respectively. The pore water pressure difference is 1 MPa, 2 MPa, and 3 MPa, respectively. The axial pressure loading rate is constant at 0.02 mm/min.

3.3.1 Analysis of Sandstone Deformation and Damage Under Mining Stress

This part mainly analyzes the whole process of rock changes from the original state to the instability and failure through the change of stress–strain curve of rock. The characteristics of each part in the stress–strain curve reflect the morphological characteristics and physical significance of different stages of rock. This section mainly shows the stress–strain (axial strain and annular strain) curves of the sandstone under the penetration pressure of 1, 2, and 3 MPa, respectively (Fig. 3.9). According to the overall analysis of the stress–strain relationship in the figure, the whole process from the original state to the final instability can be divided into four stages:

(1) Fissure compaction stage: This stage is mainly near the starting point of the curve in the early stage of the triaxial test. In this stage, the original cracks and pores in the rock are gradually closed under the action of axial compression. The pore space in the rock is gradually reduced, and the whole rock becomes more homogeneous and denser. At the same time, the stress–strain curve at this stage shows a slightly concave trend.

(2) Linear elastic deformation stage: As the native fissure and pore gradually close, the rock specimen becomes homogeneous and compact. This phase corresponds to the linear phase of the initial loading phase in the Fig. 3.9.

(3) Development stage of rock fissure initiation: With the continuous loading of axial stress, the rate of rock strength increase begins to decline. The sandstone begins to produce cracks and constantly develop, expand and penetrate.

(4) Sandstone instability and failure stage: With the further increase of axial stress, the failure cracks in sandstone continue to expand and connect. When the ultimate strength of sandstone is not enough to offset the axial stress, the cracks of sandstone are connected under axial loading and accompanied by stress drop.

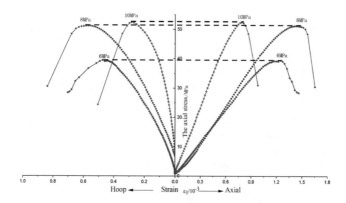

(a) Water pressure difference 1 MPa

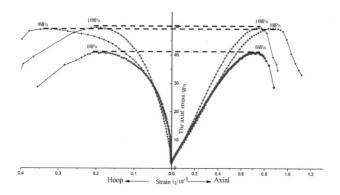

(b) Water pressure difference 2 MPa

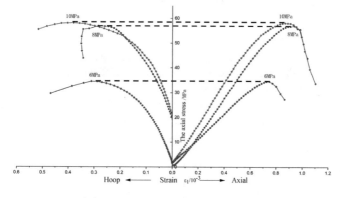

(c) Water pressure difference 3 MPa

Fig. 3.9 Stress–strain curves of sandstone under different stress paths

From the horizontal comparative analysis of stress–strain (axial and annular) under the osmotic pressure condition in the following Fig. 3.9: With the penetration sandstone axial stress increase, the annular strain and axial stress should become basically the same trend. However, the axial strain is more sensitive to stress change before sandstone failure. The hoop strain sensitivity of sandstone is strengthened after peak strength failure. The overall dispersion of sandstone is strengthened after peak strength failure, and a large number of fracture spaces will be formed.

Through the longitudinal comparative analysis of the stress–strain curves under the infiltration pressure of 1, 2, and 3 MPa, it can be concluded that the infiltration pressure difference gradually increases, gradually reducing the rate of reaching the peak strength. Taking the 10 MPa condition as an example, when the axial strain is 0.75, 0.78, and 0.9, the axial stress reaches the peak value and is destroyed. When the seepage pressure increases from 1 to 2 MPa, the peak stress decreases by 2 MPa. When the seepage pressure reaches 3 MPa, the peak strength increases by 5 MPa, The results indicate that when the seepage water pressure is low, with the increase of water pressure, the development, expansion and penetration of cracks will be promoted to a certain extent. The process accelerates the failure of rock and reduces the peak strength. However, when the water pressure is too large, there will be some high-pressure water in the internal cracks. This will give some resistance to the compression process, and then increase the strength of the rock rupture. When the sandstone reaches the peak damage, as the osmotic pressure increases, the change rate of the hoop strain after the peak gradually decreases. When the osmotic pressure is 3 MPa, the hoop strain change rate is particularly slow after the sandstone failure. When the sandstone is damaged, the water pressure in the fracture plays an important role in supporting the further shear displacement of the whole.

3.3.2   Analysis of Sandstone Permeability Characteristics Under Mining Stress

In order to further analyze the change pattern of the sandstone hydraulic conductivity along with the mining stress loading process under different surrounding pressures and different osmotic pressures. In this section, the hydraulic conductivity evolution curves under the conditions of osmotic pressure of 1 MPa, 2 MPa and 3 MPa without confining pressure loading are drawn, as shown in Fig. 3.10. Meanwhile, in order to compare the influence law of the osmotic pressure on the sandstone hydraulic conductivity, the evolution curve of the coefficient without osmotic pressure under the ambient pressure of 6 MPa, 8 MPa and 10 MPa is drawn, as shown in Fig. 3.11. This section mainly describes the features of Fig. 3.10c and Fig. 3.11c.

According to Fig. 3.10a–c, the sandstone remains stable before the initial loading under the action of permeable water pressure, and the whole sample remains in the seepage state. When the water pressure difference is 3 MPa, the permeability of the sandstone samples is 23.79×10^{-10} cm/s, 5.54×10^{-10} cm/s, and 8.57×10^{-10} cm/

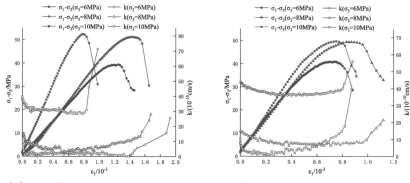

(a) Osmotic pressure difference 1MPa   (b)Osmotic pressure difference 2MPa

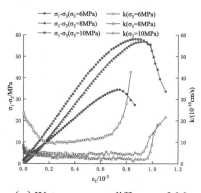

(c) Water pressure difference 3 Mpa

Fig. 3.10 Evolution curves of mechanical properties and hydraulic conductivity of sandstone under the same water pressure and different confining pressures

s, respectively, in response to the circumference pressure of 6, 8, and 10 MPa. The change of rock permeability under the condition of permeability pressure is 3 MPa. In the crack pressure stage, the sandstone permeability value decreases rapidly with the increasing of strain. When the sandstone enters the linear elastic deformation stage, the sandstone permeability is basically stable, with an average permeability of 10.39 $\times 10^{-10}$ cm/s, 2.93 $\times 10^{-10}$ cm/s, and 4.24 $\times 10^{-10}$ cm/s, respectively. The penetration rate increases very slowly near the linear elastic deformation stage at the end of the linear elastic deformation stage and increases gradually at approximately the end of the stage, indicating that the fissure development and expansion rate of the rock are also accelerating near the peak destruction. When the test specimen is damaged, the permeability of the sandstone rises sharply until the test specimen is completely penetrated. Through the analysis of the curves of different confining pressure conditions under the water pressure difference of 3 MPa in Fig. 3.10c, it can be seen that the influence of sandstone permeability surrounding pressure is very obvious. As the surrounding pressure increases from 6 to 8 MPa, the hydraulic conductivity decreases

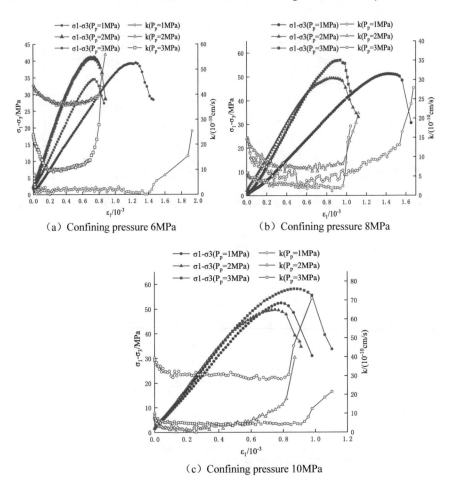

Fig. 3.11 Evolution curves of mechanical properties and hydraulic conductivity of sandstone under the same confining pressure and different water pressure differences

from 23.79×10^{-10} cm/s to 5.54×10^{-10} cm/s. The initial permeability decreases by 78.2%, indicating that the surrounding pressure is compacting the internal cracks and pores of sandstone. For the sandstone in the research area, the compaction effect of the sandstone basically reaches the limit when the surrounding pressure reaches 8 MPa. When the pressure increases to 10 MPa, the initial hydraulic conductivity is slightly increased from 5.54×10^{-10} cm/s to 8.57×10^{-10} cm/s, indicating that the increase of the circumference pressure causes further extrusion and crushing of the original sandstone fissure. Through the analysis of the sudden increase of the hydraulic conductivity after the peak, the hydraulic conductivity of sandstone after instability and failure reaches 42.65×10^{-10} cm/s, 18.01×10^{-10} cm/s and 21.52×10^{-10} cm/s under the condition of seepage pressure of 3 MPa and confining pressure of 6 MPa, 8 MPa and 10 MPa respectively. The values are 32.25×10^{-10} cm/s, 15.08

$\times\ 10^{-10}$ cm/s and 17.28×10^{-10} cm/s higher than the average hydraulic conductivity of sandstone in elastic stage. The change range of post-peak hydraulic conductivity shows the degree of stable failure of the test. When the confining pressure is 6 MPa, the change of post-peak hydraulic conductivity of sandstone is the most obvious, indicating that the failure deformation of sandstone with small confining pressure is more obvious. The change of hydraulic conductivity is basically stable after the increase of confining pressure, indicating that high confining pressure has a certain stabilizing effect on the whole rock after instability and failure.

Through the overall analysis of the permeability changes of sandstone under different confining pressures under the water pressure difference of 1, 2, and 3 MPa, it can be seen that when the confining pressure is 6 MPa, the initial permeability of sandstone increases with the increase of water pressure difference. When the confining pressures is 8 and 10 MPa, with the increasing water pressure difference, the initial permeability of the sandstone decreases continuously. When the confining pressures is 6 MPa, the water pressure difference is the main influencing factor in the process of sandstone compression damage, whereas when the confining pressures reaches 8 MPa or even 10 MPa, the main influencing factor affecting the sandstone permeability is the lateral confining pressures.

The sandstone failure characteristics and permeability changes under different water pressure differences according to Fig. 3.11. Taking the surrounding pressure of 10 MPa as an example, as the water pressure difference of sandstone increases from 1 to 2 MPa, the peak strength of sandstone decreased by 2.724 MPa. When the water pressure difference increased to 3 MPa, the stress peak increased by 8.417 MPa. Meanwhile, the post-peak hydraulic conductivity increased by 40.678×10^{-10} cm/s, 35.524×10^{-10} cm/s and 17.082×10^{-10} cm/s, respectively over the average hydraulic conductivity of 10 cm/s. Therefore, when the pressure is large, the degree of rock instability is mainly controlled by the pressure, and the influence of water pressure difference is very limited. According to the hydraulic conductivity and the trend of stress strain curve, the increase of water pressure under high circumference pressure will cause certain damage to the circumference pressure. However, with the increase of water pressure to 3 MPa, the high-pressure permeable water in sandstone will block the damage of sandstone and weaken the damage of sandstone. At the same time, the slow decrease of hydraulic conductivity after 3 MPa peak also confirms this understanding.

Through Fig. 3.11a–c, different permeability pressure sandstone mechanical characteristics and permeability changes help with comparative analysis. With the increasing of surrounding pressure from 6 to 8 MPa, the sandstone permeability increases in the stability stage. When the pressure increases from 8 to 10 MPa, the stable stage of sandstone decreases instead, which proves that low confining pressure can play a certain role in stabilizing sandstone, and excessive confining pressure will accelerate the destruction of rock. Through the change rate of hydraulic conductivity after different confining pressure peaks, it can be found that the increase rate of sandstone permeability after peak decreases slowly with the continuous increase of confining pressure during the whole experiment, indicating that confining pressure will have a certain supporting effect on the secondary stability of sandstone after destruction.

According to the above test results, the permeability change of sandstone in the triaxial seepage test is influenced by many factors. The internal is directly affected by the closure of primary fractures and the expansion of shear fractures, and the external is mainly affected by high pressure and high water pressure difference. The water pressure difference and surrounding pressure have different influences on sandstone strength. The increase of low water pressure difference will reduce the strength of sandstone, and the high water pressure difference will strengthen the destructive strength of sandstone. At the same time, the increase of confining pressure will reduce the crushing rate of rock after the failure of sandstone, which has a certain supporting effect on the secondary stability of rock.

3.4 Summary

This chapter describes the mining rock cover rock fissure on the basis of the permeability partition. Through the bench-scale triaxial mechanical test analysis of the first barrier under the influence of mining stress, the deformation damage characteristics and permeability evolution characteristics, for the prevention and control of coal seam roof water–sand inrush disaster and protect ecological fragile mining area to provide theoretical guidance, have the following main observations:

(1) Under the premise of determining the water rich area and water rich strength of the roof of the working face, the main inducing factor of the roof water damage accident in the working face is that the mining stress leads to the fracture expansion of the rock mass and changes the permeability of the rock mass. The rock layer with weak permeability increases the water conductivity, increases the water inflow of the mine, and then increases the risk of roof water damage accidents in the mine.

(2) Sandstone stress-seepage coupling tests under different surrounding pressure sand water pressure differences result in the conclusion that the sandstone permeability changes are mainly affected by internal and external factors. The interior is directly affected by the closure of its original cleft and the expansion of the shear cleft. The exterior is affected by the change of the circumference pressure and the water pressure difference. Under the condition of the same surrounding pressure, the local transformation of brittle to ductility occurs under the action of permeable water pressure, which affects the strength and permeability change of the sandstone itself.

References

1. Chen H, Chen Z, Zheng C et al (2018) Damage evolution process of cement-stabilized soil based on deformation and microstructure analysis. Mar Georesour Geotechnol 36(1):64–71
2. Liang YK (2017) Experimental study on water and sand transport characteristics of rock mass in the caving zone of shallow coal seam mining in the west. China University of Mining and Technology, Xuzhou
3. Liu HL, Wang HF, Zhang DS et al (2014) Study on the development law of water flowing fractured zone in extremely thick coal seam mining at oasis mining area. Legislation Technol Pract Mine Land Reclam 28(01):321
4. Shi XC (2016) Research on the deformation, damage and permeability evaluation of overlying strata in coal mining. China University of Mining and Technology (Beijing), Beijing
5. Wang LY, Liu XF, Tang AT et al (2002) Damage evolution of metallic materials during high temperature plastic deformation. Trans Nonferrous Metals Soc China 03:379–382
6. Wang YG (2016) Study on failure characteristics and mechanism of overburden rock in high-strength mining. Henan University of Technology, Jiaozuo
7. Zhang SC, Li Y, Li J et al (2020) Experimental study on the change characteristics of physical parameters in the process of water inrush and sand burst from mining fractures. J Coal 45(10):3548–3555

Chapter 4
Research on the Influence of Coal Mining on the Evolution of Groundwater Circulation

4.1 Hydraulic Properties of Roof Clay Aquiclude

4.1.1 Mechanical Index of Roof Clay Aquiclude

The clayey soils in the mining area include the Lishi Formation loess and Baode Formation red soil. The clay of Lishi Formation and Baode Formation together constitutes the aquiclude of the mining area. Based on the data of 276 boreholes in the mining area, the thickness of clay is between 0 and 89.78 m. The average is 23.05 m with a standard deviation of 21.39 m.

The thickness of the Lishi loess ranges from 0 to 65.55 m, generally between 5 and 20 m. The lithology is silty clay with a small amount of sandy soil. Columnar joints are developed, and scattered calcareous nodules can be seen. Gullies are developed in the outcropping area. The gully sides often collapse to form steep slopes. The latent erosion can be seen in gully heads. According to the geotechnical test data (Table 4.1), the porosity is 0.754. The plastic limit is 16.86%, while the liquid limit is 28.09%. The natural water content is 12.98%. The liquid index is less than 0.25. The soil is in the state of hard or hard plastic. The mechanical test showed that the loess soil had a certain shear strength and compressive strength. The compression coefficient was 0.163 MPa, whereas the compression modulus was 11.31 MPa. The hydraulic conductivity measured in the laboratory is only 4.47×10^{-5} cm/s. Such a value suggests that the clay has good water barrier ability. The collapsibility coefficient is 0.008, which is non-collapsible. Loess has large porosity, loose structure, and develops upright columnar joints. The loess is easily eroded by surface water flow, resulting in lapses.

The thickness of clay in Baode formation ranges from 0 to 89.55 m, often from 10 to 20 m. The lithology is silty clay. The soil is relatively fine, containing a small amount of sub-sandy soil, the upper part of the ancient soil, and local calcareous tuberculosis stratified distribution. According to the geotechnical test data, the porosity

Table 4.1 Soil geotechnical test results of Jinjie Coal Mine

Soil layer	Soil sample number	Sample depth (m)	Moisture content W (%)	Porosity e_0	Saturation Sr (%)	Liquid limit W_L (%)	Plastic limit W_P (%)	The plastic index I_L	Liquidity index I_L	Water ratio w/W_L	Collapsibility coefficient δ_s	Compression coefficient a_{1-2} (Mpa⁻¹)	Compression moduli Es_{1-2} (Mpa⁻¹)
Loess Q_{2L}	J1302-1	2.40–60.19	13.4	0.957	38	28	16.8	11.2	−0.3	0.48	0.01	0.21	9.23
	J1302-2		12.8	0.923	38	27.4	16.6	10.8	−0.35	0.47	0.005	0.12	16.07
	J1302-3		13.6	0.936	39	28	16.8	11.2	−0.29	0.49	0.008	0.24	8.07
	J1302-4		13.8	0.952	39	27.9	16.8	11.1	−0.27	0.49	0.008	0.13	15.02
	J1302-5		13.4	0.957	38	28.2	16.9	11.3	−0.31	0.48	0.005	0.17	11.51
	J1602-1	0.80–22.90	12.2	0.567	58	28.2	16.9	11.3	−0.42	0.43	0.011	0.19	8.25
	J1602-2		12.7	0.582	59	28.5	17	11.5	−0.37	0.45	0.01	0.13	12.17
	J1602-3		12.5	0.563	60	28.1	16.9	11.2	−0.39	0.44	0.01	0.21	7.44
	J1602-4		12.5	0.555	61	28.6	17.1	11.5	−0.4	0.44	0.008	0.12	12.96
	J1602-5		12.4	0.562	60	28	16.8	11.2	−0.39	0.44	0.007	0.16	9.76
	J1602-6		12.6	0.549	62	28.5	17	11.5	−0.38	0.44	0.004	0.15	10.33
	Average		12.98	0.754	49.33	28.09	16.86	11.23	−0.344	0.463	0.008	0.163	11.31
Red clay N_{2b}	J1302-7	60.19–69.30	12.5	0.789	43	31.9	18.5	13.4	−0.45	0.39	0.01	0.14	12.78
	J1302-8		12.3	0.776	43	31.7	18.4	13.3	−0.46	0.39	0.008	0.1	17.76
	J1302-9		12.6	0.76	45	31.7	18.4	13.3	−0.44	0.4	0.006	0.15	11.73
	J1302-10		12.4	0.757	45	31.5	18.4	13.1	−0.46	0.39	0.007	0.1	17.57
	J1302-11		12.4	0.747	45	31.4	18.3	13.1	−0.45	0.39	0.006	0.13	13.44
	J1302-12		12.8	0.743	47	31.7	18.4	13.3	−0.42	0.4	0.004	0.15	11.62
	Average		12.5	0.762	44.67	31.65	18.4	13.25	−0.45	0.393	0.007	0.128	14.15

is 0.762. The liquid limit is 31.65%, whereas the plastic limit is 18.4%. The natural water content is 12.5%. The liquid index is less than 0.25. Clay is in hard or hard plastic state and has a certain shear strength and compressive strength. It has good water insulation ability.

4.1.2 Water Isolation Analysis of Roof Clay Waterproof Layer

Loess and laterite aquicludes are widely distributed in northern Shaanxi Province. Therefore, the research results of the water-isolation of loess and laterite in the Shennan mining area and the influence of coal mining on their water-isolation performance have direct guiding value to the Jinjie mining area [5, 6, 8].

The Lishi Formation loess and Baode Formation red clay are distributed in the Shennan mining area. The results of clay mineral composition analysis show that the sample contains more smectite, illite, kaolinite, some illite and smectite mixed layer, chlorite and other clay minerals, which determines its good natural water resistance. The hydraulic conductivity of variable head was tested on two layers of clay. The loess was in hard state, and the hydraulic conductivity of unconfined variable head was between 0.04 and- 0.13 m/d. The laterite was in hard plastic state, and the hydraulic conductivity of unconfined variable head was smaller than that of loess, which was between 0.0016 and- 0.017 m/d. The results show that the two layers of loess have certain water isolation under natural conditions. There are many factors affecting the clay permeability, and the variable head test can only reflect such factors as mineral composition and compactness degree. The stress redistribution caused by coal mining causes the additional stress to compress and shear the clay layer, which changes the clay structure and results in the great change of its permeability.

The pre-mining water pressure test was carried out before the first coal seam mining in Hongliulin, Zhangjiamao and Ningtiaota Coal Mines. Field pressure water test was carried out on loess layer and red soil layer, respectively. The test results are shown in Figs. 4.1, 4.2 and 4.3. In the pressure water test of loess, when the pressure is increased from 0.4 to 0.6 MPa, the back flow phenomenon occurs in the borehole, while the leakage of laterite is stable when the pressure is pressurized to 1 MPa, and there is no water return phenomenon. that the test results indicate that the laterite has stronger water isolation than the loess.

When the first mining area of Hongliulin (borehole HL3) and Zhangjiamao Coal Mines (borehole ZJ3) in the Shennan mining area was advanced approximately 100 m, the field piezometric tests were conducted on two layers of clay at the surface of the mining area (fissure penetration area) and at the tension area approximately 50 m from the cutting hole, respectively. In the case of no pressure in the drilling holes of the connected fractured zones in Hongliulin and Zhangjiamao Coal Mines, the leakage is always greater than 100 L/min (Fig. 4.4, the curve of water level change in 1 h after 20 m of pressure less water injection in loess layer). i.e. The loss of stability of the waterproof clay layer due to shear failure is manifested by a dramatic change in hydraulic conductivity of several orders of magnitude. However,

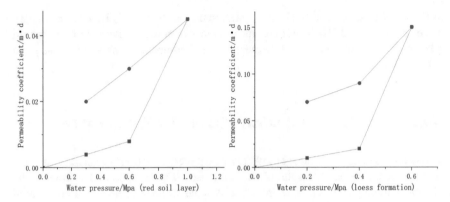

Fig. 4.1   Test results of pre-mining pressure water in Hongliulin Mine

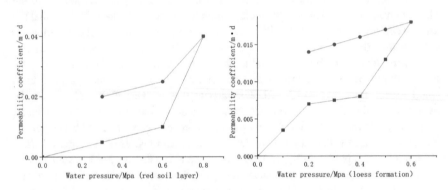

Fig. 4.2   Test results of pressure water before mining in Zhangjiamao Mine

Fig. 4.3 Results of
pre-mining laterite water
pressure test in Ningtiaota
mine

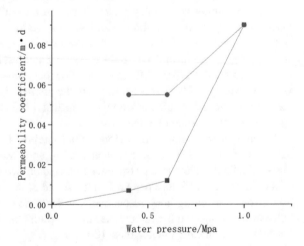

in the compressive water test of loess and laterite in the tensile area, the borehole leakage is greater than 50 L/min in the case of the loess without water pressure, while the laterite maintains good water isolation. The hydraulic pressure test results are shown in Fig. 4.5, indicating that the laminar hydraulic conductivity of the laterite after mining is slightly lower than that before mining.

The results of water pressure test above the goaf of caragana tower (NT1 borehole) show that (Fig. 4.6), the loess and laterite have obvious changes compared with those before mining. However, both of them maintain a certain water insulation capacity. Due to the thick overlying bedrock and soil layer on the first mining face of Ningtiaota, the water pressure test shows that there is no penetrating fracture. The laterite water pressure test shows that the hydraulic conductivity is lower than that before mining, and the water isolation is stronger. The hydraulic conductivity of the loess also increased significantly, and the water isolation was weakened.

In addition to the pressure water test, the triaxial full stress–strain permeation test was carried out on the undisturbed loess and laterite. The confining pressure is

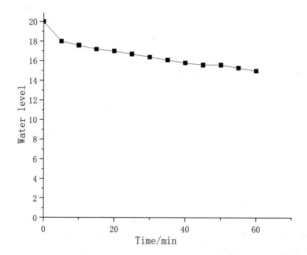

Fig. 4.4 Change curve of loess seepage water level after mining in Zhangjiamao mine

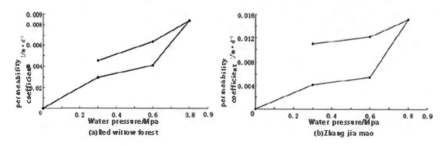

Fig. 4.5 Post-harvest tensile zone laterite water pressure test results

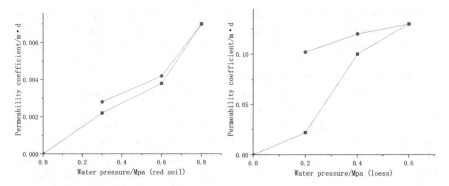

Fig. 4.6 Experimental results of upper pressure water in mined-out area of coal seam in Korshinskaota Mine

reduced to the original stress state of soil, which is 0.4 MPa for loess and 0.6 MPa for laterite. The total stress–strain loading test of the loess and laterite shows that with the increase of axial stress, the clay mainly undergoes plastic deformation. The permeability gradually decreases until the failure of clay sample. The permeability decreases more obviously after the sample failure.

The variation characteristics of the water resistance of the fracture penetration zone are that the mining of the shallow coal seam leads to the overall cutting of the bedrock, and the overlying clay layer also has shear failure, resulting in penetrating cracks. Based on the above laboratory and in-situ tests, the variation characteristics of the water-isolation of clay layer in the Shennan mining area are shown in Table 4.2.

(1)   Variation characteristics of water-resistance in fractured t zone

The mining of shallow coal seam leads to the overall cutting of bedrock. The overlying clay layer also appears shear failure, resulting in penetrating fractures. Based on previous experimental studies, the water isolation characteristics of the clay layer can be summarized as follows:

The clay layer in fissure penetrating zone is in complete unloading unconfined state. The field water pressure test shows that the clay is broken, and the permeability is several orders of magnitude higher than that before mining. The laboratory triaxial permeability test shows that the soil sample has a dramatic increase in flow at the shear failure when the confining pressure is smaller than the water pressure. The formation of the water-conducting zone after 3–1 coal mining is not conducive to the protection of the upper sand aquifer in the Jinjie mining area and is not conducive to safe production.

Previous laboratory experiments on the unloading of damaged samples show that the confining pressure has a certain control on the permeability of shear-damaged clay samples. However, when the water pressure is greater than the confining pressure, the instability of the clay layer occurs. The field water pressure test shows that the

Table 4.2 Clay hydraulic conductivity in fracture penetration zone

Soil sample		Variable head determination of K value/(m/d)	K value determined by pressure water test (m/d)		K value determined by triaxial permeation test (m/d)
			Laminar flow	Turbulent flow	
Loess	Before mining	0.04–0.13	0.00643–0.01369 (water pressure ≤ 0.4 MPa)	0.14532 (water pressure 0.6 MPa)	0.019 (confining pressure 0.4 MPa)
	After mining	–	Flow greater than 100 L/min (water pressure < 0.1 MPa)		Flow surges beyond range (confining pressure < 0.1 MPa)
Laterite	Before mining	0.0016–0.017	0.00326–0.00705 (water pressure ≤ 0.6 MPa)	0.047 (water pressure 1 MPa)	0.0061 (confining pressure 0.6 MPa)
	After mining	–	Flow rate greater than 50 L/min (water pressure < 0.1 MPa)		Flow surges beyond range (confining pressure < 0.15 MPa)

obvious erosion after the water pressure exceeds the confining pressure can be used as evidence.

The loading test of clay shows that the soil sample mainly undergoes plastic deformation in the whole stress–strain process. The hydraulic conductivity decreases with the increase of strain. Even if the clay sample is damaged by shear, as long as the confining pressure does not completely disappear, the hydraulic conductivity will continue to decrease, indicating that the clay has the characteristics of self-healing in the confining pressure control range. It is therefore meaningful for the self-healing of the aquiclude in the Jinjie mining area.

(2) The variation of water isolation of clay layer before and after mining in tensile area

According to previous results, in the stretching area approximately 50 m away from the open-cut of shallow coal seam (Hongliulin and Zhangjiamao first mining face), the variation characteristics of clay layer's water-isolation before and after mining are summarized in Table 4.3.

The field water pressure test shows that the variation characteristics of hydraulic conductivity of the loess and laterite in the tensile area are different. The loess exposed by drilling was in shear fracture state, indicating that it was subjected to shear failure. The interface between the fractures and the red soil layer was pinched out. The red soil layer was intact and located in the overall subsidence zone. After coal seam mining, the loess permeability increases by several orders of magnitude, while the laterite permeability decreases. In the case of laminar flow, the hydraulic conductivity of laterite decreases slightly, while in the case of turbulent flow, when the

Table 4.3 Clay hydraulic conductivity in tensile zone

Soil sample		Variable head determination of K value/(m/d)	K value determined by pressure water test (m/d)		K value determined by triaxial permeation test (m/d)
			Laminar flow	Turbulent flow	
Loess	Before mining	0.04–0.13	0.00643–0.01369 (water pressure ≤ 0.4 MPa)	0.14532 (water pressure 0.6 MPa)	0.019 (confining pressure 0.4 MPa)
	After mining	–	Flow greater than 100 L/min (water pressure < 0.1 MPa)		Flow surges beyond range (confining pressure < 0.1 MPa)
Laterite	Before mining	0.0016–0.017	0.00326–0.00705 (water pressure ≤ 0.6 MPa	0.047 (water pressure 1 MPa)	0.0061 (confining pressure 0.6 MPa)
	After mining	–	0.00298–0.00412 (water pressure ≤ 0.6 MPa	0.00841 (0.8 MPa)	0.0058–0.0163 (confining pressure ≥ 0.15 MPa)

water pressure decreases by 0.2 MPa, the hydraulic conductivity of laterite decreases by an order of magnitude.

Indoor loading experiments on intact red soil samples show that when the axial pressure increases, the red soil exhibits strong plastic deformation, which reduces its hydraulic conductivity. When the clay is not unstable within the confining pressure control range, the consolidation compression of the clay in the tensile area is significant, i.e., it shows plastic deformation. The reduction of permeability is conducive to coal mining with water conservation.

(3)   Variation characteristics of water isolation during excavation of near shallow coal seam

In the case of the first mining face of Ningtiaota Coal Mine, the fractures did not develop to the soil layer. The aquifuge did not lose stability. After mining, the water isolation performance became weaker, and the hydraulic conductivity increased (Table 4.4). The field water pressure test shows that even if the water flowing fractured zone does not pass through the loess and laterite on the goaf of aquiclude, the variation characteristics of hydraulic conductivity are quite different. The loess exposed by the borehole is in a loose state, indicating that it is subjected to unloading. When the coal seam is in laminar flow state after mining, the hydraulic conductivity of loess increases by one order of magnitude. However, the erosion space caused by the increase of water pressure after mining is equivalent to that before mining, which shows that the loess is in a loose and unbroken state. The red soil layer is completely located in the overall subsidence zone. There is obvious consolidation compression phenomenon. Its permeability decreases and with the increase of water pressure, the hydraulic conductivity does not change much.

Table 4.4 Hydraulic conductivity change of clay layer above mined-out area near shallow coal seam

Soil sample		Variable head determination of K value/(m/d)	K value determined by pressure water test (m/d)		K value determined by triaxial permeation test (m/d)
			Laminar flow	Turbulent flow	
Loess	Before mining	0.04–0.13	0.00643–0.01369 (water pressure ≤ 0.4 MPa)	0.14532 (water pressure 0.6 MPa)	0.019 (confining pressure 0.4 MPa)
	After mining	–	0.0242–0.121 (water pressure ≤ 0.4 MPa)	0.1313 (water pressure 0.6 MPa)	0.0055–0.0077 (confining pressure ≥ 0.1 MPa)
Laterite	Before mining	0.0016–0.017	0.0064–0.0134 (water pressure ≤ 0.6 MPa)	0.087 (water pressure 1 MPa)	0.0061 (confining pressure 0.6 MPa)
	After mining	–	0.00298–0.00412 (water pressure ≤ 0.6 MPa)	0.00841 (0.8 MPa)	0.0058–0.0163 (confining pressure ≥ 0.15 MPa)

(4) Influence and change on water isolation performance of clay layer before and after mining in Jinjie Mine

Compared with the experimental results in the Shennan mining area, the following understanding can be obtained:

In the thin bedrock area of the second disk area, the water-isolated clay layer in the crack through area is in the state of complete shear failure and unconfined after coal seam mining. After the clay layer loses water isolation capacity, the sand water is communicated. The river water in the Heze gully and Qingcaojie gully is closely related to the sand water, resulting in the increase of mining drainage flow. Such a condition is not conducive to safe production and surface water protection.

The two aquifuge clay layers in the tension zone show different mining variation characteristics. The loess is in a fractured state due to the shear action of the ground tensile fractures. The hydraulic conductivity shows several orders of magnitude changes after mining. The laterite is located in the overall subsidence zone and exhibits large plastic deformation under the influence of accessory stress. Its permeability decreases after mining, which is in favor of water-preserving coal mining. Because of the absence of laterite at 218–223, 501–502 and 505–508 working faces in the thin bedrock area of Hezegou Basin, the damage on pore water is permanent at these working faces after mining.

The test shows self-healing of the damaged clay layer, i.e., permeability decrease, within a certain confining pressure control range, when the clay layer damaged by shearing is stable, i.e. Therefore, the working face of the thin bedrock area of the second disk area should not be mined concurrently by multiple working faces, and the production of a single working face should be rapidly advanced.

The technology of backfilling goaf should be adopted to realize safe water-conserving coal mining in Jinjie Mine.

4.2   Characteristics of Competent Bedrock Waterproof Layer in Thin Bedrock Area

According to the degree of weathering, the bedrock above 3^{-1} coal in the second panel area is divided into weathered bedrock and competent bedrock. The weathered bedrock is located at the top of bedrock. The thickness varies from 3.6 to 44.5 m in the Hezegou area, generally between 10 and 20 m. The thickness is between 0 and 26 m in the Qingcaojiegou area, generally less than 10 m. The weathering results in fissures in the weathered rock layer, whereas the weathering degree gradually weakens from top to bottom. Compared with the original rock, the strength of the weathered bedrock is reduced. Unstable minerals such as feldspar are weathered into clay minerals, resulting in a decrease in rock bulk density and a permeability increase, porosity increases, water content increases, and strength decreases. Most rocks disintegrate or segregate along the cracks in a short time when they encounter water. According to the rock quality classification standard, it belongs to poor quality (grade IV), which is defined as a rock mass with poor integrity. Under saturated conditions, the regolith belongs to the aquifer and does not have the ability of water isolation.

The thickness of competent bedrock in the thin bedrock area of the second panel area is relatively small. The thickness is 1.1 m in the hezegou area, while the thickness is generally less than 10 m in the Qingcaojiegou area. According to the rock mechanics test (Table 4.5), the average uniaxial saturated compressive strength is 19.20 MPa. The roof of 3^{-1} coal seam in the mining area is divided into three categories according to rock types: argillaceous rock roof, siltstone roof, and sandstone roof. Most of the direct roof of the coal seam in the thin bedrock area of the second panel area is the siltstone roof.

The direct roof of 3^{-1} coal seam is composed mainly of siltstone. According to the observation of roof rock fractures, there are two groups of fractures in the mining area. The average strike orientation of fractures in Group I is N81 °E, and the average fracture spacing is 0.485 m. The average strike direction of group II fractures is N50.5°W, and the average fracture spacing is 0.44 m. The cracks are not well developed, and the saturated uniaxial compressive strength of roof rock is 26.8 MPa. The first collapse step of the direct roof is 12.19 m, as measured at 93103 working face. According to the Standard for Classifying the Top of Coal Mining Face in Slowly Inclined Coal Seams (MT554-1996), the direct top of 3^{-1} coal seam is determined to be the type 2 intermediate stable roof. Combined with the observation results of mine pressure on 93103 working face, the first step distance

Table 4.5 Test results of physical and mechanical parameters of intact rock above coal in typical borehole 3-1

Sample	Formation	Sample depth/m	Natural density g/cm³	Dry density g/cm³	Porosity	Specific gravity	Moisture content (%)	Compressive strength		Softening coefficient	Natural shear		Tensile strength/MPa
								Drying/MPa	Saturation MPa		Φ	C/MPa	
J913	Medium sandstone	87.00–88.11	2.38	2.36	8.19	2.57	0.52	19	12.9	0.68	38.25	1.77	0.93
	Medium sandstone	88.12–89.31	2.33	2.31	8.2	2.52	0.45	20.3	15.5	0.76	35.42	2.9	0.98
	Gritstone	93.56–94.66	2.38	2.33	9.35	2.57	0.31	27	20.7	0.77	30.33	3.48	1.11
	Gritstone	98.27–99.08	2.36	2.31	9.37	2.55	0.28	16.2	12.9	0.8	37.95	2.24	0.82
	Medium sandstone	103.78–104.46	2.39	2.35	8.96	2.58	0.26	34.8	26.8	0.77	31.74	4.83	1.91
	Siltstone	109.51–110.13	2.44	2.42	8.17	2.64	1.56	24	17	0.71	36.24	4.6	1.36
	Fine sandstone	113.00–113.61	2.29	2.26	8.62	2.47	0.86	30.5	21.4	0.7	35.26	5.53	1.58
	Fine sandstone	129.52–130.34	2.31	2.29	8.21	2.49	1.54	26	17.4	0.67	36.64	6.27	1.44
	Fine sandstone	136.16–137.00	2.33	2.32	7.8	2.52	2.12	30.9	24.2	0.78	34.86	4.21	1.69
	Siltstone	139.40–139.80	2.36	2.33	8.58	2.55	2.27	27.6	19.8	0.72	35.15	5.11	1.48
Sample	Rock	Sample depth/m	Natural density g/cm³	Dry density g/cm³	Porosity	Specific gravity	Moisture content (%)	Compressive strength		Softening coefficient	Natural shear		Tensile strength/MPa
								Drying/MPa	Saturation/MPa		Φ	C/MPa	
J1212	Gritstone	114.80–116.38	2.39	2.38	7.79	2.58	0.55	30.7	26.5	0.86	30.95	4.92	1.45

(continued)

Table 4.5 (continued)

Sample	Rock	Sample depth/m	Natural density g/cm³	Dry density g/cm³	Porosity	Specific gravity	Moisture content (%)	Compressive strength		Softening coefficient	Natural shear		Tensile strength/MPa
								Drying/MPa	Saturation/MPa		Φ	C/MPa	
	Siltstone	132.10–133.55	2.46	2.39	10.04	2.66	1.72	27.3	19.2	0.7	37	3.46	1.36
	Fine sandstone	143.00–143.90	2.44	2.34	11.2	2.64	3.79	31	20.6	0.67	36.6	4.31	1.62
J911	Siltstone	103.96–105.00	2.4	2.37	8.56	2.59	1.63	21.4	16.1	0.75	27.91	4.44	1.6
	Fine sandstone	105.40–106.62	2.37	2.33	8.97	2.56	0.75	18.5	15.3	0.83	28.04	4.54	1.15
	Siltstone	107.14–107.84	2.35	2.31	8.98	2.54	1.7	21.3	18.4	0.87	34.63	3.2	1.46
	Fine sandstone	122.00–122.99	2.42	2.34	10.47	2.61	2	24.1	19.8	0.82	35.09	5.56	1.73
	Fine sandstone	129.30–130.10	2.33	2.31	8.2	2.52	0.46	24.9	21.1	0.85	36.13	4.56	2.24
Minimum			2.29	2.26	7.79	2.47	0.26	16.2	12.9	0.67	27.91	1.77	0.82
Maximum			2.46	2.42	11.2	2.66	3.79	34.8	26.8	0.87	38.25	6.27	2.24
Average			2.37	2.34	8.87	2.56	1.27	25.31	19.20	0.76	34.34	4.22	1.44

of basic roof is 32.09 m. The 3^{-1} coal seam should be the obvious roof of grade II weighting.

According to the field observations on 31202 working face, the ground elevation of the driving section is approximately between 1194.4 and 1200 m. The terrain is flat. The surface is covered by aeolian sand. There are ancient gullies in the lower part. The ancient gullies are loose and thick. The bedrock is thin. The stratum from top to bottom is:

(1) Sand layer: the layer thickness varies from 15 to 90 m. It gradually thickens along the excavation direction. The end of the designed roadway is located in the ancient gully with a layer thickness of 90 m. The sand layer aquifer is approximately 30 m thick with high water yield.

(2) Soil layer: the layer thickness varies from 0 to 10 m and gradually becomes thinner along the excavation direction. There is a "skylight" area at the end of the designed roadway.

(3) Weathered bedrock: the layer thickness is approximately 10 m. The lithology is composed mainly of medium and fine sandstone. The fissures are developed. The whole layer contains bedrock pore and fissure water, with strong water richness and sufficient water supply. In the "skylight" area, it is connected to the sand layer aquifer, which enhances the water richness of the formation.

(4) 3^{-1} Coal competent bedrock: The layer thickness varies from 10 to 30 m. It gradually becomes thinner along the advancing direction. The thinnest part is located in the ancient gully. The lithology is composed mainly of fine, silt-stone and mudstone. The direct roof of the coal seam is mostly mudstone, with developed fractures and slip surface, which is prone to fall. As of February 24, 2011, the 31202 transportation gateway (31203 return air gateway) of Jinjie Coal Mine has been tunneled for approximately 5200 m in total, and the 31202 working face has been tunneled for approximately 283 m. Multiple roof falls have occurred between the 110 lane of the gateway and the working face. See Table 4.6 and Fig. 4.7 for details.

A large number of studies have shown that mining-induced cracks in roof are either upward or downward. The upward fracture is the water-conducting fracture formed after mining, whereas the downward fracture is the downward tensile fracture formed by surface tension. The upward fractures and downward fractures determine the water isolation of the roof strata. Mining height has obvious influence on the development height of upward fractures. Studies by Zhang Wenzhong and Huang Qingxiang show that the relationship between the development height of upward fractures and the mining height of coal seam is positively correlated and linear [4, 9] (Fig. 4.8).

As the mining stratum bends and sinks, there is a tension area in the clay layer. Therefore, under the action of head pressure, the phreatic water in the downward fissure will permeate to both sides of and below the fissure. The clay aquifuge has strong plastic deformation and water-induced swelling, resulting in the expansion of the soil on both sides of the fissure. The fissure will be closed when the expansion is greater than the width of the fissure. Because the downward fissure expands when

Table 4.6 31202 working face crossheading caving observation results table

Serial number	Position	Caving range			Falling lithology and thickness (from top to bottom)	Remake
		Lenth/m	Width/m	Heightm		
1	31203 Huishun 110 roadway mouth	7.5	5.4	0.7	Mudstone 0.35 m Coal line 0.05 m Mudstone 0.30 m	Lane height of 2.8 m, top coal is approximately 0.4 m, cutting bottom approximately 0.4 m
2	31203 Huishun 110 lane west 20 m to 111 lane mouth	25	0.4–1.03	0.3	Top coal 0.3 m	Coal seam cracks are more developed, and the roof is not exposed
3	31203 Huishun 111 lane west 11 m to the tunnel face	7.5	5.4	1.5	Siltstone 0.60 m Mudstone 0.45 m Coal line 0.05 m Mudstone 0.40 m	Lane height of 2.8 m, tunneling along the bottom, roof caving roadway tunneling after cutting approximately 1.5 m
4	31202 Yunshun 111 lane west 2 m to the tunnel face	5	5.4	0.8	Mudstone 0.35 m Coal line 0.05 m Mudstone 0.40 m	Lane height of 2.8 m, tunneling along the bottom, roof caving roadway tunneling when cutting approximately 0.8 m
5	31202 Yunshun and Pipe Link Lane (111 Link Lane)	12	5	0.4	Mudstone 0.4 m	Lane height of 2.8 m, tunneling along the bottom

(continued)

Table 4.6 (continued)

Serial number	Position	Caving range			Falling lithology and thickness (from top to bottom)	Remake
		Lenth/m	Width/m	Heightm		
6	31202 face pipe 111 lane to the tunnel face	5	5.4	1	Mudstone 0.5 m Coal line 0.05 m Mudstone 0.45 m	Lane height of 2.8 m, tunneling along the bottom, roof caving roadway tunneling when cutting approximately 1 m

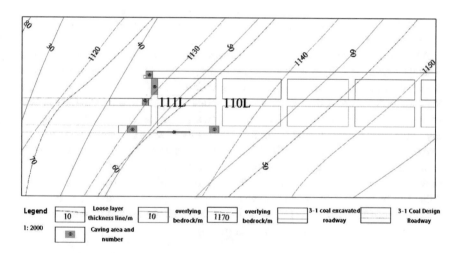

Fig. 4.7 31202 working face roof caving distribution map

encountering water, it has a bridging property and a positive effect on the stability of the aquiclude.

According to the relevant literature [1–3, 5, 7], for silty clay with large plasticity, cracks are generally generated when the tensile deformation value exceeds 6–10 mm/m. For sandy clay with poor plasticity, cracks occur when the tensile deformation reaches 2–3 mm/m. The downward fissure is generally wedge-shaped. Its width becomes smaller with the deep until it pinches out at a certain depth. According to a large number of measurements from trenching, the downward fissure pinches out within 10 m in depth when the surface alluvium is thick. The extension depth of the fissure will not exceed twice the mining height in general. Therefore, reducing the

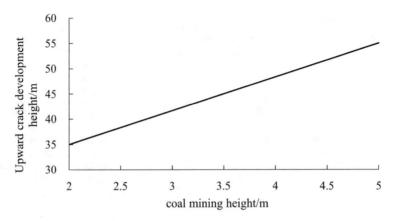

Fig. 4.8 Relationship between mining height and upward fissure height

mining height is beneficial to the stability of the water resisting layer. If the mining height is 3 m, the thickness of the water resisting rock group should be greater than 54 m to prevent the pore water damage.

As sand layer is widely distributed on the surface of Jinjie Coal Mine, the thickness of sand layer in the thin bedrock area is generally greater than 8 m. The mining height of the working face is 3.5 m. According to the calculation of 2 times the mining height of the downward crack depth, the downward crack depth of Jinjie Coal Mine after mining of coal seam 3^{-1} is 7 m. Even if the crack enters the soil layer, the water will randomly enter the crack, and the crack will be closed after encountering water.

According to the "Code for Coal Pillar Retention and Coal Pressing Mining for Buildings, Water, Railways and Main Shafts", the condition under which the overlying sand layer water is not filled with water in the 3^{-1} coal seam of Jinjie Mine is that the water-resisting layer is greater than 61 m. As a consequence, the water resisting property of the intact bedrock in the thin bedrock area of the ancient gully in the second panel area will be lost after mining.

4.3   Stratigraphic Combination Characteristics of Roof Waterproof Layer of Coal Seam 3^{-1} in Jinjie Mine

In the thin bedrock area of ancient gullies in the second panel, there are Yan'an formation, Zhiluo formation, Baode formation, Lishi formation, Salawusu formation, aeolian sand and local Quaternary alluvial proluvial sand layer above coal seam 3^{-1}. The Yan'an formation and Zhiluo formation are distributed in working face 218–223 and all working faces in Wupan district. The sand layer is mainly distributed in the low-lying areas.

According to the lithological characteristics of borehole logging, the strata in the thin bedrock area are combined and stratified. The strata above coal seam 3^{-1}

can be divided into competent bedrock, weathered bedrock, cohesive soil and sand. Figure 4.9 shows the thickness and distribution of each layer.

The general characteristics of 3^{-1} coal and aquifer in the thin bedrock area of Jinjie Mine are as follows. The thickness of aquifer is greater than that of aquiclude. The height of water conductive zone is greater than that of the aquiclude. And water and coal coexist. According to the contact relationship among complete bedrock, weathered bedrock, cohesive soil and sand layer, the combination form of 3^{-1} coal roof stratum is divided into the following four modes:

Mode 1: Fissure water filling type. The aquiclude of the Baode formation and Lishi formation divide the pore water and weathered fissure water. Although the mining fissure affects the integrity of the cohesive soil, the cohesive soil has healing ability, and the drainage capacity of the working face is small. This mode is mainly distributed between 505 and $-$ 519 working faces.

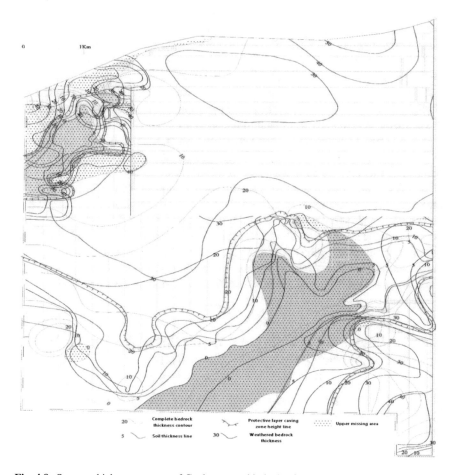

Fig. 4.9 Stratum thickness contour of Guchonggou thin bedrock area

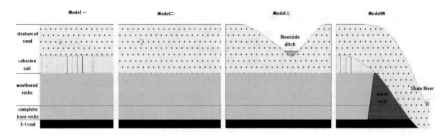

Fig. 4.10 Stratum combination model of Guchonggou thin bedrock area

Mode 2: Pore water and weathering fracture water co-fill type. The Baode group aquiclude and Lishi group aquiclude are missing between the pore water and weathered fissure water aquifers. The distribution area is at the skylight. Although the mining fissure destroys the integrity of bedrock, weathering fracture water and sand layer pore water enter the working face, and the working face drainage is large. It is mainly distributed in 502 -and 503 working faces and 218–223 working faces.

Mode 3: River leakage type. A river passes over the working face. The river water is closely related to the pore water. The lack of the Baode group and Lishi group aquiclude between pore water and weathered fissure water aquifers occurs in the skylight distribution area. When the mining-induced fissure destroys the integrity of the bedrock, the river water seeps into the sand layer to supply the pore water, the weathering fissure water and the sand layer pore water, ultimately enter the working face. The working face has a large drainage volume and a long drainage time. This mode is mainly distributed in 218–220 working faces.

Mode 4: pore water and bedrock fissure water composite filling type. Located in the burnt rock distribution area on both sides of the Qingcaojiegou, the thickness of the cohesive soil is thin or missing, and the burnt rock is closely related to the pore water hydraulic connection. The burnt rock fissure water, pore water and weathered fissure water aquifer together constitute the water filling sources of the working face in the mining area. The working face has a large drainage volume and a long drainage time. This mode is mainly distributed in 507–519 working faces (Fig. 4.10).

4.4   Variations in the Hydraulic Properties of the Top Plate Water Barrier Before and After Mining

4.4.1   Groundwater Level Change in the Case of Evacuation and Drainage

4.4.1.1   Working Faces 101–110 in the First Panel Area

The surface of 31101-104 working face is covered by wind-deposited sand in most of the area . The local soil layer is exposed. The top-down structure of the stratum is as follows. The thickness of the sand layer varies from 0 to- 18 m with most areas less than 5 m. The soil layer has a thickness between 30 and- 60 m, and the overall trend is thinning from the middle of the area to the perimeter. The thickness of the weathered bedrock ranges from 10 to 34 m. The overall trend is thickening from 31101 to 31104 surfaces. The competent bedrock is approximately 30 m thick and relatively stable.

The Jinjie Coal Mine 93103 working face is very close to 93101 and 93102 working faces. The geological and hydrogeological conditions are similar. The stratigraphic structure consists of approximately 10–20 m of wind-bound sand, 40–60 m of loess, and weathered bedrock and competent bedrock below the loess. A total of 66 water release boreholes were arranged, with a total of 6459.85 m of drilling footage. The boreholes are constructed to the end of 2 m inside the loose layer. After the completion of the construction in the orifice down to place 2 m long $\varphi 108$ mm sealing pipe, a $\varphi 108$mm gate valve and $\varphi 89$ mm water hose are installed. There are 19 boreholes without water at the beginning. Other boreholes with water have large water gushing in the range between 0.8 and 45.8 m^3/h. The largest water gushing borehole 9310 in 93104 back to the Shun 15 joint lane. Larger water inflow occurs in boreholes from cutting-eye, 45–55 joint lane of 93104 return air gateway, and 10–17 joint lane of 93104 return air gateway. The water inflow from a single hole is greater than 5 m^3/h. The total water gushing of all boreholes is 398.55 m^3/h. By 10 December 2007, 8 boreholes had reduced water inflow to zero; 8 boreholes had the same water inflow as the initial water inflow, and the rest had water inflow Reductions. The total water inflow of all boreholes was 245.4 m^3/h, and the cumulative water discharge was 1,230,997.2 m^3. The 93101 working face had a typical surge of 50 m^3/h during the actual recovery process, with a maximum of 360 m^3.

The 93104 working face of Jinjie Coal Mine is 3,401 m long and 369 m wide. The back mining started at the end of May 2008. The working face is overlain by the Quaternary Holocene wind-deposited sand (Q_4^{eol}). The thickness of the sand layer varies from 0 to 10 m whereas the soil layer has a thickness between 30 and 60 m. The soil thickness gradually thins from the middle to both sides of the working face. The weathered bedrock gradually thin from 35 m in the middle of the working face to 10 m both sides. The competent bedrock of 3^{-1} coal has a thickness between 25 and 35 m, gradually thinning from the return channel to the cutting-eye. The 3^{-1} coal thickness varies from 3.20 to 3.51 m. The elevation of the coal seam floor varies from 1144 to 1109 m. The coal seam is inclined to the west, and the direction of excavation is mainly negative slope advance. The No. 15–17 observation boreholes reveal that the thickness of the sand layer ranges from 0 to 10 m. The soil layer ranges from 30 m thick to 60 m thick. the thickness of the bedrock overlying the 3^{-1} coal varies from 39.3 to 90 m. The thicker area is located between 29 and 35 joint lanes in the working face back to the Shun, whereas the thinner area is located between 45 and 47 joint lanes in the working face to the Shun. The elevation of bedrock surface varies from 1160.12 to 1210 m. The lower bedrock surface is in the return channel area, and the higher surface is between the working face cut-hole area and the back-shun 30–35 joint lane. The 93104 working face started to construct water release borehole on December 7, 2007, and ended on May 19, 2008. A total of 55 water release boreholes were constructed, and each water release borehole was drilled at 50–60° angle to the working face. The final locations of the water release boreholes are in the soil layer. The water release layers are the pore-fracture water of weathered bedrock of Zhiluo Group (J2z) and the fracture water of Yan'an Group (J1-2y). The data from observation boreholes 15–17 shows that the loose sand layer of the 93104 working face does not contain water, and the main water-bearing layer is the pore fracture water of the weathered bedrock of Zhiluo Group, which is confined. After the working face is mined, the aquifer is the direct water filling source. Its recharge mode is to receive regional lateral recharge, from high to low transport along the top surface of competent bedrock to exudation of the Shamu River Ditch. When the face was mined to 36 m away from the cutting-eye, the old roof was pressurized for the first time. There were many water sprays in the working face, and the water inflow amounted to 130 m^3/h. When the large area of the working face was pressurized, the water inflow increased to 300 m^3/h. According to the past experience, after the first mining face is mined, other subsequent work surface will gradually reduce the amount of water gushing. However, the water volume of the 93102 working face is 210 m^3/h, and the maximum water volume of the 93103 working face is 501 m^3/h. This indicates that there is a strong water-rich area in the first mining area. By May 28, 2008, the cumulative water discharge was 1,306,539.6 m^3. Some of the boreholes with large cumulative water discharge were in the initial mining area of the working face, and these boreholes were completed early and the cumulative water discharge time were more than 200 days. The initial gushing volume of each water release borehole varies from 2.4 to 45 m^3/h, and the total gushing volume of all drill holes is 731.9 m^3/h.

By July 15, 2008, the water inflow of a single borehole ranged from 0 to 28.8 m^3/h. When the working face advanced 406 m, the water inflow of the working face was between 150 and 200 m^3/h. The total water inflow of all boreholes was 209.6 m^3/h, whereas the water inflow of 15 boreholes, numbered T1–T14 and T40, was reduced to zero. Borehole T40 had an initial gushing water of 5.1 m^3/h. The gushing water was reduced to 0 after 5 days of water discharge. The aquicludes in this area are soil layer and siltstone, sandy mudstone and mudstone in 3–1 competent bedrock. According to the statistics of drainage borehole data, the top interface elevation of the overlying bedrock ranges from 1160.12 to 1196.1 m. Although the observation borehole 15 (102.9 m from the cutting-eye) and the observation hole 17 (2957.7 m from the cutting-eye) have no water, there is still water in observation borehole 16, which is 1445.1 m from the cutting-eye. On August 27, 2008, the water level was 1184.07 m, which was higher than the lowest elevation of 1160.12 m on the bedrock roof interface on the working face. The fracture water was still confined, indicating that the water isolation of the water-resisting strata was not completely destroyed (Fig. 4.11).

Geological overview of the 105–106 working faces: The surface is covered by aeolian sand. The stratum structure from top to bottom consists of a sand layer of 5–50 m thick with an overall thinning trend from the middle to both sides of the working face. The thickness of the soil layer varies from 0 to 65 m. There are "skylights" in some places. The weathering bedrock has a thickness ranging from 5 to 55 m thick.

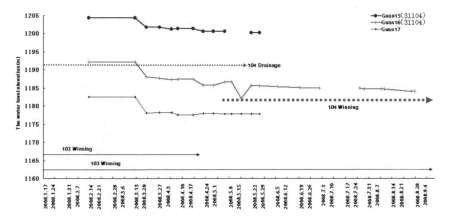

Fig. 4.11   104 comparison curve of drainage and water level of working face

The whole layer contains water, and the overall trend is thickened from the middle of the working face to both sides. The competent bedrock has a thickness between 25 and 32 m with an average of 30 m. The first working face of 31105 was tested in April 2009 and formally mined in July 2009. The maximum water inflow during mining was 1000 m^3/h. The water inflow increased with the increase of goaf area on the whole. The water inflow was over 900 m^3/h when passing the Guchong gully in the south.

The 31105 working face of Jinjie Coal Mine is 5253 m long and 203 m wide. The surface of Jinjie Coal Mine 31105 is covered by the Quaternary aeolian sand. The terrain is relatively flat with little fluctuation. From top to bottom, there are Quaternary Holocene eolian sand (Q4eol), Series Salawusu Formation (Q3S), Series Lishi Formation (Q2L), Pliocene Baode Formation (N2b), Jurassic Middle Zhiluo Formation (J2Z), and Middle Lower Yan'an Formation (J1-2y). The Quaternary Holocene eolian sand layer has a thickness ranging from 5 to − 50 m. The sand layer at the initial mining position of the cut hole is stable in thickness with a variation from 20 to − 22 m. The sand aquifer is approximately 10 m thick. Two ancient gullies (called North Gully and South Gully, respectively) are developed at 400–800 and 2000–2600 m away from the cut hole. The thickness of the eolian sand in the North Gully varies from 30 to 40 m. The thickness of the aquifer ranges from 5 to 20 m. The thickness of the eolian sand in Nangou varies from 20 to − 50 m. The thickness of the aquifer is between 10 and − 42 m. Except for the two ancient gullies, the thickness of the aeolian sand in other sections is thin, ranging from 0 to − 10 m. The thickness of the overburden soil layer of working face 31105 varies from 0 to − 60 m. The overall trend is from cut hole to back track channel. The thickness of the soil layer at the initial mining section of the cut is between 3 and − 4 m. The thickness of the soil layer at the bottom of the north ditch is 18 m, and the thickness becomes 8 m towards both sides. The thickness at the bottom of the south ditch is 9 m, and it becomes thinner to 0 on both sides. The thickness of the weathered bedrock of Zhiluo Formation on the roof of 31105 working face varies from 5 to − 55 m, which is the main aquifer of the working face and distributed throughout the area. The aquifer thickness at the initial mining section of the cut is approximately 54 m. The stratum in the two ancient gullies is thin due to denudation. The thickness of the aquifer in the north gully is varies from 14 to − 20 m, whereas the thickness in the south gully varies from 5 to − 15 m. The thickness of the aquifer between the two ancient gullies is between 15 and 50 m. The thickness of the aquifer from the south gully to the back channel varies from 30 to 55 m. The thickness of the competent bedrock of coal 3–1 varies from 25 to − 32 m with an average thickness of 30 m.

The construction of the 147 exploratory drainage boreholes at 31105 working face started on June 27, 2008 and completed on April 15, 2009. The water inflow of some boreholes exceeded 30 m^3/h with the maximum of 70 m^3/h, The flow affected the safety of construction. The final borehole is in the weathered bedrock. The initial water inflow of the borehole was between 7 and − 78 m^3/h, and the total initial water inflow was 4,050 m^3. On July 9, 2009, during the initial weighting of the basic roof, the water inflow of the boreholes varied from 0 to − 26 m^3/h. The total water inflow of all drainage boreholes was 1164 m^3. The cumulative water outflow of exploratory drainage boreholes was approximately 7.3128 million m^3. Eighteen of the water exploration and drainage boreholes were dried out, including 14 boreholes located in the goaf and four boreholes (T133, T134, T135 and T136) adjacent to goaf in the 31106 Huishun 100–101 cross roadway. Of the exploratory drainage boreholes in the 31105 working face, the boreholes with water inflow of less than 20 m^3/h account for 20%. The boreholes with water inflow between 20 and 40 m^3/h account for 69%, while the boreholes with water inflow of greater than 40 m^3/h account for 11%.

As the final positions of all boreholes are in the weathered bedrock, the number of boreholes and water inflow of single holes in different areas of the working face are quite different. The working face can be divided into two strong water rich areas according to the water inflow of the boreholes. A total of 34 exploratory drainage boreholes were constructed within 200 m of the initial cut of mining area. The initial water inflow of a single hole varied from 11 and 78 m^3/h with a total initial water inflow of 928 m^3/h. The cumulative drainage volume before the initial drainage of the main roof is 1.26 million m^3. Among them, the T7 hole (101 lane of the exploratory roadway) is the largest borehole with a water inflow of 78 m^3/h in the whole working face. In the ancient gully area, the north gully is 400–800 m away from the cut. Between the 89–96 water exploration lane and the 89–93 Yunshun water exploration lane, 16 water exploration and drainage boreholes were constructed. The initial water inflow of a single hole was between 19 and − 69 m^3/h, and the total initial water inflow was 594 m^3/h. Before the initial drainage of the main roof, the cumulative drainage volume was 1.17 million m^3. The south gully is 2000–2600 m away from the cut between the water exploration roadway 52–63 and Yunshun 53–65. A total of 25 exploratory drainage boreholes were constructed in this area. The initial water inflow of a single hole was between 9- and 51 m^3/h. the total initial water inflow was 708 m^3/h. The cumulative drainage volume was 1.54 million m^3 before the initial weighting of the basic roof.

During the water exploration and drainage, the water level at boreholes 21, 11, 10, 9 and 8 were observed regularly. Table 4.7 presents the borehole water level observation records.

According to the initial water level data at boreholes 21 and 11, the loose sand layer and weathered bedrock in the initial mining area of 31105 face contain water. The thickness of the sand aquifer is approximately 10 m. The thickness of the weathered bedrock aquifer varies from 50 to 55 m and is slightly confined. Once the sand layer water in the pre-mining cutting-eye area of the working face has been completely drained, the water level of the weathered bedrock aquifer decreased by 10 m. The total aquifer thickness in the cutting-eye area is reduced by approximately 23 m by

Table 4.7 Depth to water (m) during drainage of 31105 working face

Date	Borehole 21 (50 m from 31105 face cut)	Borehole 11 (31105 cutting-eyes above)	Borehole 10 (105 face Huishun 98 joint north approximately 15.4 m)	Borehole 9 (31105 face Huishun 67 Lianxi 99.1 m)	Borehole 8 (105 side back to 55 alliance west 42.8 m)
2008.6.15		10.72	10.88	15.75	11.86
7.25		10.72	10.88	15.88	12.83
8.27		10.72	10.88	16.49	13.72
9.25		10.98	11.07	16.73	14.61
10.16		10.98	11.83	16.93	15.07
11.21		12.57	12.67	17.37	15.83
12.3	23.6	15.51	15.63	17.72	16.82
2009.1.23	24.75	16.94	16.57	17.86	16.96
2.18	27.22	19.39	17.81	17.86	17.93
3.16	29.48	21.07	19.18	Mud sand plugging hole	18.65
4.15	32.52	22.77	20.23		19.02
5.15	34.33	23.03	23.6		19.49
6.18	36.03	Mud sand plugging hole	25.02		19.56
7.2	36.62		25.82		20.14
Speed-down (m/d)	0.06	0.036856	0.04	0.01	0.02
Sand floor depth/m	21.4	20.13	20.70	22.47	53.88
Buried depth of clay floor/m	24.54	23.33	26.8	0	63

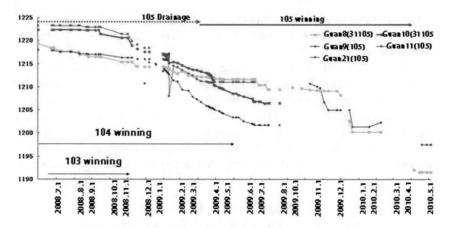

Fig. 4.12 105 working face drainage and water level comparison curve

the water exploration and drainage program. On July 9, 2009, there was no water in borehole 21 after the initial collapse of the main roof. The initial collapse step was approximately 90 m, and the water inflow of the working face was 95 m³/h. After May 2009 (Fig. 4.12), borehole 11 was drained, indicating that the water isolation of clay was damaged and pore water of sand layer entered the working face after mining of 31105 working face.

The stopping time of 31105 working face is from April 9, 2009 to April 8, 2010. There are 7 hydrological observation boreholes numbered Guan 1 to Guan 7 between Working Face 31105 and Qingcaojigou. All water level observation boreholes are in the sand layer. According to the comparison between the pre mining and post mining water levels of 31105 working face, the water levels of the seven hydrological observation boreholes decreased continuously (Table 4.8). The water level at Guan 1 borehole, Guan 5 borehole, and Guan 7 borehole dropped 7.42 m, 9.69 m, and 8.41 m, respectively. The water level changes of Guan 1, Guan 5, and Guan 7 boreholes indicated that the pore water in the skylight area has not been drained. However, the water level drop of Guan 2 borehole in the non-skylight area is obviously large. The clay thickness at Guan 2 borehole outside the skylight is 5.68 m, and the water level drop is 6.57 m. The clay thickness atGuan 3 borehole is 14.04 m, and the water level drop is only 3.87 m. The clay thickness at Guan 6 borehole is 38.63 m, and the water level drop is up to 14.05 m. It can be seen that the clay thickness at Guan 6 borehole near the 31105 working face is large, and the water level drop is the largest. According to J405–J406 data, the thickness of the bedrock in this area is between 67 and − 92 m. It can be concluded that the water resistance of the clay layer at Guan 6 borehole was damaged by 31105 mining.

The following observations can be made based on the long-term water level data at Guan 1, 2 and 6 over eight years from 2007 to 2015 (Fig. 4.13):

Table 4.8 Table of water level change of hydrologic observation hole west of 105 working face in the first panel Unit: m

Hole number	Position	Elevation of borehole	Thickness of sand	Depth of soil layer	Buried depth of soil roof	Initial water depth before 2007.4	2008.8 buried depth (before water exploration)	2009.5 buried depth (before mining)	Buried at the end of August 2010	Static water level elevation	Lower than initial water level	Sand aquifer thickness
Observation 1	940 m west of 31105, 245	1225.4	70.00	0	Mix	8.83	9.54	10.41	16.25	1209	7.42	53.75
Observation 2	1910 m from Yangjiagouzhang, 1650–1690 m	1225.2	64.35	5.68	64.35	12.80	13.54	14.24	19.37	1205	6.57	44.98
Observation 3	from Cuijiagouzhang	1222.7	43.96	14.04	43.96	20.12	20.40	20.91	23.99	1198	3.87	19.97
Observation 4	470 m west of 31105,	1243.6	14.30	23.00	14.3	39.20	39.27	39.84	Unwatering	Unwatering	Unwatering	Unwatering
Observation 5	1190–2380 m from	1233.8	31.80	0	Mixed water	14.39	14.93	17.31	24.08	1209	9.69	7.72
Observation 6	Yangjiagouzhang from Cuijiagou 837–2670 m	1229.2	27.65	38.63	27.65	12.69	13.40	16.58	26.74	1202	14.05	0.91
Observation 7		1233.0	28.28	0	Mix	16.09	16.55	18.49	24.50	1208	8.41	3.78

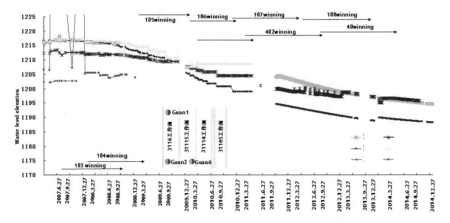

Fig. 4.13 105 comparison curve of drainage and water level of working face

- Before 2008, the drainage of working face 31105 did not start. The water level at observation holes Guan 1 and 6 was higher than that of observation hole Guan 2. At this time, the groundwater runoff direction was from northeast to southwest.
- After the stopping of 31105 working face started, the water level at Guan 6 observation hole decreased close to the water level elevation at Guan 2 observation hole by December 2009.
- With the production of 31106, 31107 and 31108 working faces, the water level at observation borehole Guan 6 continued to decrease and was lower than the water level elevation at observation borehole Guan 2.
- Since December 2011, the water level at Guan 2 and Guan 1 was higher than that of Guan 6. The groundwater flow direction has also changed, pointing to the northeast.

The overall flow direction of the Qingcaojie ditch in the well field is from northeast to southwest. The groundwater is confined and flows in the direction of the ditch valley and finally discharges to the Qingcaojie ditch. With the east side of the basin 31105-110 workings back to mining, the north side of the 31401-404 workings back to mining, and 3110 to the east of the workings back to mining, it is anticipated that the Qingcaojie ditch recharge water will be reduced. As the ditch water is further reduced, more ditch palm springs face the threat of disconnection.

The 31106 working face of Jinjie Coal Mine is 5390 m long. The geological and hydrogeological conditions of the 31106 face are similar to those of the 31105 face. The groundwater flow direction is from the 31106 face to the 31105 face. Therefore, the construction of drainage borehole on the 31106 face can intercept part of the upstream catchment to the goaf of the 31105 face. The 31106 working face is covered by the Quaternary eolian sand, with little variation. According to the drilling data, the stratum consists of the Quaternary Holocene eolian sand (Q4eol), Upper Pleistocene Salawusu Formation (Q3S), Middle Pleistocene Lishi Formation (Q2L), Tertiary Pliocene Baode Formation (N2b), Jurassic Middle Zhiluo Formation

(J2Z) and Middle Lower Yan'an Formation (J1-2y). The thickness of the sand layer varies from 5 to 50 m. The thickness of the sand layer in the initial mining section of the cut between 20 and − 25 m. The sand layer does not contain water. Two ancient gullies are respectively developed 300–850 m and 1800–2800 m away from the cut hole (the upper reaches of the grassy boundary gully are called Beigou and Nangou, respectively). The thickness of the eolian sand in the north gully is between 30 and 40 m, while the thickness of the aquifer varies from 0 to 5 m. The thickness of the eolian sand in Nangou is between 20 and 40 m, while the thickness of the aquifer is between 10 and 20 m. Except for the two ancient gullies, the thickness of the aeolian sand in other sections is relatively thin and does not contain water. The thickness of the soil layer ranges from0 to − 65 m. The overall trend is to thicken from the cut hole to the return track channel. The thickness of the soil layer in the initial mining section of the cut hole is approximately 10 m. There may be skylight areas near the 35–45 cross roadways of the return air gateway of the working face and the 40 cross roadways of the transport gateway. The weathered bedrock of the Zhiluo Formation has a thickness between5 and − 55 m. and is the main aquifer of the working face. The thickness of the initial cut section is approximately 50 m. The stratum in the two ancient gullies is relatively thin due to erosion. The thickness of the Zhiluo Formation in the north gullies is between 30 and − 50 m. The thickness of the Zhiluo Formation in the south gullies varies from 5 to -15 m. The competent bedrock thickness of coal seam 3-1 is between 25 and − 32 m, with an average thickness of 30 m. The thickness is relatively stable.

The exploratory drainage boreholes are arranged at the cut hole, cut hole connection roadway and return air chute of 31106 working face. The construction of exploratory drainage drilling holes started on July 15, 2008 and completed on September 29, 2009. A total of 40 boreholes were drilled, with a total drilling footage of 3130 m. Among them, 9 boreholes were drilled for the construction of cut hole and cut hole connection roadway, and 31 boreholes were drilled in return air gateways for the 31107 working face. The diameter of the exploratory drainage hole is $\varphi75$ mm. Most of the final borehole layers are in the weathered bedrock, while a few of the final borehole layers penetrate into the soil layer. The purpose was mainly to drain the water in the weathered bedrock aquifer. The water inflow of T22 and T18 boreholes is relatively small due to the fact that the gravel drill is not constructed to the aquifer level during the construction. However, the water inflows of other holes areas relatively large when they are finished. The initial water inflow of the boreholes range from 3 to- 60 m^3/h, and the total initial water inflow is 1490 m^3/h. As of October 9, 2009, the water inflow of the boreholes varied from 0 to − 25 m^3/h, and the total water inflow of the boreholes was 541 m^3/h. The cumulative water discharge of the exploratory drainage boreholes is approximately 780,000 m^3, relieving the water pressure of the 31105 surfaces. Among them, the boreholes with water inflow in the range between 30 and 50 m^3/h account for 50% of the total, and are distributed in all working faces. The water inflow less than 30 m^3/h accounts for 25% of the total, most of which is distributed between the south gully and the back tracking channel of the working face, indicating that the water yield in this area is relatively weak. The boreholes with water inflow of more than 50 m^3/h account for 25% of the total. Most of them are distributed in the cut outs of the 31105 working face, 92–96 alleys in the working face and 50–65 alleys in the working face. Many exploratory drainage boreholes have been constructed in these three areas, with large water inflow and long drainage time. During the mining period from the 31105 working face to the north gully, the water inflow of the working face increased significantly, indicating that the water yield is strong in the cut hole, north gully and south gully areas.

Based on the water level observation data of Kongguan No. 24 hole of the weathered bedrock near the cutting hole of 31105 working face, the bottom elevation of the sand layer is 1223.9 m. The top elevation of the bedrock is 1207.85 m. The depth to the groundwater is 30.04 m. The thickness of the weathered bedrock aquifer is 41.56 m, and the water head height is 44.42 m. The weathered bedrock water is still pressurized during drainage. The water level of fissure water in the 31105 working face immediately dropped after mining. The water level dropped to 1204.93 m after 2 days of mining, which was lower than the elevation of bedrock roof. The water level observation of the sand layer in the south and north ancient gullies at Kongguan 32 and Kongguan 33 revealed that the floor elevation of the sand layer was 1210.39 m and 1205.07 m, respectively. Before mining, the water drainage did not completely drain the sand layer aquifer (Fig. 4.14). After mining, the water level of Kongguan 33 decreased immediately, with an increase in the decline. Two days later, it fell to 1207.768 m on April 6, 2010, and to 1206.018 m on October 19, 2010.

The working face 31107 of Jinjie Coal Mine is located at the upstream of Qing-caojigou. The surface is covered by the aeolian sand, and the terrain is relatively flat. According to the drilling data, the stratum consists of the Quaternary Holocene eolian sand (Q4eol), Upper Pleistocene Salawusu Formation (Q3S), Middle Pleistocene Lishi Formation (Q2L), Tertiary Pliocene Baode Formation (N2b), Jurassic Middle Zhiluo Formation (J2Z) and Middle Lower Yan'an Formation (J1-2y) from top to bottom. The thickness of the sand layer varies from 1 to − 35.m. In general, the thickness become smaller from the cut hole to the back track channel. The thickest

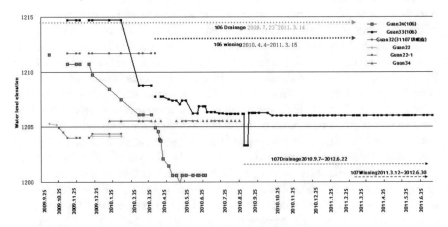

Fig. 4.14 106 working face drainage and water level comparison curve

part is located at the vicinity of No. 32 hole in the 95 lane of the 31107 transport gateway. Two ancient gullies are developed 450–950 m and 2100–2900 m away from the cut hole, respectively. The thickness of the eolian sand in the north gully is approximately 35 m, whereas the thickness of the eolian sand in the south gully is approximately 30 m. Except for the two ancient gullies, the thickness of the aeolian sand in other sections is relatively thin. The thickness of the weathered bedrock of Zhiluo Formation is between 5 and − 55 m. It thins from the cut hole and back track channel to the middle of the working face (ancient gully). The thinnest part is located in hole J307 (close to the transportation gateway 63 lane). The competent Competent bedrock of coal seam 3-1 is relatively stable. Its thickness ranges from 23- to 40 m with an average of 30 m. According to the hydrological observation hole data, the sand layer of the 31107 working face does not contain water. The main aquifer is the weathered bedrock aquifer. The thickness of the weathered bedrock aquifer is between 5 and − 55 m. The thickness becomes thinner from the cut hole and back tracking channel to the middle of the working face (ancient gully). The thickness of the cut hole initial mining area is between 40 and − 50 m. The thickness of the south gully is the thinnest, approximately 10 m. The weathered bedrock is mainly coarse and medium grained sandstone, with conglomerate at the bottom. The weathering degree of the whole layer is from medium to strong depending on argillaceous cementation, local calcareous cementation, and fissure development. The water drainage program was started on September 7, 2010 and completed on January 11, 2011. A total of 57 boreholes were drilled. Most of the final boreholes were in the weathered bedrock. A few boreholes were drilled into the sand layer in the ancient gully area. The initial water inflow of the boreholes varied from 12 to 60 m^3/h with an average initial water inflow of 35 m^3/h. The total initial water inflow was 1971.5 m^3/h. The boreholes with water inflow more than 30 m^3/h accounted for 72% of the total, and were distributed in most areas of the working face, indicating that the weathered bedrock of the whole working face has strong water yield. The boreholes with water inflow of more than 40 m^3/h accounted for 35% of the total, which were distributed between the middle of the working face and the cut hole, mainly in the two ancient gullies with strong water yield. Due to the stopping of the 31401 working face near the cut hole, the water yield was reduced. The water inflow less than 30 m^3/h accounted for 25% of the total, which was mainly distributed between the middle of the working face and the return track channel, indicating that the water richness near the return track channel was relatively weak. As of January 13, 2011, the water inflow of the drilling holes ranged from 5 to − 18 m^3/h. The total water inflow of drilling holes was 621 m^3/h, and the cumulative water discharge of drilling holes was approximately 1.26 million m^3. With water drainage of the 31106 working face the water head pressure of the 31107 working face, the Guan 24, Guan 32, Guan 33 and Guan 34 boreholes near this working face before the mining of the 31107 working face could no longer be observed.

The maximum water inflow of the 31108 first face is 218 m^3/h. The water inflow increases with the increase of goaf area. In the section close to the withdrawal channel, the water inflow increases obviously compared with the initial mining area.

Most of the surface of Working Face 31109 is covered by the Quaternary eolian sand. According to the drilling data, the stratum consists of Quaternary Holocene eolian sand (Q4eol), Upper Pleistocene Salawusu Formation (Q3S), Middle Pleistocene Lishi Formation (Q2L), Tertiary Pliocene Baode Formation (N2b), Jurassic Middle Zhiluo Formation (J2Z) and Middle Lower Yan'an Formation (J1-2y) from top to bottom. The thickness of the sand layer is 0–20 m, and the thickness of the sand layer varies greatly. The thickness of the 80–105 connecting roadway is almost 0, and the maximum thickness of the 50–60 connecting roadway is approximately 20 m. The thickness of the soil layer is between 10 and- 62.4 m. The thinnest 50–60 connecting roadway at the return air gateway of 31109 working face is approximately 10 m. The thickest layer is near the return passage. The thickness is generally thin in the north and south, and thin in the middle. The weathered bedrock of Zhiluo Formation is 10–60 m thick, and the middle part becomes thinner and thicker from north to south. The thickness of 85–100 connecting roadway at the return air gateway of 31109 working face is 50–60 m. The competent bedrock thickness is relatively stable, generally approximately 30 m. The 31109 working face is not threatened by surface water hazards. The overlying aquifer of the working face includes sand layer and weathered bedrock aquifer. According to the observation hole data, the aquifer in the 31109 working face is mainly weathered bedrock aquifer, with a thickness of approximately 10–30 m, generally approximately 20 m. The sand aquifer may only contain a small amount of water locally, and most sections have no water. The water drainage drilling of the 31109 working face was started on August 2012, and completed on December 16, 2012. There were 46 actual boreholes. All the final boreholes were in the weathered bedrock, mainly used to drain the water in the weathered bedrock aquifer. The drainage time is approximately 6 months, and the initial water inflow from the cut hole of the 31109 working face to the connecting roadway 55 of 31109 transport gateway is 3.5 m^3/Between h and 52.9 m^3/h, the water inflow of individual boreholes at 55 connecting roadways of 31109 transport gateway exceeds 50 m^3/h, and the total initial water inflow is 1081 m^3/h; By January 14, 2013, the water inflow of boreholes was between 0.9 and 11.3 m^3/h, with some exceeding 10 m^3/h. The average water yield of a single borehole is 5.23 m^3/h, the total water inflow of the borehole is approximately 240 m^3/h, and the cumulative water discharge is approximately 970,000 m^3. The maximum water inflow during stopping at 31109 working face is 217 m^3/h. The water inflow generally increases with the increase of goaf area. Working face 31110 is located in panel 1, adjacent to the goaf 31401 of panel 4 in the north. There is no ponding in this area, and adjacent to Huangtulmiao Coal Mine in the south. Coal mine 31 has been closed; it is adjacent to the 31109 working face being mined in the west.

The 31110 working face is 243 m wide and 3792.5 m long, and the surface is covered by Quaternary eolian sand. According to the drilling data, the stratum consists of Quaternary Holocene eolian sand (Q4eol), Upper Pleistocene Salawusu Formation (Q3S), Middle Pleistocene Lishi Formation (Q2L), Tertiary Pliocene Baode Formation (N2b), Jurassic Middle Zhiluo Formation (J2Z) and Middle Lower Yan'an Formation (J1-2y) from top to bottom. The thickness of the sand layer is 0–10 m, and the overall trend is gradually thicker from both sides of the working face to the middle. The thickness of the cut area is 0–5 m, and the maximum thickness of the withdrawal channel is approximately 10 m, basically without water. The thickness of the soil layer is 10–70 m, and the overall trend is gradually thicker from the withdrawal channel to the cut hole. The thickness of the withdrawal channel is approximately 10 m, and the thickness of the cut hole area is 60–70 m. The weathered bedrock of Zhiluo Formation is 18.45–41.61 m thick, and the thickness of the whole working face is relatively stable. The competent bedrock thickness of 3-1 coal seam is 24.92–33.65 m, and the thickness of the whole working face is relatively stable. The water drainage project was commenced on May 8, 2014 and completed on June 3, 2014. There are 16 boreholes were actually drilled, and most of the final boreholes were located at the top interface of weathered bedrock or 1–2 m into the soil layer. Initial water inflow of borehole is 3.1–24.3 m^3/h, the total initial water inflow is 156.9 m^3/h. The average initial water inflow of the borehole is approximately 9.8 m^3/h. Water inflow is greater than 9.8 m^3/h boreholes of account for 44% of the total (more than 20 m^3/h accounts for 6.3% of the total), which is mainly distributed near the 61 connecting roadway at the edge of the water rich area of the working face, indicating that the water rich property of the aquifer in the section of the whole working face's cut hole and withdrawal channel is strong. By November 5, 2014, the water inflow of the borehole was 0.2–1.64 m^3/h, the total water inflow of borehole is 14 m^3/h. The cumulative water discharge of the borehole is approximately 154,000 m^3. The overlying aquifer of the working face includes sand layer and weathered bedrock aquifer. The initial aquifer of the sand layer is 0–3 m thick. The mining and long-term drainage water of the 31105-31109 working face has almost no water in the sand layer. The weathered bedrock aquifer is 3–7.6 m thick and generally distributed around 5 m in the whole area. There is no ground observation borehole in 31109 working face and there is no dynamic change of water level.

4.4.1.2 Working Face 201–210 in the Second Panel Area

Most of the surface area of 31201 working face of the Jinjie Coal Mine is covered by wind-bound sand. The stratigraphic structure from top to bottom is: the sand layer is 0–33 m thick, the overall trend is gradually thickening from the middle of the working face to both sides, the sand layer near the watershed is the thinnest, the thickest is in the initial mining area of the cutting-eye, the layer thickness is approximately 30; the soil layer is 4–25 m thick, the overall trend is gradually thickening from the cutting-eye to the return rutting channel, the cutting-eye area is the thinnest, the layer thickness is approximately 5 m, the other areas of the soil layer is generally approximately 20 m thick. The weathered bedrock of straight Luo group is 40–74 m thick, gradually thickening from the rutting passage and cutting eye to the middle of the working face, the cutting-eye is approximately 46 m thick, the thickest place is near borehole J909 in the middle of the working face, the lithology is mainly coarse and medium-grained sandstone, the bottom contains conglomerate, the whole layer is strongly-medium weathered, mud cementation, local calcareous cementation, fissure development. The competent bedrock is approximately 30 m thick and the thickness is relatively stable. All boreholes were drilled in the weathered bedrock to release water from the weathered bedrock aquifer. The initial surge volume of the boreholes is between 22 (31201 back to Shun 9 lane) and 57.6 m³/h (31201 back to Shun 88 lane), the total initial surge volume is 4957 m³/h, until December 8, 2010, the surge volume of the boreholes is between 7 and 23 m³/h, the total surge volume of the boreholes is approximately 1744 m³/h, the accumulated water release is approximately 3.6 million m³. The sand layer water is mainly recharged by atmospheric precipitation and flows along the top surface of the soil layer from high to low, i.e. from east to west, and is discharged in the form of spring in Hezegou, and there is a "skylight" area near the cutter, so the sand layer water is directly recharged downward to the weathered bedrock aquifer water. The lateral recharge of the weathered bedrock aquifer is mainly from high to low runoff along the top surface of the competent bedrock of 3^{-1} coal, i.e., from east to west, and discharged in the form of springs in Hezegou on December 8, 2010, the amount of water gushing from each borehole at the cutting eye is still large, between 9 and 17 m³/h, indicating that the fissures of the weathered bedrock in the area are developed and the recharge water source is sufficient.

According to the observation of the water level of the No. 25 borehole, the No. 25 borehole is protected by the solid pipe under the loose layer, but the soil layer of the borehole is only 4 m thin, and there is a "skylight" on the west side. The sand layer water supplies the weathered rock water, and the pore water is closely related to the fissure water. The construction time of the borehole is 2010.6.5–13. On November 2, 2010, the water level elevation of 1173.38 was close to the sand layer floor (Fig. 4.15). On November 11, 2010, the water level elevation was lower than the sand layer floor, and the drainage time was approximately 4.5 months. The water level decreased by 21.53 m, with an average monthly decrease of 4.8 m. On December 22, 2010, the water level is below the clay aquiclude floor. The dynamic of observation borehole No. 25 experienced two stages: water sparing and releasing decline, and back mining decline.

Most of the surface area of the 31202 working face is covered by the wind-deposited sand of the fourth system. The sand layer is 0–35 m thick, and the overall trend is gradually thicker from the middle of the working face to both sides, and the sand layer near the watershed is the thinnest; The thickest is located in the initial mining area of the cut, with a thickness of approximately 30 m. The thickness of the soil layer is 4–25 m, the overall trend is gradually thickening from the cutting-eye to the rutting channel, the cutting-eye area is the thinnest, the layer thickness is approximately 5 m, there is a "skylight" area, except for the cutting-eye, the other areas of the soil layer thickness is generally approximately 20 m. The weathered bedrock of straight Luo Group is 40–74 m thick, gradually thickening from the rutting passage and cutting-eye to the middle of the working face, the cutting-eye is approximately 46 m thick, the thickest near borehole J909 in the middle of the working face, the lithology is mainly coarse and medium-grained sandstone, the bottom contains conglomerate, the whole layer is moderately weathered, muddy cementation, local calcareous cementation, fissure development. The competent bedrock is 25–45 m thick, generally 30–35 m, with a relatively stable thickness. The underground water

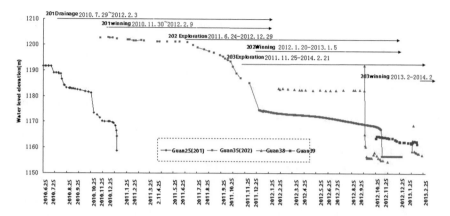

Fig. 4.15 Contrast curve between drainage and water level of working face 201 in the second panel area

release drilling started on June 21, 2011, and was completed on October 1, 2011. All the boreholes were drilled with φ75 mm borehole diameter ending in the weathered bedrock, mainly releasing water from the weathered bedrock aquifer. The surface of working face 31202 belongs to Hezegou watershed, approximately 1400 m to the west of Hezegou, and the sand aquifer is 0–25 m thick, mainly distributed in the cutting-eye area of working face, and the thickest part is located near the open-cut area. Due to the drainage of underground water, the sand aquifer is approximately 5 m thick in the open-cut area on November 1, 2011, and the sand water is mainly supplied by atmospheric precipitation. It flows from east to west along the top surface of the soil layer, and is discharged in the form of spring in the Heze gully. There is a "skylight" area near the eye cut, so the sand layer water directly supplies the weathered bedrock aquifer water. The thickness of the weathered bedrock aquifer is 20–60 m, the thickness of the cutting-eye is approximately 25 m, and the thickness of the middle of the working face is approximately 60 m. The weathered rock water is mainly supplied laterally, with sufficient water supply and strong water enrichment. The fissure water flows from east to west along the top surface of competent bedrock of 3^{-1} coal and drains out in the form of spring in Hezegou. The initial water inflow from boreholes is between 27.3 (31202 Huishun 99 lane) and 66.3 m^3/h (31203 Huishun 101 lane), and the total initial water inflow from boreholes is 2875 m^3/h. By November 1, 2011, the total water inflow from boreholes is between 12.3–29.8 m^3/h, and the total water inflow from boreholes is approximately 1271 m^3/h. The cumulative water discharge is approximately 2.41 million m^3. The initial water inflow of 4 drilling boreholes with auxiliary cutting-eye and 30 drilling boreholes near the initial mining area of cutting-eye was mostly between 3045 m^3/h, some of which exceeded 60 m^3/h, with a total of approximately 11,004 m^3/h. On November 1, 2011, the water inflow of most drilling boreholes was between 12 and 184 m^3/h, with a total of approximately 480 m^3/h. It indicates that fissures are developed in weathered bedrock in this area.

On November 29, 2010, No. 35 observation borehole was constructed in the middle of the working face at approximately 888 m away from the withdrawal channel. The water level of No. 35 observation borehole decreased by 8.7 m during the underground drainage period (4 months), with an average monthly decrease of 2.2 m. On November 1, 2011, the sand layer was no longer water-bearing, and the weathered bedrock aquifer was approximately 47 m thick. The borehole has experienced three decline stages: the first stage is affected by the drainage of 201 working face, the second stage is affected by the drainage of 201 working face and 202 working face, and the third stage is the rapid decline stage affected by the drainage of working faces 201, 202 and 203 (Table 4.9).

The surface of the 31203 working face is covered by the aeolian sand, and the stratum structure is from top to bottom: the thickness of the sand layer is 12.25–65.5 m, and the thickness of the sand layer is 65.5 m in the initial mining area of the hole cutting. The thickness of the soil layer is 0–26.5 m. The thickness of the cutting-eye area is the thinnest and there is a skylight. The thickness of the soil layer in other areas is 10–20 m. The thickness of weathered bedrock in Zhiluo Formation is 10.26–70 m, which gradually becomes thinner from the middle of the working face to the cutting-eye, and there is little change from the middle of the working face to the retreat channel. The thickness of competent bedrock is mostly approximately 30 m, the thickness around the hole cut is approximately 42 m, and the thickest is approximately 61.5 m near the 40 borehole, which is relatively stable. The surface of 31203 working face belongs to Hezegou Valley, approximately 1000 m to the west of Hezegou, and the overlying aquifer includes sand layer and weathered bedrock aquifer. The sand aquifer is 10–46.57 m thick, and the thickest is near the cutting-eye (46.57 m thick). The sand layer water is mainly supplied by atmospheric precipitation and flows from high to low along the top surface of the soil layer, i.e., from east to west, and is discharged in the form of spring in Heze gully. When the soil layer is missing near the eye cut, sand water directly supplies the weathered bedrock aquifer water downward. Weathered bedrock aquifer thickness of 10–60 m, cutting-eye near the thickness of approximately 10 m, the thickest working face in the middle of approximately 60 m, weathered fissure water is given priority to with

Table 4.9 Watch 35 borehole water level observation record table

Location	Elevation of bore hole (m)	Drainage water before water bit depth (m)	Current water depth (m)	The amplitude of decrease compared with that before undredging (m)	Current static water level (m)	Downhole water discharge date
31202 midplane 888 m from return channel	1245	44.19	52.89	8.7	1192.1	2011.06.21–2011.11.01

lateral recharge, in the cutting-eye near the "skylight" also has vertical recharge, weathered bedrock fissure water along the 3-1 coal competent bedrock top surface from east to west, in the river ditch in the form of spring discharge. Underground drainage boreholes in November 24, 2011 to July 09, 2012 completed, drainage time of approximately 7 months, all boreholes with aperture φ75 mm final borehole in the weathered bedrock, the main drainage weathered bedrock aquifer water. The initial water inflow of the borehole is between 0 (31203 cutting-eye) and 65.6 m^3/h (31203 cutting-eye), and the total initial water inflow is 2529m^3/h. Up to July 14, 2012, the total water inflow from the borehole is between 0 and 29.15 m^3/h, and the total water inflow from the borehole is approximately 848 m^3/h, and the cumulative water discharge is approximately 2.3 million m^3. On December 31, 2011, the observation No. 38 borehole was constructed at a place approximately 82 m away from the hole cut within the working face, and the water level of the observation 38, J1008 and J808 boreholes was observed regularly while the underground water was detected and released. The solid pipe retaining wall under the loose layer of No. 38 borehole, the initial water-bearing medium thickness is 75.76 m, the initial water level burial depth is 18.93 m, the initial aquifer thickness is 56.83 m, of which the sand layer aquifer is 46.57 m thick, and the weathered bedrock aquifer thickness is 10.26 m. On July 9, 2012, the thickness of the aquifer was 49.95 m, of which the thickness of the sand aquifer was 39.69 m, and the thickness of the weathered bedrock aquifer was 10.26 m. The borehole drained water for approximately 4.5 months, and the water level dropped by 6.9 m. Although the water level of observation borehole No. 38 has decreased significantly, the aquifer in the sand layer has not been completely drained. The water level of observation borehole No. 38 is 1131.44 m higher than the bedrock top interface. Similarly, the floor elevation of the sand layer in observation borehole No. 39 is 1150.784 m. The water level of observation borehole No. 39 decreased from 1164.46 m on October 29, 2012 to 1162.434 m on February 8, 2013, and the groundwater level was still in the sand layer. Due to the absence of clay layer in observation borehole No. 38 and No. 39, it can be seen from the water level decline rate of observation borehole No. 39 that the pore water and fracture water in this part have a unified water level, and the hydraulic relationship is close.

Due to the serious uneven distribution of sand layers on the whole working face, the sand layer in observation borehole No. 38 (near the cutting-eye) near Hezegou is very thick, and the rest of the sections are all less than 10 m. Except for the sand layers near observation borehole No. 38 and other sections that contain water, the rest of the sections have no water. For example, in the outer hole J1008, the elevations of sand floor and clay floor were 1194.099 m and 1182.999 m, respectively. The water level of fracture decreased from 1186.239 in February 2012 to 1175.859 m on March 3, 2013, and the water level of fracture was lower than that of clay floor (Fig. 4.16).

In the duration curve of 31203 working face water inflow observation data (Fig. 4.17), when the working face advanced 12 m on January 20, 2013, the water inflow was 133 m^3/h; when the working face advanced 55.5 m on January 25, 2013, the roof collapsed and the water inflow increased greatly, and the water inflow reached 330 m^3/h at this time. It indicates that the water-conducting fractures have communicated with pore water in the sand layer, and it can be seen from the observation

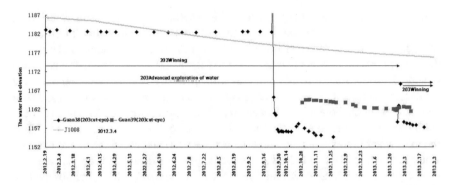

Fig. 4.16 Correlation curve between mining and water level of working face 203 in the second panel area

of drainage water in working face 203 that the hydraulic relationship between pore water and fracture water is close.

On March 16, 2014, the water drainage project of the 31205 working face has completed 22 boreholes, and the maximum water inflow of a single borehole is 114 m³/h, indicating that the area near the borehole cut of the 31205 working face is rich in water, overlying the Quaternary Salawusu Formation and the strong water-rich area of aeolian sand, and the water supply of weathered bedrock is better, and the aquifer near the "skyligh" is directly communicated. The thickness of the sand aquifer is approximately 28–30 m, the soil layer is 0–5 m, the thickness of the weathered bedrock aquifer is approximately 40–45 m, and the thickness of the competent bedrock is 22–31 m. The soil layer revealed by the drilling is not very obvious, and the exposed soil layer is all fine and medium grained sand with good permeability. The water inflow from a single borehole of the drilling is 71.7–114.1 m³/h. The

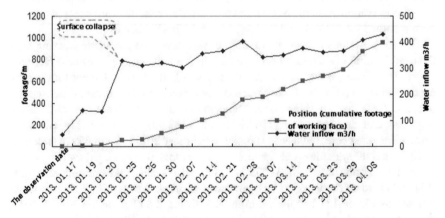

Fig. 4.17 Duration curve of water inflow observation data of 31203 working face in the second panel area

water inflow of single borehole in the 31205 route area is 1166.6 m³/h. Cutting-eye water inflow anomaly drilling decay faster, no sand inrush phenomenon. As can be seen from the data of the water level observation borehole on the working surface 204 and 205 (Fig. 4.18), with the influence of the mining on the working face 203 and the water exploration and drainage on the working face 204, the water level of the observation borehole No. 40 and No. 41 has a great influence on the water level reduction and drainage, especially in the observation borehole No. 40, the water level drop value is greater than 20 m. The water level of observation borehole No. 40 is 1161.8 m lower than the soil floor elevation, and the water level of observation borehole No. 41 is 1157.7 m lower than the soil floor elevation. However, the dynamic variation range of water level in observation borehole No. 42 and No. 43 is small. Only the water level in observation borehole No. 42 decreased by 0.56 m, the floor elevation of sand layer in observation borehole No. 42 was 1161.01 m, the soil layer thickness was 38.17 m, the weathered bedrock roof elevation was 1122.84 m, the floor elevation of sand layer in Guan 43 borehole was 1155.60 m, and the soil layer thickness was 0 m. The water level of observation borehole No. 42 and No. 43 is higher than that of the sand floor, indicating that the sand layer near observation borehole No. 42 and No. 43 has not been drained, and the sand layer is relatively water-rich. Especially, Guan 43 borehole is in the skylight area, and the water level decreases the least.

It can be seen from the water level dynamics of observation borehole No. 40, 41, 42 and 43 that the water level decreases from small to large is in the order of observation borehole No. 43 < observation borehole No. 42 < observation borehole No. 41 < observation borehole No. 40, i.e., the skylight area has the smallest decrease, which reflects the close hydraulic relationship between pore water and fracture water,

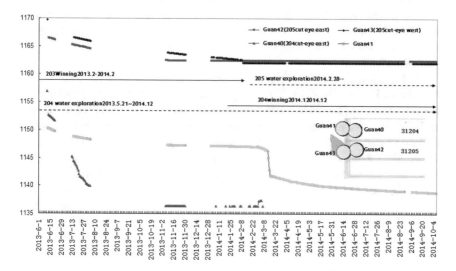

Fig. 4.18 Comparison curve of relation between stopping and water level of working face 204 in No. 2 panel area

good water wealth, and the water level decreases greatly away from the skylight area, and the pore water is drained.

The surface of the working face 31208-210 is covered by aeolian sand, and the top-down structure of the stratum is: the thickness of the sand layer is 0–40 m, and the thickness of the sand layer aquifer is 0–10 m. The soil layer is 0–40 m thick, and the overall trend is thickening from the middle of the working face to both sides, and there are some "skylights". The weathered bedrock is 5–46 m thick, the whole layer contains water, and the overall trend is thinning from the cutting-eye to the Return channel. The competent bedrock is 10–32 m thick, generally 30 m thick, the thickness is relatively stable. In Jinjie Coal mine, the 31208 faces have been mined for the first time and the press step distance is approximately 65 m, and the maximum water inflow during the first time is 205 m³/h. The initial step distance of surface 31209 is approximately 73 m, and the maximum water inflow during the initial emergence is 100 m³/h. The thickness of the sand layer on the 31210 working face near Shamu River is 0–34 m, and the sand layer is missing in the Return channel of the working face, and the thickest is in the middle of the working face, and the sand layer is generally water-free. The soil layer thickness is 0–34.9 m, with a stable thickness of approximately 10 m in most sections. The thickness of weathered bedrock is approximately 2.12–40 m, slightly thin in the north and thick in the south. The competent bedrock is approximately 20–32.29 m, and the thickness is stable. In the cutting-eye area, the competent bedrock is approximately 30 m, and the weathered bedrock is approximately 40 m, and the whole layer contains water. On January 27, 2010, in the normal mining process of 31210 working face in the four-point shift of the second fully mechanized mining team, when the cumulative advance of approximately 40 m (head 38.5 m, tail 41 m), the working face pressure began to be abnormal, and then a small amount of old pond water gushing appeared, to 18:00 at night, the working face water inrush gradually increased to approximately 200 m³/h. In the later period, the water inrush increased unsteadily, and the water inrush in the working face reached approximately 500–600 m³/h at approximately 23:00 at night.

The Jinjie Mine 93208 working face is 2168 m long, 294 m wide and began mining at the end of January 2009. The hydrogeological condition of the 93208 working face is similar to that of the 93101-104 working face in one panel area. The loose sand layer does not contain water, and the main aquifer is the weathered bedrock aquifer of Zhiluo Formation. The working face is covered by quaternary aeolian sand, and the terrain is relatively flat with little fluctuation. From top to bottom, there are Quaternary Holocene eolian sand (Q_4^{eol}), Upper Pleistocene Salawusu Formation (Q_3s), Middle Pleistocene Lithi Formation (Q_2l), Tertiary Upper Neoocene Bade Formation (N_2b), Middle Jurassic Zhiluo Formation (J_2z) and Middle Lower Yan'an Formation ($J_{1-2}y$). The sand layer is 1–10 m thick, generally approximately 5 m thick. The soil layer is 21–61 m thick and thickens from the Return channel to the cutting-eye. The weathering bedrock of Zhiluo Formation is 23–45 m thick and thickens from the Return channel to the cutting-eye. The 3^{-1} coal competent bedrock thickness of approximately 30 m, the thickness is stable. According to the data of observation borehole No. 18, No. 19 and No. 31, the working face aquifer is the weathered bedrock aquifer of Zhiluo Formation, with a thickness of 26–30 m. The construction of water exploration and drainage boreholes began on December 1, 2008 and was completed on February 11, 2009. A total of 53 boreholes were Except for a few boreholes in the weathered bedrock, most of the boreholes were in the soil layer. The initial water inflow of the borehole is between 4 and 65m^3/h, and the total initial water inflow is 1421 m^3/h. By the end of the water exploration and drainage project on February 11, 2009, the working face had advanced 140 m, and 16 boreholes had become waterless (14 boreholes in the goaf, and the other two boreholes are boreholes T34 and T38).The water inflow of other drilling boreholes is between 6–32 m^3/h, and the total water inflow of drilling boreholes is 587 m^3/h. The 93208 working face began mining on January 31, 2009, the old top was first pressed on February 3, and the water exploration and drainage project was completed on February 11. Up to February 11, the cumulative water discharge volume of the water exploration and drainage borehole was 620,806 m^3, and a large amount of standing water reserves of the weathered bedrock aquifer were released. The water inflow of most of the water exploration and drainage boreholes in this working face is greater than 20 m^3/h, so it can be considered that the weathered bedrock of the whole working face is water-rich. However, the water inflow of boreholes in different areas of the working face is quite different. For example, the borehole cut is a water-rich area, and the length of the borehole cut is 294 m. The initial water inflow of nine water exploration and drainage boreholes is between 13 and 65 m^3/h, among which borehole T47 (65 m^3/h) is the borehole with the largest water inflow of the whole working face. Yunshun 34–38 lane and its corresponding Hui Shun (1070–1370 m away from cutting-eye) are strong watery zone. A total of 7 water exploration and drainage holes were constructed, and the initial water inflow of a single borehole was between 30 and 60 m^3/h, and the total initial water inflow was 330 m^3/h. The average water inflow per borehole is approximately 50 m^3/h.

The observation borehole No. 31 (4,294,223.07, 37,428,293.95, 1244.99) was completed on December 17, 2008, with a depth of 106.60 m, sand layer thickness of 1 m, soil layer thickness of 61 m, and φ133 mm solid pipe wall protection under the loose layer. Then the final borehole of the competent bedrock top interface of 3^{-1} coal is constructed with φ104 mm aperture. According to the water level data of Guan31 hydrological observation borehole, the loose sand layer in the cutting-eye area of the 93208 face does not contain water, and the aquifer is weathered bedrock. Before the construction of water exploration and drainage project, the weathered bedrock aquifer is 26.60 m thick, and before the initial caving of the old top of the working face on February 2, 2009, the weathered bedrock aquifer is 18.30 m thick, and the water exploration and drainage project reduces the aquifer in the open-cut area by 8.30 m. On February 3, after the initial collapse of the old roof, the thickness of the aquifer was 12.40 m, the initial collapse distance was 51 m, and the water inflow of the working face was 205m^3/h. When the roof collapses, the retaining pipe in the borehole is deformed, and the buried depth of the borehole water level cannot be measured (Table 4.10 and Fig. 4.19). As can be seen from the figure, the water level drops immediately after the drainage of water at the 208 working face, and the decrease rate of the 208 working face increases after mining. When the old roof collapses at the beginning, the water level drops substantially.

The 93209 working face in the second panel of Jinjie Coal Mine is 1605 m long and 300 m wide. The hydrogeological conditions of the 93209 working face are the same as those of the 93208 working face. The stratigraphic structure is: the thickness of the sand layer is 0–10 m without water; soil layer 10–30 m, thinning from the reverse direction to the forward direction; Zhiluo group weathered bedrock thickness of 20–47 m, the initial mining area is thick, back to the Return channel thinning,

Table 4.10 208 water level dynamics of observation borehole No. 31 during water exploration and discharge of working face

The observation date	2008.12.19	2008.12.21	2009.1.1	2009.1.11	2009.2.2	2009.2.3
Water depth (m)	75.40	76.36	78.71	80.01	83.70	89.60
Aquifer thickness (m)	26.60	25.64	23.29	21.99	18.30	12.40
For note	(Preliminary survey)	Cutting-eye drilling construction is completed (9borehole)		Completion of 31 boreholes, of which the cutting-eye within 200 m drilling completed (17borehole)	Water level before initial collapse, 43 boreholes completed	The old roof collapsed at first

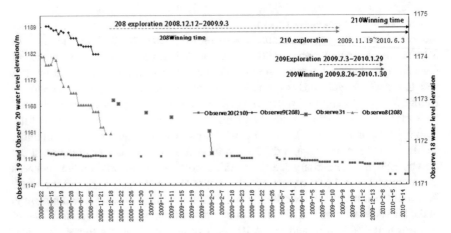

Fig. 4.19 Comparison curve of the relationship between mining and water level of 208 face in the second panel area

lithology is given priority to with grayish yellow medium and coarse sandstone, fissure development, as the overlying aquifer on the working face. The competent bedrock thickness of 3^{-1} coal is approximately 35 m, and the thickness is stable. The construction of water exploration and drainage boreholes began on June 22, 2009 and was completed on July 10, 2009. A total of 8 boreholes were constructed. The initial water inflow of boreholes was between 2 and 45 m^3/h, and the total initial water inflow was 107 m^3/h. By September 1, 2009, there was no water in the four boreholes located in the cutting-eye, and the water inflow of the other boreholes was between 2 and 4 m^3/h, and the total water inflow of the boreholes was 10 m^3/h.

The Jinjie Coal Mine 31210 working face is adjacent to the 3^{-1} coal fire boundary in the west. It is pointed out in the *Summary of the Detection Project of the thickness of Overlying bedrock in the second Panel Area* and the 3^{-1} coal fire area that there is a 2^{-2} coal fire area near borehole Jb8 and Jb10. The thickness of the burned rock is approximately 20 m, which forms a unified aquifer with the surrounding weathered rock aquifer, enhancing the water richness of these areas. The surface of the working face 210 is covered by quaternary aeolian sand. The stratum structure is as follows: the sand layer is 1–40 m thick, thickening from the cutting-eye and Return channel to the middle of the working face, and the sand layer is water-free. The soil layer is 0–40 m thick, the thinnest at the Return channel and the thickest at borehole Jb9 in the middle of the working face. The thickness of weathered bedrock is 2–40 m with the thinnest at borehole Jb4 in the middle of the working face and the thickest at cutting-eye and Return channel. The competent bedrock thickness of 3^{-1} coal is 16–35 m. The overlying bedrock in the area between Jb9 and Jb4 in the middle of the working face is the thinnest, and the thickness of other areas is stable, approximately 30 m. The overlying bedrock in some boreholes is above 50 m. The 93208 working face of the first mining face had 53 boreholes for water exploration and drainage before mining, which had been drained and lowered to the overlying aquifer by approximately 10 m. The construction of the boreholes for water exploration and drainage of the 31210 working face began on October 30, 2009, and was completed on March 15, 2010. A total of 16 boreholes were constructed. The water exploration and drainage boreholes were constructed toward the roof of the working face at an elevation angle of 45–50°. The detection borehole in the 3^{-1} coal fire area was constructed horizontally to the west of the working face. The final borehole of the 75 mm water exploration and drainage borehole is located in the weathered bedrock, and the single detection borehole in the 3^{-1} coal fire area is 50 m deep. There is no fire area in the construction, and the final borehole is still in the 3^{-1} coal. The initial water inflow of boreholes was between 0 and 60 m^3/h, and the total initial water inflow was 468 m^3/h. By March 18, 2010, the cumulative water discharge was 154,715 m^3. There is no water in 6 boreholes, among which 5 boreholes are located in the goaf. The water inflow of other boreholes was between 6 and 25 m^3/h, and the total water inflow of boreholes was approximately 109 m^3/h. At this time, the cumulative mining of the working face is approximately 650 m. On January 27, 2010 when the total working face mining was 42 m, the old roof initially collapsed, and the water inflow of the working face increased from 200 m^3/h before the initial collapse to 530 $m^3/$h. The water inrush area was mainly between the Yun Shun and 80 supports (140 m away from the Yun Shun). On January 28, the water inflow of the working face fell to 245 m³/h, and on March 18, 2010, the water inflow of the working face was still around 240 m³/h. There are two main reasons for water inrush in the cutting-eye area: first, the weathering bedrock in the cutting-eye area is 25–40 m thick, and the weathering rock is rich in water. Second, there are few water drainage boreholes in the cutting-eye area, and the water in the weathered bedrock aquifer near the 31209 face goaf is discharged, while the water in the weathered bedrock aquifer near the Yunshun area is not discharged.

Comparison of water level observation data and borehole stratigraphic structure indicates that after April 2008, there was no water in No. 18, No. 19, No. 20 and No. 31 boreholes of the sand layer. Before water discharge at 208 working face, the water level of each observation borehole was in the order of view 19 > View 31 > View 18 > view 20. The water level at borehole No. 19 has pressure property and began to decline since April 2008. However, it is always higher than the weathering bedrock top interface elevation of 1165.07 m. There is no water in the sand layer of observation borehole No. 31. Since the water discharge of the 208 working face, the water level of the fissure water level at observation borehole No. 31 is lower than the weathering bedrock top interface elevation of 1182.99 m. The water level decreased from 1169.59 m on December 19, 2008 to 1155.39 m on February 3, 2009, and the fissure water above 3^{-1} coal was close to dry. The lowest water level of Guan 18 borehole occurred on December 18, 2008, and the water level elevation of 1172.21 m was still higher than that of the weathered bedrock top interface of Guan 18 borehole, 1155.79 m. The fracture water still had pressure property. The water level of observation borehole No. 20 has been continuously decreasing since the discharging of water at the 31208 working face where the clay thickness is only 2.5 m. The water level of the borehole has been reduced from 1155.593 m on May 9, 2008 to 1152.343 m on February 8, 2010. The decrease and speed of the water level are far lower than those of observation boreholes No. 31, No. 18 and No. 19 in the upstream. This indicates that the water richness of the working face near the Qingcaojie Gully is enhanced, and the water level in observation borehole No. 31 is reduced, which also indicates that the supply from the south to the Shamu River is reduced.

4.4.1.3   Four Panel Area 401–404 Working Face

Most of the surface area of the 401 working face is covered by the Quaternary aeolian sand. The stratum structure is from top to bottom: the thickness of sand layer is 5–60 m, and the general trend is gradually thickening from the cutting-eye to the return channel. The sand layer is the thinnest near the watershed, and the thickness of the sand layer in the initial mining area of the cut hole is approximately 30 m, and the thickness is approximately 60 m near the return channel. The soil layer was 0–75 m thick, and the general trend was gradually thickening from the return channel to the cutting-eye. The soil layer between Huishun 30–47 and Yunshun 27–45 was zero in the 31401 working face. The thickest was in the area of opening cut. The weathering bedrock of Zhiluo Formation is 13–54 m thick and gradually thickens from the Return channel and the cutting-eye to the middle of the working face, and the thinnest part is in the cutting-eye. The thickest part is located near borehole 28 in the middle of the working face. The competent bedrock thickness of 3^{-1} coal is 14–60 m, which gradually thickens from the rut passage and the cutting-eye to the middle of the working face. The thinnest place at the cutting-eye is approximately 14 m thick, and the thickest place is near Yunshun 50 lane and Huishun 40 lane with a layer thickness of approximately 60 m. Overlying aquifers on the work surface include sand beds and weathered bedrock aquifers. The sand aquifer is 0–50 m thick, distributed between the middle of the working face and the Return channel, and the thickest part is located near the return channel. The weathered bedrock aquifer thickness of 13–54 m, cutting-eye the thinnest, view the thickest borehole 28, fissure water with micro pressure. The thickest part of the total aquifer is also at observation borehole No. 28, with a layer thickness of approximately 70 m and no soil layer. The sand layer water directly supplies the weathered bedrock water. Second, the total aquifer at the return channel is approximately 60 m thick, mainly the sand aquifer, and the soil layer is thin. This working face belongs to the Qingcaojie gully watershed, and the sand layer water of the working face is mainly replenished by atmospheric precipitation. It flows from high to low along the top surface of the soil layer, i.e, from east to west, and is discharged in the form of spring in the Qingcaojie gully, and directly replenishes the weathered bedrock aquifer water in the skylight area near borehole No. 28. The weathered bedrock of the working face is mainly supplied laterally and is supplied by water from the upper sand layer near the observation borehole No. 28. The runoff along the top surface of the competent bedrock of 3^{-1} coal is from high to low, i.e., the runoff is from the middle of the working face to the direction of the cutting-eye and return channel and is discharged in the gully area. The 31401 face excavation for the fixed rturn channel position, in June 2009 construction T1 (31402 Huishun to 6 lane), T2 (31401Huishun to 7 chamber) two boreholes, the rest of the water exploration drilling on December 26, 2009, to May 1, 2010 completion, a total of 95 boreholes. All the boreholes in the initial mining area were drilled through the weathered bedrock to the final borehole layer in the soil layer, and the final borehole layers of the other boreholes are in the weathered bedrock, mainly to drain the weathered bedrock aquifer water. The initial water inflow of the borehole is between 15 (31402 Huishun 113 lane) and 70 m^3/ h (31401 cutting-eye), and the total initial water inflow is 3462 m^3/h. By May 4, 2010, the water inflow from the borehole was between 3–23 m^3/h. The total water inflow from the borehole was approximately 1121 m^3/h, and the cumulative water discharge was 2.36 million m^3. At the same time of water exploration and release, the water levels at No. 26, No. 24, No. 28, No. 27 and No. J608 boreholes were observed regularly. The water level observation records are shown in Table 4.11 and Fig. 4.20.

Boreholes Guan 26 and J608 are water level observation boreholes of weathered bedrock. Observation borehole No. 6 is located near the groundwater watershed, and the sand layer is not water-bearing. On May 4, 2010, the weathering bedrock aquifer of the borehole is 26.35 m thick, the head height is 36.42 m, and it is confined. After

Table 4.11 Surface borehole water level observation record table in Sipan area

Borehole number	Location	Elevation of bore borehole (m)	Depth of initial water level (m)	2010.5.4 water level		Decrease from initial water level (m)	Sand floor elevation
				Depth of embedment (m)	Elevation (m)		
Guan 26	31401 Mianyunshun 92 alley	1290.17	52.50	60.06	1230.11	7.56	1285.77
Guan 27	31401 face Yunshun 12–13 joint roadway	1221.41	5.60	8.86	1212.55	3.26	1181.39
Guan 28	31401 Yunshun 29 united lane	1227.58	9.03	10.88	1216.70	1.85	1202.58
J608	31402 Surface Yunshun 9th lane	1222.84	7.40	8.84	1214.00	1.44	1192.07

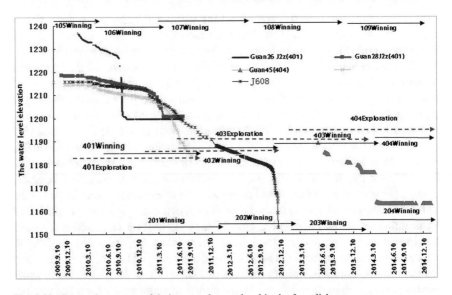

Fig. 4.20 Comparison curve of drainage and water level in the four disk areas

the 401 mining, the decline rate of Guan 26 borehole increases sharply, and the water level drops to 1221.67 m below the weathering bedrock top interface after August 23, 2010. There is no pressure.

The sand aquifer of borehole J608 is 30.7 m thick, the weathered bedrock aquifer of borehole J608 is 47.25 m thick, and the water head height is 79.88 m. On July 6, 2011, the water level of the 401 working face dropped to 1199.49 m after the end of mining. With the influence of drainage of the 402 working face, the water level dropped to 1192.07 m below the sand floor on October 25, 2011, and the water level dropped to 1181.84 m below the clay floor on July 10, 2012. After the implementation of water exploration and drainage in the 403 working face and the mining of working face, the water level drop rate of Guan 608 borehole was aggravated. On November 16, 2012, the water level dropped to 1152.84 m, which was approximately 30 m lower than the roof of the weathered bedrock.

The observation borehole 27 is a sand water level observation borehole. The buried depth of the sand floor is 40.02 m, the elevation is 1181.39 m. The clay thickness is 30.25 m, and the elevation of the weathered bedrock roof is 1151.14 m. On May 4, 2010, the sand layer aquifer at Guan 27 borehole was 31.99 m thick, while the aquifer of the weathered rock was approximately 35 m thick. Drainage of working face 401 caused the water level of Guan 27 borehole to continuously decrease. After the water exploration and drainage of the working face 402 was superimposed after January 19, 2011, the water level of Guan 27 borehole decreased more rapidly, while the water level of Guan 27 borehole was still higher than the sand floor when the mining of the working face 401 was completed. On August 25, 2011, the water level was 1183.795 m, indicating that water richness is better near borehole 27. For example, the maximum water inflow at the first 31401 working face during mining is 1576 m^3/h, and the water inflow increases with the increase of goaf area on the borehole. The water inflow reaches more than 650 m^3/h in the section near the return channel to the 30th roadway, and gradually increases in the later section, which is significantly higher than that in the initial mining area.

The missing clay in borehole No. 28 was located in the well-forming section of "skylight" area in the weathered bedrock, and the water level of fracture water was consistent with that of sand layer. The buried depth of sand layer floor was 25 m, and the elevation was 1202.58 m. On May 4, 2010, the thickness of aquifer in the sand layer of observation borehole No. 28 was 14.12 m, and the thickness of aquifer in the weathered rock was 54.30 m. Drainage and drainage of the working face 401 caused the water level of observation borehole No. 28 to continuously decrease. After the water exploration and drainage of the working face 402 was superimposed after January 10, 2011, the water level of observation borehole No. 28 decreased more rapidly. The mining of the working face 401 had not been completed, and the water level at borehole No. 28 was observed to be 1200.76 m around April 20, 2011, and the sand aquifer had not been completely drained. According to the rate and decrease of water level of observation borehole No. 28, the water is rich near the skylight area of Guan 28 borehole.

The surface of the 31404 working face is covered by the Quaternary aeolian sand. The stratum structure is from top to bottom: the thickness of sand layer is 4.39–27.3 m, which is the thinnest from the middle of working face (at Yunshun Lane 60) and the thickest at Yunshun lane 85 m. The thickness of the soil layer is 1076 m, among which the thickness is 76 m near borehole J210, and it gradually thins to approximately 10 m in the direction of the return channel. The thickness of the weathered bedrock is 10–45 m, the thinnest thickness around the cutting-eye is approximately 10 m, and it gradually thickens to 45 m towards the return channel. The thickness of the competent bedrock of 3–1 coal is approximately 15–70 m, the thinnest thickness of hole cutting is approximately 15 m, the thickest thickness of Huishun 73 link lane in the middle of working face is approximately 70 m, and most sections are above 30 m. The water drainage program started on December 22, 2012 and was completed on April 27, 2013. 62 boreholes were constructed, and most of the final boreholes were 1–2 m in the weathered bedrock top interface or soil layer. The initial water inflow from the boreholes was between 4.09 and 100 m^3/h, and the total initial water inflow is 2172 m^3/h. By June 21, 2013, the water inflow from the borehole is between 0.8 and 15 m^3/h, and the total water inflow from the borehole is 409 m^3/h. The cumulative water discharge from the boreholes is approximately 136.25 m^3/h, which greatly dredges the standing water reserves of the weathered bedrock aquifer at the working face. The initial water inflow of boreholes is approximately 35.03 m^3/h on average. The boreholes with water inflow greater than 30 m^3/h account for 52% of the total (and the boreholes with water inflow greater than 40 m^3/h account for 33.9% of the total), which are mainly distributed in the middle of the working face and the part of the withdrawal channel, indicating that the aquifer from the middle of the whole working face to the withdrawal channel is rich in water. The amount of water inflow less than 30 m^3/h accounts for 48% of the total, and the working face is distributed in the whole area. Most of the boreholes in the area near the cutting-eye of the working face are between 15 and 20 m^3/h. It is consistent with the fact that the cutting-eye area itself is a ground watershed and the water abundance is relatively weak. Drainage of the working faces 403 and 404 caused the water level at observation hole No. 45 to continuously decrease. On April 2, 2014, the water level of observation borehole No. 45 dropped to 1176.45 m, and the aquifer of the sand layer has not been drained, and the water level is still higher than the elevation of sand layer floor at 1171.5 m. After mining of the working face 404, on April 11, 2014, observation borehole No. 5 dropped sharply to 1163.6 m, lower than the floor elevation of the impermeable clay layer at 1167.750 m. The clay layer thickness of observation borehole No. 45 was only 3.75 m, indicating that the water resistance performance of the clay layer had been destroyed.

The surface of the 31405 working face at Jinjie Coal Mine is covered by Quaternary wind-deposited sand. The stratigraphic structure from top to bottom is as follows: sand layer thickness 4.39–50 m, the thinnest near the cutting-eye (borehole JB12) and the thickest near transporting Shun 20 Union Lane; soil layer thickness 10–76 m, with the thickest 76 m near borehole J210, gradually thinning to approximately 10 m in the direction of the retreat passage. The thickness of the weathered bedrock is 7.03–49.91 m, the thinnest thickness is approximately 7.03 m near the cutting-eye, and gradually thickens to 31.15–44.88 m towards the return channel; the thickness of the competent bedrock is 10.54–44.96 m, the thinnest thickness is approximately 10.54 m at the cutting-eye, and the thickest thickness is approximately 45 m at the working face of the tunnel of Yunshun 11. The overlying aquifer of the working face includes the sand layer and the weathered bedrock aquifer. According to the observation borehole of No. 45, the sand layer in the return channel area contains a small amount of water, and the sand layer in the rest of the area almost does not contain water; the weathered bedrock aquifer is 10–45 m thick, and the distribution of the whole area is generally 20–30 m, among which the cut-hole area is thinner at approximately 10 m, which is the direct source of water filling when the working face is retrieved. The water drainage project started on 9 June 2014 and was completed by 16 November 2014. 69 boreholes were actually constructed, with most of them ending up at the top interface of weathered bedrock or 1–2 m within the soil layer. The initial surge volume of the boreholes ranged from 3.1 to 87.6 m^3/h, with a total initial surge volume of 2164.7 m^3/h. By 14 January 2014, the surge volume of the boreholes ranged from 0 to 26.4 m^3/h, with a total borehole surge of 388 m^3/h. The cumulative volume of water released from the borehole is approximately 2,191,200 m^3. The average amount of initial gushing water in the boreholes is approximately 31.37 m^3/h, with 53.6% of the total number of boreholes with gushing water greater than 30 m^3/h (including 48.6% of the total number of boreholes with gushing water greater than 40 m^3/h), mainly distributed in the central part of the working face and the retreat channel area. 46.4% of the total number of boreholes with gushing water less than 30 m^3/h are distributed throughout the working face, with the area near the working face eye cutting. Most of the boreholes are in the range of 3.1 and 13.4 m^3/h, which is in line with the fact that the area of the eye cut itself is a ground watershed, with a thin aquifer thickness and relatively weak water-richness.

The maximum gushing water amount was 1576 m^3/h when the 31401 first mining face was retrieved. The gushing water amount reached more than 650 m^3/h in the section from the return channel to the 30 Union Lane. The maximum gushing water amount was 992 m^3/h when the 31402 working face was retrieved (near the return channel), and the gushing water amount was approximately 200 m^3/h in the initial mining area of the eye cutting. The maximum gushing water amount was 843 m^3/h when the 31403 working face was retrieved. The maximum gushing water amount was 482 m^3/h when the 31404 working face was retrieved as of January 14, 2015. The gushing water gradually increased in the working face near the water-rich area of the return channel.

From the above each working face drainage and water level dynamic change can be seen:

(1) From the dynamic observation of pore water, it can be seen that the water level in the sand layer at the working face is substantial, indicating that the pore water in the mine area is closely linked to the fracture water hydraulics, especially the large volume of evacuation water at the first mining face, and the evacuation water includes a large amount of pore water in the sand layer.

(2) After mining, the clay water barrier is destroyed and the pore water level drops, but the pore water drop in the sand layer and the drop is smaller than the fracture water, reflecting that the Baode Formation and the Lishi Formation have a certain degree of water barrier in the mine area.

(3) The pore water and fissure water of the sand layer in the skylight area has a uniform water surface morphology and dynamic change pattern with good

water-richness, and the pore water of the sand layer has not been completely evacuated and dried after the recovery of some observation boreholes due to the healing of the water barrier.

(4) Watertightness damage of complete bedrock in mining area after mining.

4.4.2 Hydraulic Properties of Roof Aquiclude

4.4.2.1 Problems Revealed by Changes in the Amount of Water Gushing from the Working Face

The field measurement shows that there is a close relationship between the mine water inflow, the roof water outflow and the roof pressure, which indicates that the change of the roof waterproof layer changes the circulation way of groundwater. When the working face of 93101 advances to 57.5 m, the maximum water inflow of the working face is 360 m^3/h. The water volume gradually decreases and tends to normal. When the working face advanced to 799.1 m, the water inflow of the working face appeared to be a large anomaly, and the water inflow reached 155 m^3/h. According to the analysis in Table 4.12 and Fig. 4.21, the maximum water inflow of the working face appears near the initial pressure position of the old roof, and the relationship between the water inflow and the advancing distance of the working face is not obvious thereafter, which reflects that the water inflow in the later period is the supply quantity available within the influence range of the working face, namely, the water-resisting layer of the roof is conductive to intensify the external supply.

When the 93102 working face advanced to 27.5 m, the maximum water inflow of working face was 210 m^3/h. Due to the large water inrush, the working face stopped production for 7 days when it advanced to 28.5 m. On July 3, 2007, the maximum water inrush of the working face was 210 m^3/h, and then the water began to gradually decline to the normal water inrush. Since then, there have been several anomalies (Table 4.13). As can be seen from Fig. 4.22, the maximum water inflow of the working face appears near the initial pressure position of the old top. The water

Table 4.12 93101 working face water inrush anomaly, advance distance, seasonal relationship statistical table

Date of advancement	Advance distance m	Inflow of water m^3/h	Remark
2006.10.4	57.5	360	Near initial pressure
2007.5.13	529	105.5	
2007.6.3	671.25	132	
2007.8.3	799.1	155	Rainy season
2007.9.7	895	103	Rainy season
2007.9.24	965.5	130	Rain season, the end of mining

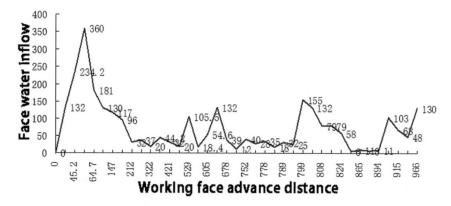

Fig. 4.21 93101 face water inflow and advance distance relation curve

	Date of advancement	Advance distance/m	Inflow of water m³/h	Remake
Table 4.13 93102 working face water inrush anomaly, advance distance and seasonal relationship statistical table	2007.6.30	27.5	210	Near initial pressure
	2007.7.3	28.5	210	Rainy season
	2007.08.09	187.6	141	Rainy season
	2007.11.14	596.3	146	Winter
	2008.02.29	797.5	190	Winter
	2008.04.20	1107.8	180	The end of mining

inflow of the working face shows an obvious increasing trend with the increase of the mining area, and the periodic conduction of the water-barrier shows abnormal water inflow.

When the 93103 working face advanced to 23.9 m, the water inflow of the working face began to increase. When the working face advanced to 42 m, the water inflow reached the maximum value of 222 m³/h in this mining stage. Then the water inflow gradually decreased and returned to normal. When the working face advanced to 65.5, 283 and 289 m, the water inflow appeared abnormal, and the water inflow was much larger than the water inflow near the initial pressure. When the working face advanced to 610.5 m, the water inflow of the working face reached the maximum of 501 m³/h in the whole mining process. After that, the water inrush was abnormal at 863.5 m, 1073.5 m and 1310.8 m, respectively (see Table 4.14). As can be seen from Fig. 4.23 of the relationship between water inrush and advancing distance of the working face, the water inrush of the working face was abnormal near the initial pressure position of the old top, but it was not the maximum water inrush. Different from the water inrush characteristics of the 93101 and 93102 working faces, the water inrush of working faces shows an overall increasing trend with the increase of

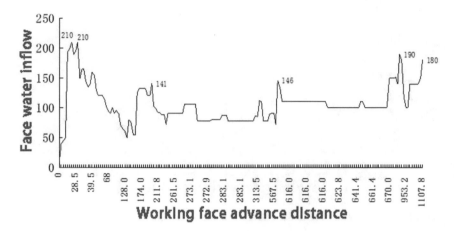

Fig. 4.22 93102 relation curve between water inflow of working face and advancing distance

mining area, indicating that the water inrush of working faces is closely related to the area of goaf.

Table 4.14 93103 working face water inflow anomaly and advancing distance relationship statistics table	Date of advancement	Advance distance (m)	Inflow of water (m³/h)	Remake
	2007.11.10	42	222	Near initial pressure
	2007.11.16	65.5	290	
	2007.12.09	283	370	
	2008.01.10	610.5	501.0	Winter
	2008.02.06	863.5	472.0	Winter
	2008.02.25	1073.5	467.0	
	2008.03.26	1310.8	387	

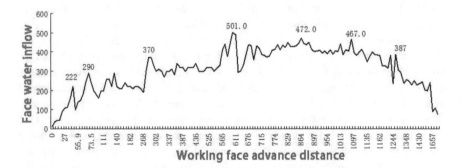

Fig. 4.23 93103 working face water inflow and advance distance relationship curve

In addition, it can be seen from Fig. 4.23 that when the working face was advanced to 1310 m, i.e., when it is advanced to the vicinity of the goaf of the 93102 working face, the water inflow of the working face decreased on the whole, and then the water inflow was stable at 338 m^3/h, indicating that drainage funnel centering on the goaf of the 93101 and 93102 was formed around the goaf. After one year of drainage, the water level of the aquifer and the water inflow of the adjacent working face were reduced.

To sum up, the different characteristics of the working face water inflow, mine water inflow and goaf area, working face advancing distance relationship is complex. Is is also one of the reasons for the low accuracy of conventional mine pit water inflow calculation results. The abnormal point of water inflow indicates that the waterproof layer of the roof leads to stimulate the upper storage into the well, while the stable displacement is only the lateral runoff supply of the working face.

4.4.2.2   Conductivity of Aquifuge

(1)   Quaternary unconfined water level variation

Guan 6 belongs to the sand layer water level observation borehole. The coal burial depth of 3^{-1} is 100.82 m, the thickness of roof bedrock is only 33.54 m, and there is 19.07 m Q2l loess and 19.56 m N2b laterite. In early December 2009, the 31105 working face passed through the east of the borehole, and the water level continued to drop. After February 8, the loess and laterite were presumed to be completely conductive, and then the diving level dropped significantly. It temporarily maintained at this level until 29 June 2010 (Fig. 4.24).

The aquifers in Guans 2, 6 and 7 are Q3s. Guan 2 is located in the middle of the mining area, the thickness of the sand layer of Salawusu formation is 58.32 m, the underlying is 8.05 m laterite and mudstone, and the buried depth of 3^{-1} coal seam

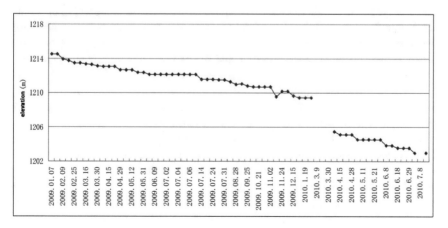

Fig. 4.24 Dynamic change curve of Guan6 borehole before and after mining

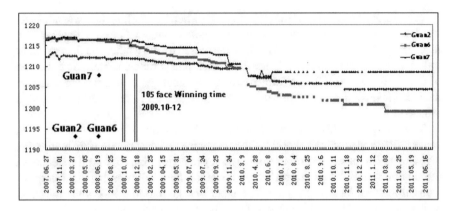

Fig. 4.25 Comparison of dynamic change curves between Guan2 and Guan7 boreholes

is 98.86 m. In November 2009, the 105 working face passed through at a distance of 940 m due east. Although the borehole is far away from the working face, the water level dynamic shows a trend of continuous decline since 2007. It indicates that the drainage of other working faces in the periphery has affected the shallow pore water. Guan 6 is located at Guan 2 east, 480 m from the working face 105, while Guan 7 is located at Guan 6 north, 480 m from the working face 105. As can be seen from Fig. 4.25, with the working face advancing from north to south, the water level in Guan 6 changed greatly. Before February 8, 2010, the groundwater flowed to the southwest direction, and the mining face advanced to the central position of the 105 working face, and the groundwater flowed to the southeast direction, i.e., the groundwater migrated to the goaf. Groundwater flow field changes under mining conditions.

(2) Variation of fissure water level in Zhiluo Formation

The observation borehole of Guan 26 belongs to the observation borehole of weathering fissure water of bedrock, which is located in 31401 transportation lane, and the buried depth of 3^{-1} coal is 137.87 m. The decrease of water level in the borehole began to increase on August 17, 2010, and the working face approached the borehole at the end of August, 2010, and then the decrease of water level in the borehole increased sharply and temporarily stopped on September 10, 2010. From February 8, 2010 to September 27, 2010, the water level of the borehole decreased as much as 37.35 m, and the depth of the water level reached 90.62 m (Fig. 4.26).

Observation borehole No. 24 is located on the west side of the coal pillar of the 31107 working face. On March 10, 2010, the working face passed through the side of the borehole. Before the working face approached the borehole, the water level in the borehole was basically stable. On February 8, 2010, the water level elevation was 1217.46 m. After the working face approached, the water level dropped sharply. On March 9, 2010, the water level elevation was 1212.85 m, and then the water level continued to decline. On May 17, 2010, the water level elevation was 1,206.7 m.

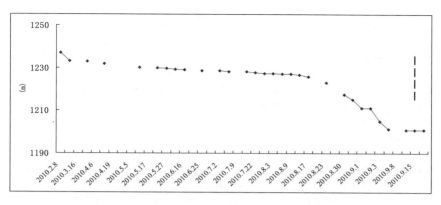

Fig. 4.26 Dynamic change curve of Guan26 borehole before and after mining

On May 27, 2010, the water level elevation rebounded to 1207.34 m (Fig. 4.27) and temporarily maintained at this level. The target layer of the borehole is J2Z, and the water is stopped by the solid pipe under the loose layer. Before the roof failure, the weathered fracture water was confined, and the pressure measuring water level is located in the Q2l loess. After the working face passed through the borehole, the fracture water changed from the confined state to the pressure-free flow.

The observation boreholes No. 15 and No. 25 are both J2Z aquifers. The observation borehole No. 15 is located at 31104 working face, 102.9 m away from the cutting-eye. The aquifer of observation borehole No. 15 is J2Z, Q4 aeolian sand does not contain water, and there is 42.86 m clay at the top of J2Z. After the mining of the working face, the water level of observation borehole No. 15 continued to decline, and the water level could not be measured on May 26, indicating that the section was basically drained. The weathering zone has been communicated and is not recharged by the upper pore water. The observation borehole No. 25 is a weathered bedrock water level observation borehole with a thickness of 32.15 m overlying the competent

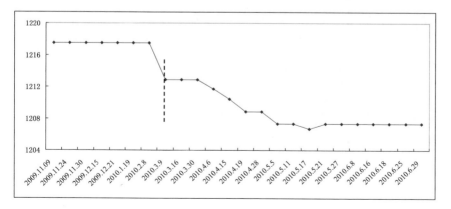

Fig. 4.27 Dynamic change curve of observation borehole No. 24 before and after mining

bedrock. The working face 31201 was close to this borehole in early December 2010. The groundwater level decreased from 1164.368 m on November 29 to 1158.668 m on December 9. For example, according to the observation of No. 31 borehole on the 208 working face, the thickness of weathered bedrock aquifer in the cutting-eye area was 26.60 m, and the thickness of the aquifer was 18.30 m before the initial collapse of the old top on February 2, 2009. After the initial collapse of the old top on February 3, the thickness of the aquifer was 12.40 m, and the water level decreased by 5.90 m.

The observation borehole No. 8 is located approximately 42.8 m to the west of Huishun 55 of the 31105 working face. The aquifer of observation borehole No. 8 is Q and J_2z, Q_4 aeolian sand is 53.8 m, and there is 9.12 m N_2b clay on the top of J_2z. The burial depth of coal 3^{-1} is 98.65 m. The overlying bedrock is only 38.65 m. The thickness of overlying bedrock is less than the height of the caving zone, and the distance of 42.8 m from working face is less than the influence area of moving basin. Before and after December 5, 2009, the working face reached the east of observation borehole No. 8, and the sharp water level drop began on November 30, and then the water level continued to decline. Mining caused bedrock fracture development (Fig. 4.28).

Based on the above dynamic characteristics of different types of groundwater observation boreholes, the groundwater level of either pore water or Zhiluo Formation fissure water has a sharp decline, indicating that the integrity of loess of Lishi Formation, laterite of Baode Formation, mudstone of Jurassic Formation and complete sandstone overlying the roof of coal seam above the goaf has been destroyed. Under mining conditions, the fracture or separation of the roof waterproof layer led to the enhancement of the overall or local permeability of the waterproof layer.

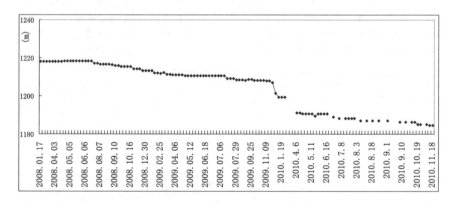

Fig. 4.28 Dynamic change curve of Guan8 boreholes before and after mining

4.4.2.3 The Bridging of Aquifuge

The observation borehole No. 22-1 is located 100 m north of the southern waist lane of 93105 face. The thickness of Q4 aeolian sand is 45.13 m, the thickness of Q_2l loess is 14.65 m, the thickness of J_2z sandstone is 26.31 m, the thickness of overlying bedrock is 38.22 m. The aquifer is Q_4 and J_2z. The working face 105 began to be mined on April 9, 2009. From the dynamic point of view of the water level of observation borehole No. 22-1, the water level of the borehole began to drop sharply from October 26, 2009 (Fig. 4.29), indicating that the aquifuge has been fractured, and the aquifuge was completely damaged for 14 days. By November 9, 2009, the water level reached a temporary dynamic equilibrium state. The water level was briefly stable for 36 days, and the groundwater level maintained at 1203.995 m level. Before and after November 24, 2009, the mining face passed under the observation borehole No. 22-1. After 21 days, the water level in the borehole began to rise on December 21, 2009, indicating that the fractures began to close. It took 42 days from the complete destruction of the aquiclude to the closure of the aquiclude. Due to the conductivity and bridging of the waterproof layer, the water inflow of the 105 working face presents the change characteristics of "small-large-small". The corresponding displacements of the working face in October, November and December are 1333 m^3/h, 1453 m^3/h and 1230 m^3/h on average, respectively.

The observation borehole No. 33 is located at the 93107 working face. The observation target layer is Q_4 aeolian sand, and the thickness of the overlying bedrock of 3^{-1} coal is 67.27 m. In early September 2010, the working face was close to G33. On September 2, the water level of the Quaternary system began to drop sharply to the lowest point, which took 18 days, and then the water level rose, and the water-barrier layer appeared to be restored (Fig. 4.30).

The drainage engineering of the Daliuta Coal Mine 201 working face in the north of Jinjie Mine also shows the conductivity and recovery of the aquiclude. In this drainage project, vertical boreholes were drilled from the surface to the working face through the loose layer aquifer to the coal roadway, and the groundwater was

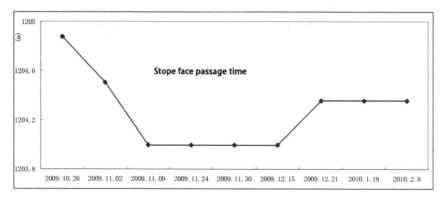

Fig. 4.29 Dynamic change curve of Guan22-1 boreholes before and after mining

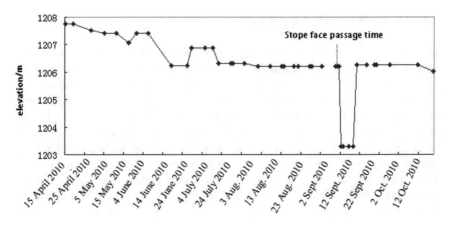

Fig. 4.30 Dynamic change curve of Guan33 boreholes before and after mining

discharged into the roadway. The water level observation system consists of Guan1, Guan2, Guan3 and two pumping boreholes, No. 1 and No. 2. The working face 201 started mining on July 10, 1995, and the old top of the working face was first pressed when the mining reached 40 m on August 8. The working face cycle to press step distance is approximately 10 m, normal advance, every day can form a cycle to press. August 11 should be the third cycle to press, then the roof fracture zone developed to the maximum. According to the water level observation results of observation borehole No. 3 and observation borehole No. 2 (Fig. 4.31), the water level of borehole 2 decreased sharply at 9 a.m. on August 11, which was 11.18 m lower than that observed 2.5 h ago, while the water level of other boreholes still decreased smoothly. Therefore, it can be inferred that the roof waterproof layer is conductive, and the groundwater penetrates down along the water conduction fracture zone. The water level of borehole No. 2 drops sharply, while other parts of the working face are not conduction, and the water level of borehole No. 3 does not drop significantly. The rock layers in Hole 2 are divided into three sections. The rock layers between 0 and 35.53m are loose sand layers, the rock layers between 35.53 and 80.62m are J_{1-2y} bedrock, and the remaining part is mining coal seams. Compared with other surrounding boreholes, the rock column near borehole No. 2 has the lowest position, so the water level change is observed first in borehole No. 2.

As shown in Fig. 4.32, the water level in borehole No. 2 raised again rapidly after the sharp drop of the groundwater level in the borehole. The rise lasted for 12 days. After the rise, the water level still did not reach the level before the rapid decline, indicating that the current stage was in a negative equilibrium state. From the dynamics after the rebound, the water level of view 3 borehole keeps the same change. The above dynamic change process shows that the water-barrier of rock mass breaks suddenly and then closes slowly.

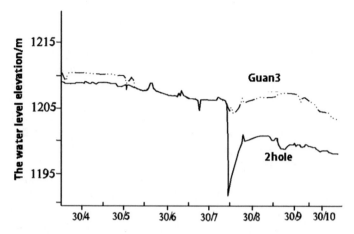

Fig. 4.31 Dynamic curve of water level in borehole No. 3 and No. 2 of 201 working face in Daliuta Mining area

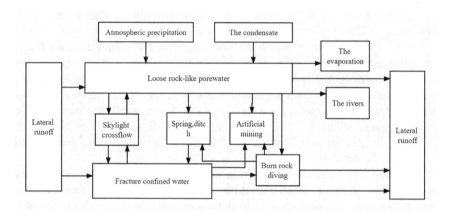

Fig. 4.32 Model diagram of water circulation under natural conditions

4.4.3 Variation of Water Circulation Pathway

4.4.3.1 Water Circulation Mode Under Natural Conditions

As mentioned above, the direct precipitation recharge is the main recharge source for sand diving in this area, while the lateral recharge and condensate recharge are weak. The unconfined runoff migrated along the loess top to the gully, Qingcaojie gully and Heze gully, and finally was excreted mainly in the form of spring, followed by evaporation and vertical excretion. In the area where the soil layer is absent, the sand layer water infiltrates into the weathered rock fissure water of Zhiluo Formation.

Most of the natural groundwater discharge points are distributed in each branch of Qingcaojie gully, which is a steep slope and wide valley, with a topographic elevation difference of 15–25 m. The groundwater mainly exudes linearly from the north and west slope feet of the gully, and the obvious sections of the exudation are different. The length of the seepage section is 50–150 m, and the seepage flow per meter is 0.118–0.179 L/S. According to the observed data before 2008, the seepage discharge of the northern branch of Yangjiagou was 47.15 L/S; the second branch of Cuijiagou was 63.32 L/S; and the Baijiawan branch was 28.73 L/S. The seepage displacement of sand layer is closely related to the distribution area of the aquifer, the thickness of the aquifer, the condition of gully cutting and other factors.

There used to be many lakes (locally referred to as Haizi) in the northern beach area of Qingcaojie Gou, but now the Haizi in the area have all dried up, and the depth of the water level is 1.2–6 m. Under the natural conditions, the natural discharge of groundwater is relatively stable, and the peak of gully discharge mostly occurs in September and October (33,041 m^3/d). The discharge in dry season is 19,987 m^3/d, with an average discharge of 21,417.70 m^3/d. The dynamic characteristics of the groundwater level are that it rises sharply in the rainy season and decreases continuously in the winter. The annual variation is generally less than 1.5 m, and the dynamic type of groundwater is meteorological.

The pore fissure phreatic water-confined aquifer of the Middle Jurassic Zhiluo Formation mainly accepts the lateral recharge of the region and the seepage recharge of the upper skylight. The runoff generally migrates from high to low along the bedrock surface to the infiltration and overflow discharge of the valley area. When the local borehole is exposed, it is artesian water. For example, the water head of the roof aquifer of the 3^{-1} coal seam at the J705 borehole is 8.72 m higher than the orifice, and the artesian flow rate is 1.5 m^3/h. The dynamic type of groundwater shows runoff characteristics.

The distribution of fissure boreholes in the burnt rock is small and receives lateral recharge from quaternary loose layer diving and bedrock weathering zone for a long time. In the terrain of low concave, the burnt rock outcrops by a spring excreted in gully, such as small gully burnt rock spring.

4.4.3.2 Mining Water Cycle Variation Under Mining Conditions

Since the completion of Jinjie Mine in 2006, there has been a general trend of annual increase in the mine's water influx, with the maximum water influx in the mine being greater than 5300 m^3/h. The increase in water influx is accompanied by a decrease in the springs' flow and a drop in the regional water level, reflecting changes in water circulation conditions. The coal seams in the Jinjie mine field are shallow, and the surface of the workings have sunk to varying degrees after mining, affecting the surface topography, groundwater, vegetation and crops. The main manifestation of surface subsidence is the appearance of temporary or permanent cracks and step downs of varying degrees one after another. The width of the cracks and the step drop can be up to 10 cm (Fig. 4.33). At present, only 3^{-1} coal is mined in the mine

Fig. 4.33 Ground subsidence in observation borehole 27 (2011-7-20)

area, and the estimated area affected by surface subsidence of 3^{-1} coal is 26.8–36.4 m^2. The maximum horizontal deformation value is 24.23 mm/m, and the average maximum subsidence value of 3^{-1} coal seam is 2.55 m.

According to the observation data from nearby mines, there are 2 types of ground collapse in terms of time: firstly, short interval collapse after hollowing, i.e., ground collapse occurred during and shortly after mining at the working face. With the expansion of the workings, the overlying rock layer is periodically damaged during the coal mining process and within a short period of time (i.e., a few minutes to a few days after the overlying rock is damaged) the overlying loose layer is affected and causes the ground to collapse. This collapse is rapid in thin bedrock areas and slow in thick bedrock areas. The second is the sudden collapse at long intervals after mining, i.e., the ground collapse occurs after six months to several years after the coal seam has been mined. According to the research results of the mining research institute of Taiyuan University of Technology and Taiyuan Yiyuan Coal Science and Technology Company in April 2008, the hydrogeological significance of the mineral pressure pattern of the 3^{-1} coal seam of the Jinjie Coal Mine 93103 is as follows:

(1) Caving step distance of working face immediate roof is 12.19 m. The initial weighting interval is 32.09 m. The periodic weighting interval varies from 3.46 m to 17.3 m, and the average periodic weighting interval is 7.82 m. Due to the destruction of the roof caused by periodic weighting, the groundwater circulation path changes.

(2) The maximum load on the working face is 42–50 MPa. During the pressure period on the working face, the apparent strength of the mineral pressure (distribution) is relatively average at 40.42 MPa, the 250 m working face length roof collapse and fall has been sufficient, and the pressure on the roof tends to be consistent. The roof slab collapse and the riser mining fissure constitute the underground water storage space and the channel of movement.

(3) Auxiliary transport roadway, working face in front of the beginning of the affected area 130–152 m, an average of 141 m. The front violent influence area is 0–37.5 m, and the influence range of roadway surrounding rock behind the working face is 129–135 m, with an average of 132 m. The average influence distance of advance abutment pressure of two stations in 3^{-1} coal seam roadway is 20.2 m. The average range of the initial area in front of the auxiliary roadway is 131 m, and the influence area behind the working face is 142 m. The permeability of strata within the influence range will change.

(4) The observation and analysis results of the roof convergence of three sections in the auxiliary haulage roadway show that under the roof condition of the 3^{-1} coal seam in Jinjie Mine, the high-level separation subsidence of the roof strata is inevitable, and the separation value can reach 32–59 mm, which is very dangerous compared with the mudstone roof with well-developed layer joints. The maximum roof-to-floor convergence of the auxiliary roadway is 378.51 mm, which belongs to the class III medium-stable roadway type and is not easy to maintain. The greater the separation value, the greater the storage capacity of groundwater.

Taking the 31101 working face as an example, the 31101 working face is the first fully mechanized mining face produced after the production of Jinjie Coal Mine. When the mining advance length of the first mining face is 36 m, the main roof is first pressed, and there are many water sprays in the working face, and the water inflow is 130 m^3/h; when the working face is pressed in a large area, the water inflow increases to 300 m^3/h. With the advancement of the working face mining and the operation of the drainage project, the water inflow of the working face and the water inflow of the drainage borehole gradually decrease. When the working face advances more than half, the water inflow has been reduced by more than half. After the end of the mining, it is only the water inflow in the goaf, and the water volume is approximately 100 m^3 /h. From the water inflow of the 31104, 31105 and 31208 working faces, it can be seen that it has the same law as that of 31101 working face, i.e., the water inflow of mining face generally shows the change rule of first large and then small, i.e., after the working face advances a certain distance (approximately half later), the water inflow decreases obviously, which explained volume storage of the discharged aquifer, reflecting weak lateral runoff. According to past experience, after the first mining surface is mined, other subsequent working surfaces will gradually reduce the amount of water gushing. However, the subsequent mining of the 31102 workings had a surge of 210 m^3/h and the 31103 workings had a maximum surge of 501 m^3/h. On the one hand, it indicates that the water-rich seam on the roof of the 3^{-1} coal seam is strong and has a large hydrostatic storage capacity. On the other hand, it indicates that with the mining of adjacent workings, the mining void area expands and there is a possibility of linking the riser fracture zones of different workings.

In addition, according to the correlation curve between rainfall and mine water inflow, there is no obvious correlation between mine water inflow and atmospheric precipitation. It shows that the main reason is that the atmospheric precipitation is only the recharge source of the loose sand aquifer on the surface. According to the investigation of the production mine, after the fracture generated after the collapse of the mined-out area is connected to the loose layer, the rainwater enters the underground goaf through the water-conducting fracture directly through the surface after the mining in the rainfall season, resulting in different changes in the mine water inflow with the season, so the precipitation is the indirect water source for the mine water filling. The first mining 31101 and 31102 working face, overlying weathered bedrock thickness of 10–20 m and 20–25 m, respectively. After the implementation of water exploration and drainage project, the working face water inflow is less than 100 m³/h, the 31103 working face overlying weathered bedrock thickness of 25–32 m. After the implementation of water exploration and drainage project, the working face water inflow was 100–300 m³/h, the 31104 working face overlying weathered bedrock thickness of 28–35 m. After the implementation of water exploration and drainage project, the working face water inflow 160–350 m³/h, the 31105 working face overlying weathered bedrock thickness of 40–50 m, up to 70 m thick. The working face water inflow is still as high as 1300–1500 m³/h. All reflect the working face overlying weathered bedrock thickness, the greater the general law of water inflow. It shows that the source of water inflow in mining face and mine is closely related to the thickness of weathered bedrock aquifer. With the increase of the thickness of weathered fissure water aquifer, the drainage volume of mining area increases accordingly.

Through the analysis of the change trend and composition of the drainage volume and the water circulation channel in the mining area, it can be seen that the circulation path of groundwater has changed significantly under the mining conditions. The recharge items of the groundwater system are still atmospheric precipitation, condensate water and lateral runoff, but the lateral runoff increases with the water level of the mining area. Among the discharge items of groundwater, lateral runoff, evaporation, gully channel and spring discharge decrease, and most of the groundwater is discharged through the mine.

The mining conditions have the greatest impact on the pore water subsystem and the fracture confined water subsystem (Fig. 4.34). Due to the decrease of the water level of the coal-bearing strata caused by the mine drainage, the increase of the induced overflow, the recharge of the unsealed borehole, the seepage of the caving zone and the small structure, the water level of the pore water subsystem is decreased, the spring flow is attenuated, the river is cut off, and the sea is disappeared. The dynamic type of groundwater presents the mining dynamic characteristics.

A large number of experimental studies have shown that the mining-induced fractures of the overlying rock and soil are mainly composed of upward fractures and downward fractures. The upward fractures are formed by the bottom-up collapse of the roof and the subsidence of the separation layer after mining. The fracture zone is usually called the water-conducting fracture zone (Fig. 4.35a). The downward fractures are the downward-developed tensile fractures formed by the subsidence movement of the stratum on the surface of the stratum (Fig. 4.35b). The communication between the upward fractures and the downward fractures causes the change of the water cycle in the mining area.

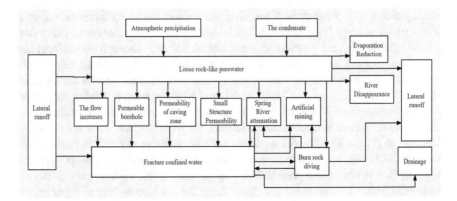

Fig. 4.34 Water circulation pattern under mining conditions

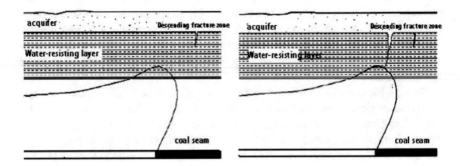

(a)The fissure zone is not penetrated up and down (b)Fissure zones communicate with the upper aquifer

Fig. 4.35 Roof fracture change diagram

4.4.4 Impact of Roof Slab Water Barrier Failure on Surface Trench Flow

(1) Yangjiagou palm, one of the sources of Qingcaojiegou, approximately 1300 m from the nearest 31105 goaf. According to the 2002 survey report, the spring flow here is 59.9 m³/h. In April 2009, there was a small amount of seepage near the palm of the ditch, and there was local water accumulation, almost no flow. At present, there is no water in the gully area, and there is a 6 m diving well at the gully.

The eastern side of Yangjiagou palm and Cuijiagou palm of Jinjie mine is 31105 working face, especially the Spring 2 and Spring 1 of Yangjiagou palm are only 941 m away from 105 working face, which started back mining in 2009 and has been back mined out, according to the field investigation, the groundwater overflow in the upstream of Yangjiagou palm and Cuijiagou palm

Fig. 4.36 Source spring cutoff in upstream of Yangjiagou

was significantly reduced on 23 July 2011, especially the spring point in the upstream of Yangjiagou palm (The groundwater type is sand diving) has dried up (Fig. 4.36), and at the time of the surveys in August 2014 and January 2015, Springs 1, 2 and 3 were all dry and waterless.

(2) The Cuijiagou palm is one of the sources of the Qingcaojie ditch, where the tributaries of Cuijiagou and Yangjiagou converge, and the flow rate of the ditch was between 487 and 578 m³/h before June 2009 and between 378 and 550 m³/h in May 2009. At the time of the survey from August 2014 to January 2015, springs 4 and 5 had been dry for many years, and there was surface flow downstream of spring 5 in August 2014 (Fig. 4.37), Spring 5 was cut off in January 2015 and there was a small amount of surface runoff downstream of Springs 7 and 8 (Fig. 4.38).

(3) Xiaogou Village East is Qingcaojiegou branch ditch, according to the nearest 31210 working face 300 m. There are two branch ditches in the small ditch.

Fig. 4.37 Spring 5 downstream surface runoff 2014.8

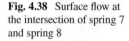

Fig. 4.38 Surface flow at
the intersection of spring 7
and spring 8

According to the 2002 survey report, the south branch ditch flow is 11.1 m³/h,
and the north branch ditch flow is approximately 42.5 m³/h. October 2008 survey
company in this ditch exploration, south branch ditch flow of approximately 13
m³/h, north branch ditch flow of approximately 40 m³/h. In March 2015, August
2014, January 2015, the south branch ditch no water, spring 9 cut-off, north
branch ditch flow of approximately 1 m³/h, spring 10 still outflow, downstream
surface flow (Fig. 4.39).

(4)   Three field investigations were carried out between March 2014 and January
2015 at the springs on the north bank of the Shamu River near the 5th panel.
Compared with 2001, the flow decreased significantly, and springs 11, 13, 15,
16, 17, and 18 all had outflows (Figs. 4.40 and 4.41).

(5)   The flow of the springs in the Hezegou watershed also decreased to some extent,
and three field surveys were conducted between March 2014 and January 2015,

Fig. 4.39 Spring 10 downstream surface flow (2014.8)

Fig. 4.40 Spring 13 flow measurement (2014.8)

Fig. 4.41 Spring 17 flow
measurement (2014.8)

which showed a decrease in spring flow compared to 2001, with some of the
upstream spring sites drying up, but the decrease in spring flow in the upper
Hezegou was not caused by mining in the Jinjie mine, for example, springs
21, 22, 23, 19 and 20 were close to cutting off (Fig. 4.42), showing a seasonal
outflow, and a significant reduction in the flow of spring 24. Other springs, such
as springs 28–32, have flowed (Fig. 4.43). The springs in the Hezegou basin
have not been affected by 3^{-1} coal mining.

It is also evident from river runoff that although precipitation has increased since
2011, river runoff has not increased, especially in the upper reaches of the Shamu
River where spring flows have gradually decreased. According to the results of the
molybdenum ion measurements of the springs in January 2015 (Fig. 4.44), the EC of
the fracture water recharged springs was higher than that of the pore water recharged
springs, and the molybdenum ion fraction changed less, and the flow rate of each
spring in the Shamu River decreased, indicating that the re-mining not only drained

Fig. 4.42 Spring 21 outflow
situation

Fig. 4.43 Spring 28 Outflow

the fracture water but also the pore water, and that the pore water and fracture water were closely related under mining conditions.

It can also be seen from the water level deceleration contour that there is a hydraulic connection between the pore water and fissure water under mining conditions. In Fig. 4.44, the area with the largest reduction rate is in the second and fourth panels. In the second panel, the water level of the Hezegou watershed has a large reduction rate in the skylight area, and the observation borehole No. 45 area in the fourth panel has a large reduction rate. There is also a small-scale skylight area, and the other areas have a relatively small decrease. First, it reflects the poor water abundance of the peripheral pore water, which is blocked by the cohesive soil of the Baode group and the Lishi group. On the other hand, due to the small thickness of the aquifer, the sand layer is drained. Superimposed on the pore water and fracture water drop rate contours, it can be seen that although there are differences in the drop rates of pore water and fracture water, their drop ranges basically overlap, indicating that the pore

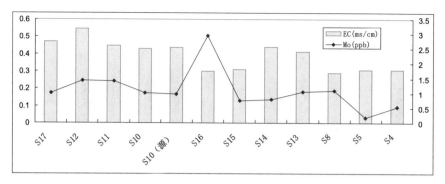

Fig. 4.44 Comparison of spring water chemical indexes in Shamu River Basin

water and fracture water are closely linked hydraulically due to roof damage under mining conditions, and the two together constitute the water-filled water source of the mine area.

References

1. Fan ZL, Fan KS, Liu ZG, Feng YT, Wei Hua, ZYS (2021) Experimental study on the mining-induced water-resistance properties of clay aquicludes and water conservation mining practices. Adv Civ Eng 2021
2. Ganiyu SA (2018) Evaluation of soil hydraulic properties under different non-agricultural land use patterns in a basement complex area using multivariate statistical analysis. Environ Monit Assess 190(10)
3. Hu YJ, Su BY, Mao GH (2004) An experimental approach for determining unsaturated hydraulic properties of rock fractures. Nordic Hydrol 35(3)
4. Huang QX, Wei BN, Zhang WZ (2010) Study on downward crack closure of clay aquiclude in shallow coal seam. J Min Safety Eng 27(01):35–39
5. Liu ZG, Fan ZL, Zhang YJ (2019) Fracture characteristics of overlying bedrock and clay aquiclude subjected to shallow coal seam mining. Mine Water Environ 38(1)
6. Sun Q, Jiang YZ, Ma D, Zhang JX, Huang YL (2022) Mechanical model and engineering measurement analysis of structural stability of key aquiclude strata. Min Metall Explor 39(5)
7. Wang SQ, Niu JJ, Liu Y, Du P, Zhang ZZ, LiangX S (2014) Hydrogeological characteristics of Jinjie coal mine and prediction of mine water filling risk. Coalfield Geol Explor 42(06):55–58
8. Yu JZ (2014) Study on failure characteristic simulation of seam floor for coal mining above confined aquifer based on fluid-solid coupling theory. Appl Mech Mater 3547(675–677)
9. Zhang WZ, Huang QX (2014) Study on the development height of upward fissure in local filling mining of shallow coal seam. Coal Mine Safety 45(04):40–42

Chapter 5
Exploration and Analysis of the Law of Overburden and Surface Ecological Damage Caused by Mining

5.1 The Height of Water Conduction Fracture Zone is Divided by Field Measurement and Bedrock Thickness

Both borehole measurement and ground measurement are used to determine the development height of roof cracks in coal seam mining. At present, the method used for the borehole measurement is the double end water shutoff measurement method. The measurement instruments used in this method are convenient to carry and highly operable. However, when this method is used for measurement, the labor required is extensive, and the measurement data cannot be automatically realized in real time [1–3]. The methods currently used in the surface measurement method include optical fiber logging, micro resistivity scanning imaging logging, borehole television logging, and borehole leakage logging. Although the optical fiber sensor logging technology is a new logging method, the logging transmission data is unstable. For example, the well cable used in the optical fiber logging is extremely easy to be broken and the signal cannot be transmitted during the process of rock falling [4–6]. Therefore, the traditional and mature method of drilling fluid leakage and drilling TV comprehensive measurement is selected for this field measurement [7].

5.1.1 Field Measurement Means and Methods

The observation method of the traditional and more mature drilling fluid leakage and drilling fluid leakage in the comprehensive measurement method of drilling TV adopted this time is mainly to comprehensively determine the top interface of the water conducting fracture zone by observing and recording the water level change in the borehole during drilling and the leakage of the flushing fluid. The borehole television logging method is to lower the high-definition camera probe with full automatic multi angle into the borehole, and comprehensively judge the fracture

© The Author(s), under exclusive license to Springer Nature Switzerland AG 2023
Y. Zeng et al., *Roof Water Disaster in Coal Mining in Ecologically Fragile Mining Areas*, Professional Practice in Earth Sciences,
https://doi.org/10.1007/978-3-031-33140-4_5

development characteristics according to the lithology characteristics and fracture characteristics in the borehole acquired in real time. Figure 5.1 shows the schematic diagram and field measurement diagram of drilling fluid leakage logging, whereas Fig. 5.2 shows the principle of borehole television logging and field measurement.

In order to verify the development height of overburden water conducting fractures after the completion of high-intensity mining in the study area, drill boreholes are designed in the middle of the goaf of the mining working face, the boundary of the transportation chute, the middle of the section coal pillar, etc. for drilling analysis. The number of monitoring boreholes for the development height of overburden water conducting fractures is 4 in total. The monitoring borehole numbers are LD-1, LD-2, LD-3 and LD-4. LD-1, LD-2 and LD-3 monitoring boreholes are arranged 2420 m away from the cutting borehole of the working face. LD-1 is arranged at the center

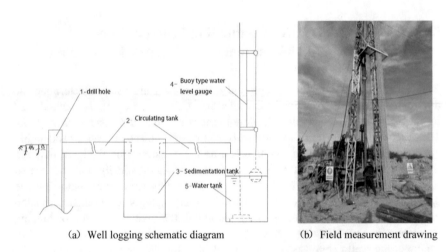

(a) Well logging schematic diagram   (b) Field measurement drawing

Fig. 5.1 Drilling fluid leakage logging principle diagram and field measurement diagram

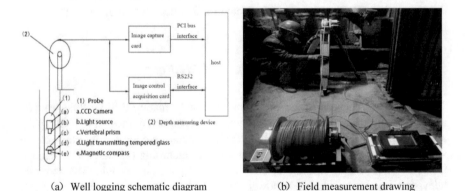

(a) Well logging schematic diagram   (b) Field measurement drawing

Fig. 5.2 Principle of borehole TV logging and field measurement chart

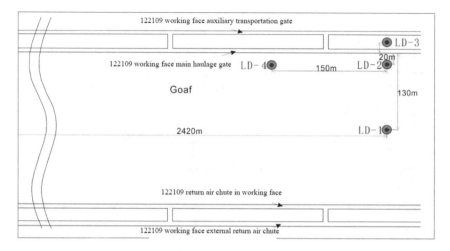

Fig. 5.3 Layout of "two belts" monitoring boreholes

line of the goaf of the working face 122,109, and the borehole is 132 m away from the middle of the transportation chute. LD-2 is arranged in the goaf and 20 m away from the transportation chute. LD-3 is arranged in the middle of section coal pillar. LD-4 is arranged at 2270 m away from the cutting hole of the working face in the goaf and 20 m away from the smooth transportation trough. Layout of four monitoring boreholes is shown in Fig. 5.3.

5.1.2 Analysis of Measured Results

5.1.2.1 Analysis of Observation Results of Flushing Fluid Leakage

The observation method of flushing fluid leakage used in this field measurement is a method to judge the water conducting fracture zone and the top interface of collapse zone in the borehole by observing the change of water level in the borehole and the consumption of flushing fluid with the drilling process. This method is a field measurement method i.e., widely used and has high reliability. The curve of the water level in the 4 monitoring boreholes during the drilling process varying with the drilling depth is shown in Fig. 5.4.

It can be seen from the curve of the water level in the borehole changing with the drilling depth in Fig. 5.4 that the changing relationship of each monitoring borehole is shown as follows: the water level in the borehole decreases with the increasing drilling depth until the water in the borehole leaks completely when drilling to the top interface of the water conducting fissure, so that the water level cannot be detected. The drilling water level of LD-1 is 91.45 m deep, and all flushing fluid is lost. The buried depth of the coal seam roof at LD-1 is 301.50 m, and the mining thickness

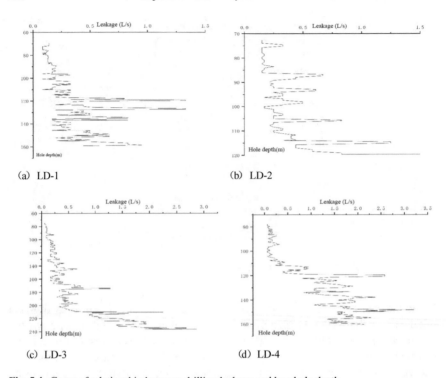

(a) LD-1   (b) LD-2

(c) LD-3   (d) LD-4

Fig. 5.4 Curve of relationship between drilling leakage and borehole depth

of the coal seam at LD-1 working face is 11 m. Therefore, the height of the water conducting fracture zone at LD-1 drilling borehole is 210.05 m. The borehole water level observation of LD-2 starts from 72.76 m, and all the flushing fluid is lost from 86.7 m. The buried depth of the coal seam roof at LD-2 is 300.07 m, and the mining thickness of the coal seam at LD-2 is 10.5 m. Through comprehensive analysis and calculation, the height of LD-2 water conducting fracture zone is 213.37 m. The observation of water level change in LD-3 borehole started from 74.95 m to 168.6 m, and all flushing fluid in the borehole was lost. The buried depth of coal seam roof at LD-3 is 309.12 m, and the mining thickness of coal seam at LD-3 working face is 11.5 m. Through comprehensive analysis and calculation, the height of the water conducting fracture zone of LD-3 borehole is 140.52 m. The observation of water level change in LD-4 borehole started from 77.76 m, and the water level could not be measured in 84.05 m borehole. The buried depth of the coal seam roof at LD-4 is 299.06 m, and the mining thickness of the coal seam at the LD-4 working face is 10.5 m. Therefore, the height of the water conducting fracture zone observed in LD-4 borehole is 215.01 m. The comprehensive detection results of the four monitoring boreholes are shown in Table 5.1.

According to the analysis of the exploration results and the hydrogeological characteristics of the study area, the overburden damage height after mining in the study area has reached the Neogene Baode Formation laterite layer. After mining, the pore

Table 5.1 Detection results of monitoring boreholes in working face

Borehole number of drilling	Mining thickness (m)	Water-conducting fracture zone	
		Height (m)	Ratio of direct roof thickness to mining height
LD-1	11.0	210.05	19.09
LD-2	10.5	213.37	20.32
LD-3	11.5	140.52	12.21
LD-4	10.5	215.01	20.48

fissure confined water aquifer of weathered rock, the pore fissure confined water aquifer of Jurassic Middle Zhiluo Formation and the pore fissure confined water aquifer of Jurassic Middle Yan'an Formation 2^{-2} coal aquifer section in the study area are all damaged. The degree of disturbance and destruction of the phreatic aquifer of the Quaternary Upper Pleistocene Salawusu Formation is controlled by the thickness of the laterite layer of the Neogene Baode Formation. However, according to the change of the thickness of the laterite in Baode Formation in the study area, the laterite in the eolian sand area is thin, even absent. Therefore, through the analysis of the development law of water conducting fractures after mining in the study area, it is easy to cause irreversible "secondary disasters" when unreasonable high intensity mining is carried out in the fragile ecological mining area.

5.1.2.2 Analysis of Borehole TV Imaging Observation Results

The borehole television system is to put the waterproof camera probe with LED and camera into the underground borehole through the cable with depth counter. The waterproof camera probe continuously collects the section information in the borehole through the cable lowering process, and the collected section information is transmitted to the ground acquisition system through the cable; The image post-processing system can identify the fracture of rock stratum at different positions according to the acquired borehole section information. At the same time, combined with the depth counter, the location of the top interface of the water conducting fissure can be determined. Figure 5.2a shows this system that consists of the host of the borehole imager, probe, depth counter, etc.

The TV logging of this borehole was completed in 30 days. The fracture interfaces of each borehole are shown in Figs. 5.5a–d. Based on the summary and analysis of the hole wall changes of the four "two belt" observation boreholes under construction, the four "two belt" observation boreholes can be divided into two categories as a whole: (1) The internal fractures of the borehole walls of LD1, LD2 and LD4 are seriously affected by mining, and the vertical and horizontal fractures are developed, the fracture surfaces are fresh, and the fracture connectivity is strong; (2) The internal fissures of LD3 borehole wall are basically not affected by mining, and there are no large fissures. The main reason is that the borehole is located in the center of the coal pillar and is less affected by mining. According to the four "two zone" observation boreholes shown in Fig. 5.5, it can be seen intuitively that boreholes and fissures of different scales are developed on the borehole wall of the observation section.

According to the distribution characteristics of fractures in the borehole at 91.2–93.9 m in LD-1 borehole in Fig. 5.5a, it can be seen that there are different sizes of boreholes, fissures and falling blocks in the borehole wall fissures affected by mining, and longitudinal fractures begin to appear at 92.15–92.55 m. The fractures are relatively short, but the fracture surface is fresh, with obvious mining damage characteristics. Therefore, it can be concluded from the analysis that the top of the water-conducting fracture zone of LD-1 borehole is 92.15 m deep, while the ground elevation of LD-1 borehole is 1,274.28 m, so the top elevation of the water-conducting fracture zone is 1182.13 m, and the coal seam roof elevation is 972.73 m. For this reason, the development height of the water conducting fracture zone is 209.40 m. The distribution of fractures at 82.9–85.6 m in LD-2 borehole in Fig. 5.5b shows that longitudinal fractures begin to appear at 85.4 m. Therefore, the top of the water conducting fracture zone is located at 85.4 m of the borehole depth, while the ground elevation of LD-2 borehole is 1272.79 m. The top elevation of the water conducting fracture zone is 1187.40 m, and the top elevation of the coal seam is 972.729 m. Therefore, the development height of the water conducting fracture zone is 214.667 m. According to the analysis of fractures in the borehole, as shown in Fig. 5.5c, it can be seen intuitively that the observation section is complete as a whole. No large fractures crisscross mining fractures occur, and only longitudinal mining fractures occur at 168.6 m of the borehole. The main reason for this phenomenon is that the borehole is located in the center of the coal pillar, and the overlying strata are less affected by mining, so there is no large mining fissure and no collapse zone. According to Fig. 5.5d, the longitudinal cracks appeared at 84.05 m and water inrush occurred at 84.25 m. According to hydrogeological analysis, the main reason is that the original cracks become crisscross after the rock stratum is damaged by tension and shear after mining.

To sum up, the detection results of the development height of the water conducting fissure in the overburden after the completion of high-intensity mining in the study area are shown in Table 5.2. The overburden failure height can be 4.88 based on the span mining ratio and 19.98 based on ratio of caving zone to mining thickness. However, according to the empirical calculation equation of the current water conducting fracture zone, after full coal thick and high-strength mining in this ecologically vulnerable area, the top interface of the fracture enters into the weathered bedrock layer. In some areas, the fractures enter into the bottom interface of laterite. Therefore, reasonable determination of mining strength is of great significance for overburden failure.

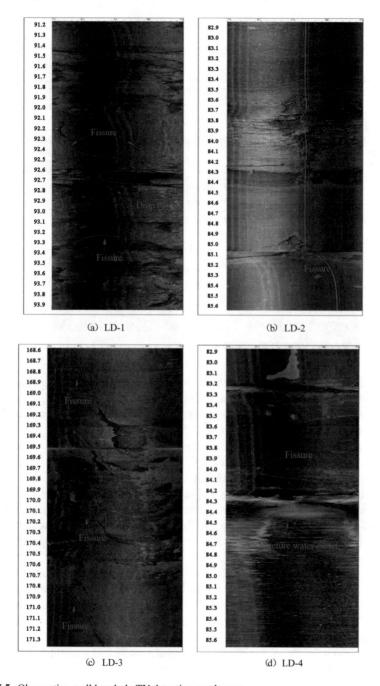

Fig. 5.5 Observation well borehole TV detection result map

Table 5.2 Borehole TV detection results of "two-zone" observation well after mining in working face 122,109

Ratio of direct roof thickness to mining height	LD1	LD2	LD3	LD4	
Coal mining method	Fully mechanized top coal caving mining method				
122109 Burial depth of working face roof (m)	301.55	300.07	309.12	299.06	
Height of water-conducting fracture zone (m)	Borehole depth at vertex position	92.15	85.40	168.60	84.05
	Observation height	209.40	214.67	140.52	215.01
	Ratio of caving zone to mining thickness	19.03	20.44	12.22	20.48
Height of caving zone (m)	Borehole depth at vertex position	247.16	249.74		247.78
	Observation height	54.34	50.33		51.28
	Ratio of caving zone to mining thickness	4.94	4.79		4.88

5.1.3 Analysis of Fitting Calculation Equation for Water Conducting Fracture Zone

According to the analysis of the above measured results, on the premise that the hydrogeological conditions are determined, the damage degree of the mine roof overburden is greatly affected by the mining intensity. However, the design of water-proof coal pillar and the upper limit of coal mining before the mining of coal seams are calculated by the existing empirical equation, but the existing empirical equation is the summary of the development height of roof fractures in coal seam mining in the central and eastern regions by scientific researchers. At present, the focus of coal mining in China is mainly concentrated in the western regions with fragile ecological environment. However, the geological characteristics of coal seams in the west are quite different from those in the middle east. If we continue to use the previous empirical equation for calculation, it is very easy to bring hidden dangers to safety production.

Therefore, this section refers to the measured data of the water conducting fracture zone in Yushenfu mining area, as shown in Table 5.3, carries out data fitting and regression, and determines the calculation equation for the development height of the roof water conducting fracture zone in the working face under the condition of high-intensity mining in Yushenfu mining area.

The measured data of the height of the water conducting fracture zone of the Jurassic coal seam mining in the Yushenfu mining area collected in Table 5.3 are fitted and analyzed, and the fitting relationship between the development height of the water conducting fracture zone of the Jurassic coal seam mining in the Yushenfu mining area and the mining thickness of the coal seam is established, as shown

Table 5.3 Measured data table of water-flowing fracture zone height in Jurassic coal seam mining in Yushenfu mining area

Measured location	Mining thickness (m)	Working face (drill borehole)	Mining coal seams	Height of water-conducting fracture zone (m)	Ratio of caving zone to mining thickness
Caojiatan coal mine	6	DZ1	2	136.10	22.63
	6	DZ2		139.15	23.19
	11	LD-1		209.70	19.06
	10.5	LD-2		214.02	20.38
	10.5	LD-3		215.01	20.48
	10.3	LD-4		172.93	16.79
	10.3	LD-5		182.69	17.74
	10.3	LD-6		133.50	12.96
Yushuwan coal mine	4.5	H-3	3	116.20	25.82
Yuyang coal mine	3.5	ZP1	2	96.30	27.51
	3.5	ZP1	2	84.80	24.23
Ningtiaota coal mine	4.8	Hole1	1	153.46	26.46
Xiaobaodang coal mine	5.80	XSD1	2	158.78	27.38
		XSD4		175.57	30.27
Jinjitan coal mine	5.5 m	JT3	2	111.49	20.30
		JT4		126.40	23.00
		JT5		146.18	26.60
		JT6		120.25	21.90
Liugang coal mine	8.8	L1	3	117.84	13.40
Yushuwan coal mine	5.0	Y3	2	130.50	26.10
		Y4		137.30	27.46
		Y5		138.90	27.78
		Y6		117.80	23.56
Hanglaiwan coal mine	4.5	H3	3	108.32	24.07
		H4		114.38	25.42
		H5		107.83	23.96
		H7		93.87	20.86
Bulianta coal mine	6	BKS6	1	157.1	23.20
Huojitu coal mine	3.5	21,201	1	81.00	23.30
Daliuta coal mine	6.6	J60	5	129.00	19.50

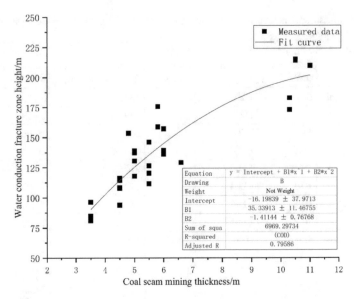

Fig. 5.6 Relationship between height of water-conducting fracture zone and coal seam mining thickness in Shenfu mining area

in Fig. 5.6. $R^2 = 0.81$ can be seen from the figure, and the fitting polynomial is significantly correlated.

5.1.4 Bedrock Thickness Division Based on the Height of Water Conducting Fissure

According to the research status at home and abroad, the main causes of water–sand inrush accidents in the shallow coal seam of Sthe henfu mining area in the past mining process are summarized as follows: the thickness of the bedrock of the coal seam roof is thin, and the height of the caving zone is highly developed. It directly cuts through the Neogene and Quaternary consolidated water-bearing sand bodies at the top of the bedrock, and directly bursts into the goaf through the large cracks in the caving zone. However, the above studies only consider that when the fracture of the collapse zone cuts through the water bearing sand layer, the water bearing sand body collapses into the goaf under the action of its own weight and water pressure. It is ignored that the extremely fine sand layer on the surface can also be substituted into the goaf through the water conducting fracture zone under the combined action of strong ground pressure display and hydrodynamic potential energy to cause water disasters. Therefore, in order to further analyze and summarize the relationship between material sources and channels in water–sand inrush, it is necessary to study and divide the water channel where water–sand inrush occurs on the premise of determining the formation material sources. However, the division of water diversion channel is related to the thickness of roof bedrock. Therefore, this chapter mainly defines and divides the bedrock thickness.

Based on the above analysis, this book proposes to divide the bedrock into four types: thick bedrock, medium thick bedrock, thin bedrock, and ultra-thin bedrock according to whether the bedrock still has the role of key stratum after being affected by the mining intensity of coal seams.

- The bedrock is completely broken under the influence of strong mining, which is called thin bedrock.
- The rock stratum between thick bedrock and thin bedrock after mining is called medium thick bedrock.
- The ultrathin bedrock not only cuts through the bedrock but also cuts through the upper stratum of bedrock after being affected by mining.

The schematic diagram of the division of bed rock thickness is shown in Fig. 5.7. Therefore, the bedrock with water inrush and sand break in the Shenfu mining area in the past can be called ultrathin bedrock. Combined with the research contents in Chap. 4 and the above definitions, the bedrock thickness in the study area bookcan be defined as thin bedrock, i.e., the bedrock is all affected by strong mining to produce cracks.

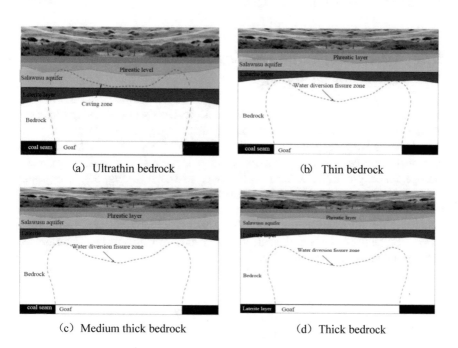

(a) Ultrathin bedrock

(b) Thin bedrock

(c) Medium thick bedrock

(d) Thick bedrock

Fig. 5.7 Schematic diagram of bedrock thickness division

According to the thin bedrock thickness map shown in Fig. 5.7b, there is a certain effective thickness of laterite layer above the water-conducting fracture zone. Therefore, the water–sand inrush of this type of stratum should not only consider the traditional source, but also consider that the soil layer loses its water isolation performance after the soil layer is subjected to strong mine pressure and the dynamic water pressure above, resulting in the surface sand layer source collapse into the underground.

5.2   Measurement and Degree Analysis of Surface Ecological Damage Under High Intensity Mining

According to the mining situation of the mine in Yushenfu mining area, a large amount of water will be collected in the surface subsidence area. It is inferred that the source of water is mainly because the discharge path of the upper phreatic water changes from the original laminar flow to the surface subsidence pit. Therefore, monitoring and determining the damage degree of surface ecology during high-intensity mining is of great significance for analyzing and summarizing and determining the reasonable mining intensity in ecologically fragile areas.

5.2.1   Overview of Surface Ecological Damage Caused by Coal Seam Mining

According to the survey of the post mining ecological landscape of the high-intensity mining area in the current Yushenfu mining area and the review of relevant literature reports, a large number of irreversible disasters have occurred in the goaf, such as ground fissures, house wall fractures, water logging in the goaf, spring water disappearance, groundwater level decline, etc., as shown in Fig. 5.8. The ground house damage and water logging occurred in the goaf after mining in a mining face. According to the following different degrees of damage to the surface after mining, if the relationship between mining intensity and ecological geomorphic bearing capacity is not coordinated, it is very easy to cause further deterioration of the fragile ecological status of the surface.

5.2.2   Measurement Method of Surface Ecological Damage Caused by Coal Seam Mining

In view of the current situation of surface ecological damage after high-strength mining of existing coal seams analyzed and summarized in the previous section, this

(a) Broken houses   (b) Ponding in subsided land

Fig. 5.8 Mining Damage phenomenon

section proposes a quantitative method to analyze and determine the degree of surface ecological damage after coal seam mining. In this book, the method used to determine the degree of surface ecological damage is the field monitoring method. The monitoring tools are level and theodolite. Figure 5.9 shows the monitoring tools and monitoring process. The layout of the surface rock movement monitoring station is mainly designed according to the relevant theory of mining subsidence. The location is on the main section of the mobile basin. There are two survey lines in the monitoring station, one along the strike direction and the other along the dip direction. The two survey lines are mutually perpendicular. The length calculation equation of observation line mainly adopts the following Eqs. (5.5) and (5.6). Figure 5.10a shows the layout of the observation station, and Fig. 5.10b shows the cross-section of the observation station.

(a) Theodolite   (b) Level measurement

Fig. 5.9 Monitoring tools and monitoring process

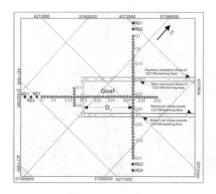
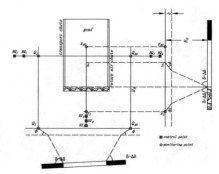

(a) Layout of observation station   (b) Cross section of observation station

Fig. 5.10 Design drawing of subsidence observation of surface movement Basin

The length of strike and dip observation in Fig. 5.10a is mainly calculated according to Eq. (5.1) and Eq. (5.2):

$$L_Z \geq 2(H_0 - h)\cot(\delta - \Delta\delta) + 2h\cot\varphi \tag{5.1}$$

$$D_1 \geq (H_0 - h)\cot(\delta - \Delta\delta) + h\cot\varphi \tag{5.2}$$

where

H_0 The mining depth of the Mining working face. It is 318 m;

δ Direction movement angle. It is 63°;

h Thickness of soil layer. It is generally 140 m;

$\Delta\delta$ Correction value of strike movement angle. It is generally 20°;

φ Movement angle of loose layer. It is 45°;

L_Z Distance from open cut of mining face;

D_1 Open cut distance from mining face.

According to the calculation, the length of strike observation line shall not be less than 661.8 m. The distance from the tilt survey line to the open cut hole shall be more than 330.9 m. Considering the accuracy of relevant parameters and the terrain, the actual tilt survey line is laid 416.8 m away from the cut hole, the length of the strike observation line is 1200 m. The location of the working face cut hole is located at the No. 15 measuring point of the strike survey line.

5.2.3 Analysis of Surface Ecological Damage Degree in Mining Area

The observation work ended when the surface settlement entered a stable state. The rock movement observation work has been conducted for 17 times. Figure 5.11 shows the measurement results, and Fig. 5.12 shows the degree of surface ecological damage in the mining area.

According to the analysis in Fig. 5.11a, during the second phase of observation, the maximum ground settlement is measured at 19 measuring point 100 m away from the cut hole, and the maximum settlement value is 1.03 m. With the advance of the working face, the degree and scope of surface damage are increasing, and finally it enters a stable state, with the maximum settlement value stable at 5.52 m.

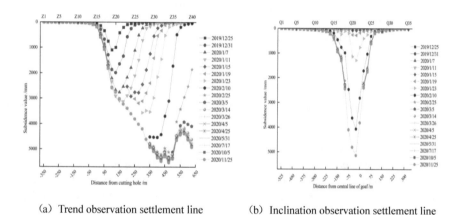

(a) Trend observation settlement line   (b) Inclination observation settlement line

Fig. 5.11 Numerical map of surface movement basin subsidence

(a) Phase I observation map   (b) The 16th observation map

Fig. 5.12 Comparison of surface subsidence and deformation in different periods

The settlement rule of each period reflected in Fig. 5.11b is the same as that in Fig. 5.11a. The maximum settlement position of the dip line is the same as that of the strike line, i.e., 32 points of the strike observation and 20 points of the dip observation are the same measuring points, and the maximum settlement value of this measuring point is 5.41 m. To further explain the surface ecological disturbance caused by high-intensity mining, while conducting surface damage observation, we recorded the surface damage of each phase through images, as shown in the first phase observation map shown in Fig. 5.12a and the sixteenth phase observation map shown in Fig. 5.12b. The locations shown in the observation maps are the same. The image data taken at the same place in the two phases show that the surface ecology has been disturbed and destroyed to varying degrees, such as waterlogging in low-lying areas, plant depletion and death, and farmland destruction.

The field monitoring and image observation and analysis indicate that the fragile ecological landform of the working face has been seriously damaged due to the long-term mining of coal seams. Therefore, when mining coal seams in ecologically fragile areas, we must consider the coordinated development of "coal resources ecological environment," otherwise it will bring great damage to the ecological environment.

5.2.4 Surface Movement Law of Coal Seam Mining

Based on the measured data of each period shown in Fig. 5.28, the surface dynamic deformation curve is drawn, which reveals the characteristics of surface movement in the mining process of the working face. Based on the data of each group, the dynamic parameters of surface movement under the mining conditions of the working face are determined as follows:

(1) The maximum surface subsidence is 5.52 m;
(2) The leading influence angle of surface movement is 56.15°;
(3) The maximum sinking speed is 43.5 cm/day, and the lag angle of the maximum sinking speed is 62.52°;
(4) The total duration of surface movement is 339 days, including an average of 24 days in the initial stage, 105 days in the active stage and 210 days in the recession stage.

Based on the above analysis of measured data, it can be seen that the active duration of surface movement is only one third of the observation period, and the decline period is two thirds of the observation period. Combined with the above drilling television detection results and the mining characteristics of the mining working face, the main reasons for the short active period of surface movement and large subsidence value are that the mining thickness of the coal seam in the mining face is large, the roof overburden is severely affected by mining. The bedrock fracture speed is fast and the damage height is high, and then the bedrock loses the ability to bear the overburden and aquifer, which ultimately leads to surface subsidence. If the water resisting layer produces large mining cracks during the sinking process, it is very easy to cause the rapid decline of the aquifer water body, which will seriously bring the surface sand layer into the goaf and cause water disasters. According to the analysis of the observed data of the strike and dip in Fig. 5.28, the surface subsidence is basically completed in a short active period. According to this phenomenon, it can be concluded that the thick seam fully mechanized top coal caving mining face in the Yushenfu mining area has a fast surface movement speed and a fast settling speed in the process of high-intensity mining.

Therefore, according to the above ground movement rules and the analysis of the characteristics of surface damage, it can be seen that if inappropriate mining methods are adopted in the mining of mines in ecologically fragile areas, not only will coordinate mining between coal resources and the ecological environment be affected, but also will cause water inrush, sand bursting and water disaster accidents in mines. According to the measured data of surface rock movement in the study area, the results of surface movement and deformation parameters are presented in Table 5.4.

5.3 Summary

This chapter summarizes the patterns of mining intensity damage to overburden and surface ecology, and explored the development height of water conducting fracture zone of overburden and the degree of surface ecological damage under the condition of high intensity mining in situ. The main conclusions are as follows:

(1) According to the analysis of the detection results obtained from the four "two belt" observation boreholes in the goaf construction of the 122,109 working face, the range of the split mining ratio is 19.07–20.38, and the range of the caving mining ratio is 4.79–4.96. Based on the statistical analysis of the development height of the existing "two zones" in the Yushenfu mining area, a prediction model of the height of the water conducting fracture zone suitable for the western Yushenfu mining area is proposed. At the same time, based on the relationship between the sedimentary thickness of bedrock and the height of fracture development after mining, the thickness standards of thick bedrock, medium thick bedrock, thin bedrock, and ultra-thin bedrock are defined.

(2) According to the analysis of the results of the exploration work on the degree of surface ecological damage under the action of high-intensity mining, it can be seen that the maximum surface subsidence after the mining of the working face is 5.52 m, and the subsidence rate is $q = 0.502$. According to the analysis of image data, there are "secondary disasters" after the mining of the working face. Therefore, the coordinated development between coal resources and ecological environment must be considered when mining coal seams in the ecologically fragile areas.

Table 5.4 Results of surface movement deformation parameters

Parameter name	Parameter value	Parameter name		Parameter value
Subsidence rate q'	0.50	Boundary angle	Strike boundary angle δ_0	57.80°
Sinking coefficient q	0.72		Uphill boundary angle γ_0	42.26°
Horizontal movement coefficient b	0.499	Moving angle	Downhill boundary angle β_0	42.26°
Maximum sinking angle θ	88.74°			
Starting distance	76 m			
Leading influence distance (l)	213.3 m			
Leading influence angle ω	56.15°	Inflection point offset	Trend — Near the incisions	52.34 m
Radius of influence — Trend γ_{trend}	95 m		Tilt — Go up the mountain	29.57 m
Radius of influence — Go up the mountain γ_{up}	125 m		Tilt — Go downhill	29.57 m
Radius of influence — Go up the mountain γ_{down}	125 m			
Tangent of influence angle — Trend $\tan\beta_{rend}$	3.4	Maximum sinking speed		43.5 cm/days
Tangent of influence angle — Go up the mountain $\tan\beta_{up}$	2.5	Hysteresis distance of maximum sinking speed L		165.4 m
Tangent of influence angle — Go up the mountain $\tan\beta_{down}$	2.5	Lag angle of maximum sinking speed Φ		62.52°
Half basin length — Go downhill L_1	475 m	Mining coefficient	Trend n_3	13.22
Half basin length — Go downhill L_2	475 m		Tilt n_1	0.57
Half basin length — Trend L_3	662.58	Influence propagation angle		84.69°

References

1. Ji WL, Tian Z, Chai J et al (2022) Multi-attribute fusion distributed water-conducting fracture zone height prediction method [J]. J Jilin Univ (Eng J) 02(14):53–60
2. Liu YF, Wang SD, Wang XL (2014) Development characteristics of water-conducting fracture zone in overlying strata of fully mechanized top-coal caving mining in deep-buried and extra-thick coal seam [J]. Acta Sinica Coal 39(10):1970–1976
3. Yi Q, Lu JL, Dai YH et al (2021) Experimental study on overburden failure height of fully mechanized top-coal caving mining in inclined coal seam [J]. Mine Surv 49(04):30–33

4. Hai FL, Duo XY, You BH et al (2015) The prediction of height of water-conducting fractured zone in overburden strata of seam using grey artificial neural networks [J]. Electron J Geotech Eng 20(13):5787–5799
5. Liu Y, Yuan S, Yang B et al (2019) Predicting the height of the water-conducting fractured zone using multiple regression analysis and GIS [J]. Environ Earth Sci 78(14)
6. Miao XX (2011) The height of fractured water-conducting zone in undermined rock strata [J]. Eng Geol
7. Ju J, Li Q, Xu J et al (2020) Self-healing effect of water-conducting fractures due to water-rock interactions in undermined rock strata and its mechanisms [J]. Bull Eng Geol Env 1:79

Chapter 6
Physical Simulation Study of Mine Water–Sand Inrush Under the Fluid–Structure Interaction

6.1 Overview

The water–sand inrush accident is a dynamic process from gestation-development-end. The water–sand inrush accidents show that when the mining fissures reach the overlying sand-bearing aquifer, the water inrush phenomenon first occurs. As the width and penetration of the cracks increase, a small amount of water–sand mixed flow appears, and finally a large amount of sand and water will emerge. Such disasters are catastrophic, secondary, and destructive, and may cause serious damage to mine drainage systems, safety production, mechanical and electrical equipment, and ventilation systems, and even cause casualties. Moreover, in the process of water–sand inrush, the structure of the loose layer of the overlying rock is unstable, resulting in a dramatic change in the pressure of the working face, which is prone to crushing accidents [1, 4, 7, 8]. Therefore, studying the characteristics of water–sand migration in the process of water inrush and sand intrusion in shallow-buried coal seams can help understand the mechanism of water–sand inrush, and then explore the main controlling factors and their influencing mechanisms.

6.2 Mechanism and Criterion of Water–Sand Inrush

If the water–sand inrush event is regarded as a system, the system consists of three key modules: water flowing fracture channel, sand-bearing aquifer, and water–sand mixed flow. The elements included are: sand aquifer, fissure channel, water–sand mixed flow. Therefore, the existing studies on the mechanism of water–sand inrush can be divided into several categories [3, 5].

Y. Zeng et al., *Roof Water Disaster in Coal Mining in Ecologically Fragile Mining Areas*, Professional Practice in Earth Sciences, https://doi.org/10.1007/978-3-031-33140-4_6

(1) Studies on the water flowing fracture channels: The influence of fracture width and fracture surface roughness on the movement of sand body particles was studied by means of theoretical calculation and similar simulation experiments.

(2) Studies on the water in the sand-bearing aquifer: It is believed that whether the sand break can occur after the formation of the fracture depends on whether the hydraulic gradient near the fracture mouth is greater than the extreme hydraulic gradient required for the start of the sand body.

(3) Studies on the main body of sand breaking—sand body: Various mechanical models are established to study the influencing factors of sand body movement and the law of sand body movement under different conditions [6].

(4) Studies on the water–sand mixed flow in the water–sand inrush accidents and flow characteristics in the vertical fracture channels: Generally, the water–sand mixed flow is regarded as a power-law fluid, and establish a complex two-phase flow mathematics model, and compile programs to solve it.

6.2.1 The Theoretical Model of "Extreme Hydraulic Gradient"

The idea of this model is: for coal mining under saturated water-bearing sand layer, when the water-conducting fracture zone touches the aquifer, the water body first migrates down along the water-conducting channel. The water level of the overlying sand aquifer near the mouth of the water inflow channel dropped sharply, and a large hydraulic gradient was rapidly formed near the mouth of the water inflow channel. Once the extreme hydraulic gradient for sand movement in the aquifer is reached or exceeded, the sand in the aquifer will collapse into the goaf on a large scale with water inflow.

The critical condition for water sand inrush can be expressed as:

$$J_r \geq J_{cr} \tag{6.1}$$

where J_r is the actual hydraulic gradient of aquifer at a point from water inflow channel and J_{cr} is the threshold hydraulic gradient.

In the application, there are several assumptions:

(1) The thickness of the aquifer is not too large, and the water-filled area is large. Once the fissure zone spreads to the aquifer, the water inflow channel can penetrate the entire aquifer.

(2) The water inflow channel can be approximately seen It can be regarded as a complete well and can be generalized as a pumping well with a radius of rw.

(3) The water in the aquifer satisfies Darcy's law.

(4) When the boundary has no obvious influence on the water head distribution in the study area, the aquifer can be regarded as an infinite aquifer without external recharge.

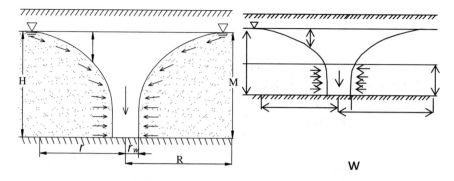

Fig. 6.1 Schematic map of complete well flow of unconfined and confined aquifers

(5) The principle of unstable movement of groundwater to a complete well is applicable.

Due to the different occurrence conditions of aquifers, they are divided into unconfined aquifers and confined aquifers (Fig. 6.1).

The final calculation equations of threshold hydraulic gradient and critical head height are:

$$\text{Unconfined aquifers: } L_1 = r_p \cdot \frac{\sin\left(\frac{3}{4}\pi - \frac{\varphi}{2} - \theta\right)}{2\cos\left(\frac{\pi}{4} + \frac{\varphi}{2}\right)\sin\theta}; \frac{\sigma_1 - \sigma_2}{2}$$

$$= \left(\frac{\sigma_1 - \sigma_2}{2} + ctg^{-1}\varphi\right)\sin\varphi$$

$$\text{Confined aquifers: } J_r = \frac{1.366K(2H - M)}{2\pi r_w(\log R - \log r_w)};$$

$$H_0 \geq \frac{\pi r_w(\log R - \log r_w)}{1.366K} + 2M \tag{6.2}$$

where H_0—initial water head of aquifer before pumping; h_w—stable water level in pumping well after pumping; r_w—radius of pumping well; K—hydraulic conductivity; R—influence radius; M—thickness of confined aquifer.

In applications, the threshold hydraulic gradient of water–sand outburst can be determined by three methods: empirical method, mechanical analysis method and experimental curve method.

Once the water head in the aquifer is greater than the critical value, there is a danger of water–sand inrush. The hydraulic gradient near the mouth of the water inflow channel is the largest. Once the extreme hydraulic gradient is exceeded, the sand body will enter the channel together with the water inflow. Therefore, the water head of the aquifer near the channel is the key research object of disaster prevention, and the corresponding effective measure is to take measures in advance to reduce the water head of the aquifer.

6.2.2 The "Pseudo-Structure" Theoretical Model

The idea of this model is that the coal seam mining causes the roof bedrock to produce penetrating cracks and spread to the bottom of the overlying sandy aquifer. When there is no water influx, according to the platts arch effect theory, part of the sand body will inevitably flow down, and the remaining sand body particles will squeeze each other to form a stable structure. The stable structure is formed by the scattered particles under the action of gravity, which is different from the structure formed by its own internal stress, so it is called "pseudo structure" (Fig. 6.2).

The theory analyzes the direction of water flow and the stress of the sediment particles after cracks and gushing occur. It is believed that the dynamic process of sediment in the process of water inrush and sand collapse presents four stages of "stabilization-instability-migration-deposition". Finally, the conditions for starting the sediment particles are:

$$v_o^2 > \frac{f(\gamma_s h_s - \gamma h)}{\rho C_D} \tag{6.3}$$

where v—water inflow velocity, which is related to the amount of underground water and water inflow fissures; f—friction coefficient of sand particles; γ_s—volumetric weight of particle; hs—distance from particle to surface; γ—unit volumetric weight of water; h—aquifer Height; C_D—drag coefficient, which is related to the flow state around the particle.

Because the flow velocity cannot be obtained by actual measurements, the horizontal flow velocity $V_{max} = \sqrt{2gh}$ is proposed. The relationship between v and the height h of the aquifer is established, and the underwater angle of repose of the sand is $\varphi = 35°$, $f = \tan\varphi = 0.7$, and CD = 0.8.The final judgment condition for the occurrence of water–sand inrush is: $h > \frac{0.44\gamma_s h_s}{\gamma(g+0.44)}$.

According to this equation, the occurrence of water–sand inrush depends on the height of the aquifer. The condition for avoiding the disaster is that the aquifer height is greater than the critical aquifer height. There is ambiguity in this statement. If the

Fig. 6.2 Force analysis of
Pseudo structure

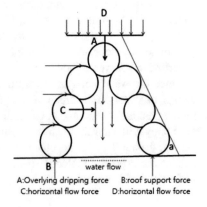

A:Overlying dripping force   B:roof support force
C:horizontal flow force   D:horizontal flow force

height of the aquifer refers to the height of the aquifer head, then the above equation can be used. If the height of the aquifer refers to the height of the aquifer itself (i.e., the thickness of the aquifer), the equation can be used for unconfined aquifers, but for the inflow of confined aquifers, because the value of the water head height of the aquifer is much larger than the value of the thickness of the aquifer, the above equation is not applicable.

6.2.3 Fluid Characteristics of "Water–Sand Mixed Flow"

This type of research focuses on theoretical analysis and believes that the water–sand two-phase fluid is a power-law non-Newtonian body. After extensive theoretical analysis and laboratory experiments, the dynamic model of water–sand mixture in the fissure seepage flow system was established. It consists of governing equations and definite solution conditions. The governing equations include basic equations and auxiliary equations, and the definite solution conditions include initial conditions and boundary conditions.

This kind of research model is relatively complex. A special program needs to be compiled to solve it. In addition, the research object focuses on the mixed fluid of water–sand, which is not closely combined with the actual conditions of the geology and mining conditions of the occurrence of water–sand inrush, and it is not realistic to apply in practice.

6.3 Similar Simulation Experimental Study of Water–Sand Inrush

The development law of coal seam mining and water conducting fracture zone is a hidden phenomenon. It is difficult to directly study the water–sand bursting process in the mining process with effective observation means, while the similar simulation experiment provides effective research means. Most of the existing studies in this area focus on the overburden failure law during mining. Because it is difficult to simulate the aquifer and control the water pressure during the experiment, there are few similar simulation experiments under the fluid solid coupling state. In this work, to solve the interaction between roof overburden failure and aquifer water under certain water pressure during coal seam mining, a targeted simulation experiment system was first developed. Based on the hydrogeological and engineering geological conditions of the 22304 working face of Shigetai Mine, the water resistance experiment material ratio experiment was completed. At the same time, the self-made solid–liquid coupling similar simulation test bed is used to study the mechanism of water sand burst and related influencing factors, so as to obtain a more intuitive and comprehensive understanding.

6.3.1 Experimental Purpose

(1) Study the overburden activity law of a specific working face under the conditions of shallow burial depth, thin bedrock, and thick and loose overburden rich in diving, and the expansion law of the water-conducting fracture zone along the advancing direction of the working face and the height of the roof in different caving stages.

(2) Study the formation process and migration characteristics of water–sand burst, the permeability characteristics and seepage flow deformation characteristics of water–sand burst channels at different stages, and analyze the relationship between migration channels, water–sand mixture and mining disturbance.

(3) Study the "solid–liquid" coupling characteristics of the overlying rock under mining conditions, and analyze the catastrophic mechanism of the surrounding rock and the flow law of groundwater in the longwall working face.

6.3.2 Experimental Method

To achieve the above experimental purpose, the solid–liquid coupling test method was adopted to solve the dynamic development process of "water body-sand body-mining fracture" and the flow characteristics of water–sand mixed flow in the fracture during the process of water–sand inrush. The application of this method in foreign countries is mainly used for similar simulation experiments of geodynamics in the lithosphere, asthenosphere, mantle convection, plate collision, etc. It is rarely used for coal mine safety. It has been more than ten years of history in China to conduct research on shallow buried coal seam rock pressure and overlying rock failure law, water-preserved coal mining, and water–sand outburst disasters by means of liquid–solid coupling similar simulation experiments. Rich achievements have been accumulated in the occurrence of water–sand inrush, fissure channels, types of water–sand outburst, critical conditions of hydrodynamics, migration characteristics of water–sand flow and influencing factors. The existing research results provide a solid foundation for the study of water–sand inrush migration characteristics of shallow buried depth, thin bedrock, and thick loose overburden rich in water.

6.3.3 Experimental System Development

The main part of the experimental system is composed of simulation device of mine pressure and automatic water injection system on the ground of the aquifer, and the auxiliary part is composed of rock layer displacement monitoring, stress pressure data monitoring system, water pressure data monitoring system, oven, heating mixer, and high-speed camera acquisition device.

Fig. 6.3 Simulation device
of mine pressure

(1) Main part

  (1) Simulation device of mine pressure: effective size is 2.0 m long, 1.8 m high
      and 0.3 m wide. Welded by No. 20 channel steel, reserved holes. On the
      left side there are vertical rows of water injection holes spaced at 70 mm.
      On the rear, various sensor installation wiring slots are reserved on both
      sides. The beam holes with an interval of 140 mm are installed on the left
      and right sides of the front and rear sides, and each hole is sealed (Fig. 6.3).

  (2) Automatic water injection system on the ground of the aquifer: It consists
      of a high-pressure tank filled with nitrogen and a cylindrical container tank
      filled with alkaline water. Using the principle of nitrogen pushing water,
      the flow and water pressure are controllable. The diameter of the water
      tank is 215 mm, the height is 1 m, and the maximum pressure is 80 kPa,
      which can simulate the real groundwater head pressure of 0–8 MPa, and
      is equipped with sensors and control instruments for water pressure, water
      volume and seepage flow time, as shown in Fig. 6.4.

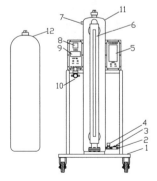

Fig. 6.4 Automatic injection device for water. *Notes* 1—flat car; 2—water outlet; 3—water inlet
valve; 4—water pressure sensor; 5—water level display; 6—water level record scale; 7—air
inlet; 8—water seepage flow time display; 9—Water pressure indicator; 10—pressure gauge;
11—container tank filled with water (alkaline water); 12—high pressure nitrogen tank

(2) Auxiliary part

   (1) Rock formation displacement monitoring: It is a southern sanding-type total station. Its angle measurement accuracy is $\pm$ 1.5″, and the ranging accuracy is $\pm (2 + 2 \times 10 - 6D)$ mm.

   (2) Stress pressure data monitoring system and water pressure data monitoring system: The main process of data acquisition of the system is: first, lay stress and pore water pressure sensors at different locations, then connect the transmission line to the acquisition system, and finally, collect data on changes in water pressure and surrounding rock through special computer software (Fig. 6.5).

(3) Oven/heated blender

It is mainly used for uniform stirring and mixing of materials. Materials such as paraffin and petroleum jelly need to be melted first and then heated. On the other hand, additive materials must be heated to a certain temperature and mixed with other aggregates, and they have certain mechanical properties after mold cooling (Fig. 6.6).

(a) CTI dynamic acquisition device

(b) Strain adaptor

(c) Computer processing software

Fig. 6.5 Test and analysis system of DH5956 dynamic signal

| Oven | heated stirrer |

Fig. 6.6 Main auxiliary equipment fort experiment

6.3.4   Similar Simulation Material Ratio Experiment

The primary difficulty in the liquid–solid coupling similar simulation is to determine the water resistance of the aquifuge material, which is the basis for the later liquid–solid coupling similar simulation experiments. In this regard, scholars have concluded that similar materials that can be used as aquifuge are: loess, sand, clay, silicone oil, vaseline, barite powder, solid paraffin, talcum powder, or hydraulic oil. Because the combination and proportion of materials selected by each person are different, the mechanical properties and water resistance of the obtained materials are different.

In this experiment, medium-coarse sand, gypsum, calcium carbonate, paraffin and vaseline were used as experimental materials. The experimental study on the ratio of similar simulated materials in liquid–solid coupling was carried out.

6.3.4.1   Experimental Purpose

The established goal of the experiment is to require the material to have a certain strength, and at the same time, during the liquid–solid coupling experiment, the material has extremely weak permeability and water absorption properties. Under the condition of water infiltration and water immersion in the aquifer, the material can still maintain a certain strength without loosening.

6.3.4.2   Experimental Scheme Design

Material experiment ideás: First, based on the columnar formation of the working face, complete the proportioning experiment of similar simulated materials in the solid phase. Secondly, select a variety of additives to carry out water resistance experiments, including compressive strength test experiments, water immersion experiments, permeability experiments and water absorption experiments. Finally, the material ratio scheme is selected.

Table 6.1 Formulation of solid in layers

Rock formation name	Sand:calcium carbonate:gypsum	Volumetric weight/g/cm^3
Medium sandstone	7:3:7	2.27
Sandy mudstone	7:6:4	2.38
Siltstone	7:5:5	2.35
Coal seam	9:7:3	1.31
Fine sandstone	7:4:6	2.31

According to the comprehensive geological histogram, the typical rock formations of Shigetai Mine are medium sandstone, sandy mudstone, siltstone, coal seam and fine sandstone. The final determination of the solid phase similar material ratio plan for each rock formation in 5 groups is shown in Table 6.1.

6.3.4.3 Experimental Scheme and Results of Water Resistance

(1) Compressive strength test: (1) According to the ratio of 5 kinds of formation materials, 0, 5, 8, and 10% of paraffin were added as cementing agent. After making different ratios of specimens, the samples were tested for compressive strength after natural curing for 7 days to understand the effect of paraffin content on the compressive strength. (2) Set the proportions of vaseline and paraffin to be 3%, 5%, and 8%, respectively, and test parameters for 7 days of natural conservation to understand the influence of paraffin and vaseline content on the compressive strength. A total of 35 sets of uniaxial compressive strength experiments were completed (Figs. 6.7, 6.8 and 6.9).

Fig. 6.7 Test samples for mechanical experiment

Fig. 6.8 Curve of paraffin content on compressive

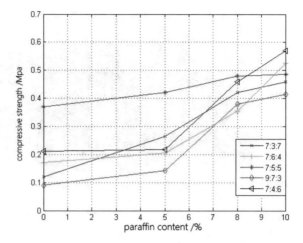

Fig. 6.9 Curve of paraffin and vaseline dosage strength on compressive strength

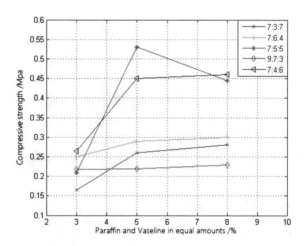

(2) Water immersion experiment: choose the proportion of paraffin wax content of 5, 8 and 10%. After the test piece is put into water for 24 h, the uniaxial compressive strength test is carried out, and a total of 20 groups of water immersion experiments are completed (Figs. 6.10 and 6.11);

(3) Permeability experiment: After preparation of a standard sample with a diameter of 2.5 cm, the permeability is tested after the sample is placed in the mold for 3 days. After conversion and calculation, the characteristic particle size value is obtained, and then the influence of paraffin on the hydraulic conductivity of different similar materials is obtained. A total of 35 sets of penetration tests were completed (Fig. 6.12).

(4) Water absorption test: determine the content of paraffin in the material ratio to be 5–8%. After sample preparation, weigh the specimen after natural curing for 7 days, measure the diameter and height of its upper and lower parts, and

Fig. 6.10 Material immersion test

Fig. 6.11 Curve of compressive strength of materials after immersion

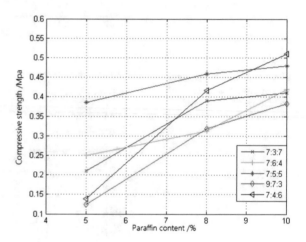

Fig. 6.12 Curve of paraffin content on material hydraulic conductivity

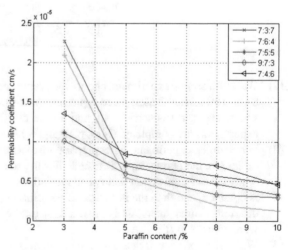

Fig. 6.13 Curve of water absorption with 8%paraffin

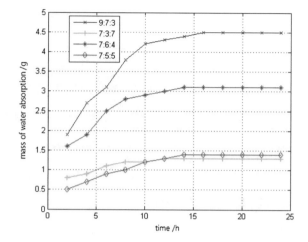

then soak it in water for 24 h. When the specimen is immersed in water, test the change in the quality of the specimen every 2 h to measure its water absorption. A total of 5 groups of samples are prepared, and 60 valid data are tested to obtain the variation relationship of the water absorption of the specimen with time (Fig. 6.13).

(5) Tensile strength test: The tensile strength characterizes the resistance of the material to the maximum uniform plastic deformation. Before the tensile specimen is subjected to the maximum tensile stress, the deformation is uniform, but after it exceeds the maximum tensile stress, the specimen begins to shrink, i.e., produces concentrated deformation. For brittle materials without (or little) uniform plastic deformation, it reflects the fracture resistance of the material (Figs. 6.14 and 6.15).

During the whole experiment, the test results of some parameters are shown in Table 6.2.

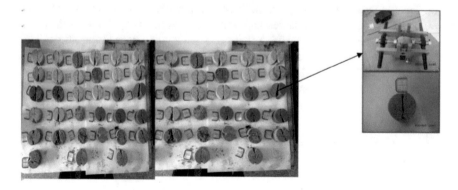

Fig. 6.14 Samples for tensile strength

Fig. 6.15 Curve of paraffin content on tensile strength

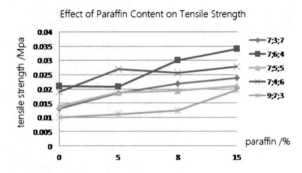

6.3.4.4   Summary of Material Tests

(1) The main materials of similar materials in the water resistance test are quartz sand, calcium carbonate, gypsum, paraffin, and vaseline. According to the proportioning scheme of paraffin and vaseline from 0 to 10%, the model material can achieve similar simulation of low strength, large deformation and water resistance of the prototype formation.

(2) Adjusting the amount of paraffin can change the compressive strength of the specimen. The higher the amount of paraffin, the higher the strength of the specimen. When the amount of paraffin reaches 8%, the plastic deformation capacity is small, the compressive strength increases, but the compressive strength increases slowly.

(3) Vaseline, as a good deformation additive, helps to achieve low strength and large deformation characteristics of similar simulated materials. Paraffin has good non-hydrophilic properties and can control the permeability of similar materials and simulate the water properties of rock formations. The reasonable collocation of vaseline and paraffin, and the complementary advantages, the simulated experimental material produced can meet the requirements of the solid–liquid coupling similar simulation experiment for the water resistance of the material.

(4) The experimental results of different proportioning schemes obtained the changing laws of mechanics and water resistance of similar materials with different proportions, which provides a basis for determining the mixing schemes of similar materials with different types of waterproof layers in the next step. The material ratio scheme for laying similar simulation experiments is finally determined as shown in Table 6.3.

6.3.5   Similar Simulation Model Design

(1) Test device setup

  (1) Model size: According to the special engineering geological conditions of the 22304 working face of Shigetai Coal Mine and the specifications

Table 6.2 Parameter results of different experimental schemes

Additive		Sand: calcium carbonate: gypsum					
		7:3:7 (medium sandstone)			7:6:4 (sandy mudstone)		
Paraffin %	Vaseline %	7d resistance MPa	Water immersion and compressi-on MPa	Penetration/md	7d resistance MPa	Water immersion and compression MPa	Penetration md
0	0	0.12	–	2.27	0.170	–	2.09
5		0.265	0.21	0.728	0.206	0.25	0.54
8		0.42	0.39	0.563	0.355	0.312	0.20
10		0.46	0.41	0.466	0.526	0.419	0.12
	3	0.165	–	1.127	0.25		1.09
	5	0.26		0.901	0.29		0.48
	8	0.28		0.739	0.3	–	0.20

Additive		Sand: calcium carbonate: gypsum								
		7:5:5 (siltstone)			9:7:3 (coal seam)			7:4:6 (fine sandstone)		
Paraffin %	Vaseline %	7d resistance/MPa	Water immersion and compression MPa	Penetration md	7d resistance/MPa	Water immersion and compression MPa	Penetration md	7d resistance MPa	Water immersion and compression MPa	Penetration md
0	0	0.370	–	1.12	0.090	–	1.35	0.212	–	1.01
5		0.42	0.3852	0.98	0.142	0.124	0.84	0.219	0.139	0.59
8		0.48	0.459	0.46	0.380	0.319	0.69	0.460	0.416	0.33
10		0.486	0.48	0.33	0.415	0.383	0.45	0.570	0.51	0.29
	3	0.209	–	0.11	0.218	–	1.13	0.265	–	0.90
	5	0.53	–	0.08	0.219	–	0.79	0.45	–	0.67
	8	0.445	–	0.02	0.23	–	0.43	0.46	–	0.47

Note When the blending amount of the conditioner is zero, 10% water is blended as the conditioner

Table 6.3 Material proportioning scheme for simulation experiment

Layer number	Lithology	Model layer thickness /cm	Number of layers/proportion number	Amount of sand per layer/kg	Calcium carbonate per layer/kg	Plaster amount per layer/kg	Water volume per layer/kg	Amount of paraffin per layer/kg	Vaselin-e per layer/kg
17	Aeolian sand	18.3	982						0
16	Mudstone	17.1	12	12.01	1.2	0.51	1.52	0.6	0
15	Medium sandstone	8.5	4	17.85	1.28	1.28	2.27	0.89	0
14	Sandy mudstone	4.9	5	8.23	0.94	0.24	1.05	0.41	0
13	Siltstone	9.9	4	20.37	2.04	1.36	2.64	1.02	0
12	Medium-coarse-grained sandstone	14.1	7	16.98	1.21	1.21	2.16	0.85	0.85
11	1 coal	0.4	1	3.41	0.34	0.09	0.43	0.17	0.17
10	Siltstone	2.91	1	23.95	2.79	1.2	3.1	1.2	1.2
9	1 coal	2.3	3	6.54	0.65	0.16	0.82	0.33	0.33
8	sandy Mudstone	4.38	4	9.2	1.05	0.26	1.17	0.46	0.46
7	Siltstone	19.2	10	15.83	1.85	0.79	2.05	0.79	0.79
6	Fine sandstone	6.2	3	17.01	1.7	1.13	2.2	0.85	0.85
5	Siltstone	9.01	5	14.83	1.73	0.74	1.92	0.74	0.74
4	Sandy mudstone	1.9	2	7.98	0.91	0.23	1.01	0.4	0.4
3	2 coal	3.51	1	29.95	3	0.75	3.74	1.5	0
2	Sandy mudstone	6.6	3	18.48	2.11	0.53	2.35	0.92	0
1	Fine sandstone	3.4	1	27.98	2.8	1.87	3.63	1.4	0

of the "solid–liquid" two-phase test bench, the similar model designed in
this experiment is 1:100. The model size is: length × width × height =
2000 mm × 300 mm × 1800 mm. The upper part is the loose aquifer, while
the lower part is the bedrock overlying the coal seam. The sand body is
saturated with water within the thickness of the aquifer, and the water in
the aquifer sand layer directly contacts the surface of the bedrock.

(2) Model sealing: First, apply waterproof paint on the surface of the model,
and use plexiglass plates on the front and back of the model. The surface
of the model and the plexiglass plate are sealed with waterproof grease.
All the connection holes are sealed with sealant.

(3) Model visibility: In order to ensure the observability of the experiment, add
baking soda to the water and phenolphthalein to the aquifer and bedrock
material. Where only the water migrates, the material is pink, which makes
the water–sand flow more visible.

(2) Determination of test parameters

The model test simulates the seepage flow situation in the underwater mining process,
and its similar conditions should include the similarity of the stress field and the
seepage flow field at the same time. According to the similarity theory, the following
similarity constants are determined:

(1) Time similarity coefficient α_t: $\alpha_t = \frac{T_m}{T_p} = \sqrt{\alpha_l} = \frac{1}{10}$

(2) Bulk density similarity coefficient α_r: $\alpha_r = \frac{\gamma_{mi}}{\gamma_{pi}} = \frac{1}{1.5}$

(3) Similarity coefficient of elastic modulus α_E: $\alpha_E = \frac{E_{mi}}{E_{pi}} = \alpha_l * \alpha_\gamma = \frac{1}{150}$

(4) Strength similarity and stress similarity coefficient $\alpha_{\sigma c}$: $\alpha_{\sigma c} = \alpha_e \times \alpha_\gamma = 225$

(5) Similar ratio of hydraulic conductivity α_k: $\alpha_k = \frac{\sqrt{\alpha_l}}{\alpha_\lambda} = \frac{1.5}{10}$ (Fig. 6.16).

Considering the boundary effect of the physical model, we used a model of 20 cm
from both sides as the excavation boundary. The length of the cutting-eye is 8 cm.
The experimental simulation of 2^{-2} coal seam fully mechanized caving working
face advances along the strike. The mining thickness is 3.5 cm. The excavation step
distance is 5 cm. During the advancement of the working face, observe the breaking of
the overlying rock structure and the dynamic distribution law of the mining fissures,

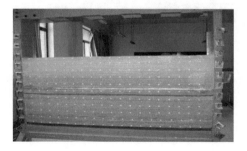
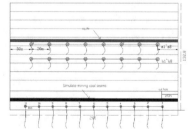

Fig. 6.16 Model for simulation experiment

study the mechanism of the mining fissures forming seepage flow channels, and the relationship between the breaking of the rock mass structure and the water seepage flow. The water inrush mechanism under the coupling action of the stress field and seepage flow field of the overlying rock during the mining operation under the water body is explained from a macro perspective.

6.3.6   Experimental Process and Result Analysis

(1) The first pressure stage of old top layer

A cutting-eye was made at a distance of 20 cm from the model boundary, which is equivalent to leaving a 20 m coal pillar, and the length of the cutting-eye is 8 cm. When the working face advanced to 33 m, the overlying rock had cracks and the direct top is bent. When the working face advanced to 38 m, the first collapse and the block-like structure of the rock block was obvious. The maximum rock block was 12.2 m long and the height of the collapsed rock layer was 1.9 m, the direct top rock layer was bent and fractured. In the goaf, the crack splits the direct top rock layer into two fractured rock masses of unequal lengths that interlock with each other, and horizontal micro-cracks appeared in the upper rock layer. When the working face continued to advance, the direct top rock stratum was separated horizontally and vertically, and the lower fault stratum gradually expanded under the action of the overlying load to form a cantilever beam structure. When the mining line reached 48 m, the direct roof collapsed again, and the rock strata under the old roof within 2 m from the coal seam occurred continuous bed separation collapse. The caving form was a flat arch with a span length of 34 m and a caving angle of 65°. The direct roof separation layer continued to caving, as shown in Fig. 6.17.

(2) The first collapse stage of old top layer

When the working face advanced by 53 m, and the direct top rock layer continued to expand along the horizontal bedding, and slowly collapses in a cantilever beam structure. At the same time, the rock layer under the old roof within 18 m from

Fig. 6.17 Cracks and initial caving in direct roof

Fig. 6.18 Continued fall and initial collapse of roof

the top of the coal seam produced bed separation cracks, and the bed separation cracks continued with the working face. The advancement had also been extended upwards, and the direct top rock layer had been continuously caving forward due to the subsidence of the roof rock layer. When the working face advanced 63 m, the overlying stratum underwent a large range of deformation and damage, resulting in the first pressure on the old roof. The first pressure step was 53 m, the caving shape was flat arch, and the horizontal span at the top of the caving body was 38 m long, the rock collapse angle was 60°, and the horizontal fissures between layers developed to 22 m from the top of the coal seam. When the old roof first collapsed, the roof collapse lagged behind the working face, and showed strong abruptness. The stratum above the stratum was only bent and deformed, and the stratum below was penetrated by fissures mainly developed vertically. The first collapse of the old roof is shown in Fig. 6.18.

(3) Periodic collapse stage of the old top

The working face advanced 73 m, and the working face experienced the first periodic caving. The caving step was 12 m. The damage developed to 45 m from the top of the coal seam, and the development height of the entire fracture zone was 6 times the mining height. The maximum gap between the roof stratum and the caving body was 0.5 m. The horizontal span at the top of the caving body was 50 m, and the caving angle of the rock layer was 60°. Horizontal cracks between layers developed to 43 m from the top of the coal seam, and the gap below 34 m from the top of the coal seam was closed under pressure. The second periodic weighting of the working face is shown in Fig. 6.19.

When the working face advanced 88 m, and the working face fell periodically for the third time. The caving step was 12 m. The damage developed to 46 m from the top of the coal seam, and the horizontal cracks between layers developed to 55 m from the top of the coal seam. The working face advanced 116 m, the first 3 cycles of pressure, the step of pressure was 12 m. The gap at 55 m from the top of the coal seam expanded to 0.5 m. The horizontal span at the top of the caving body was 44 m long, and the horizontal fissures between layers developed to 74 m from the top of the coal seam. The height of the entire breaking belt was 10 times the mining height, and the third cycle of the working face caving is shown in Fig. 6.19.

(2nd time)   (3rd time)

Fig. 6.19 The second and third times periodic collapse of roof

(4)   Fall caving stage of the old top

When the working face advanced 143 m, the working face caved for the seventh time, the caving step was 16 m. The damage developed to 94.5 m from the top of the coal seam. The height of the entire fracture zone was 12.6 times the mining height. The layer separation gap between the strata in the curving belt and the damaged rock layer on the top of the roof was closed. The layer separation space between the caving rock layers at 60 m of the roof suddenly increased, and the surface subsided significantly. In front of the working face and at the eyebrow wall of the cutting-eye, there were two obvious broken fractures extending obliquely to the top of the roof. The caving angle was 58°. The rock strata on the roof of the working face collapsed, and the changes in the abscission cracks are shown in Fig. 6.20.

When the working face advanced by 153 m, the old top had a cut-and-fall caving. The surface had a step-like subsidence. There are only "two belts" in the "upper three belts", and the height of the entire fracture belt is 16.4 times the mining height. Two broken fractures are generated 3 m behind the cutting-eye coal wall and the working face, which directly spread to the surface. The caving angle of the rock layer at the cutting-eye coal wall is 62°, and the caving angle of the rock layer at the coal wall

Fig. 6.20 Crack increase and cutting fall of caving rock layer

Table 6.4 Data of the first and period stress of roof

Name	First time pressure	Cycle to press								Period average
		1	2	3	4	5	6	7	8	
Step/m	53	11	9	14	16	13	16	11	10	12.5
caving corner/(°)	75	65	72	70	70	60	73	70	65	68.12
Crack angle/(°)	70	82	58	62	45	50	49	45	49	55

at the end of mining is 60°. The caving is shown in Fig. 6.20. Table 6.4 presents the specific initial pressure and periodic pressure data.

6.3.7 Analysis of Overlying Rock Stress and Water Pressure Change Process

(1) Development stage of overlying fissures

When the working face advanced from 8 to 38 m, there was no sharp change in the floor stress and the monitored water pressure, which is relatively regular. The No. 01 measuring point shows that the stress of the bottom plate under the coal seam before and after the cutting-eye increases. The No. 02 measuring point shows that after the working face was pushed over approximately 10 m, the stress began to change from large to small. When the working face was pushed to 58 m, the stress at No. 04 measuring point increased. When the working face advanced to 73 m, the stress reached a maximum of 0.486 MPa. When the working face advanced to 78 m, the bottom plate stress at the No. 05 measuring point began to increase. Monitoring shows that the stress-increasing area of the floor moves forward as the working face advances. Meanwhile, the floor stress in the goaf begins to decrease, and a pressure relief state appears (Fig. 6.21).

When the working face advanced to 38 m, there is no change above the cutting-eye. When it advanced from 38 to 78 m, there were changes in the measurement points of pore water pressure sensors a2, a3, a4, etc. under the aquifer. However, the water pressure at other measuring points did not change much. After the advancing distance was between 74 and 140 m, the water pressure began to show a fluctuating upward trend. When the advancing distance exceeded 140 m, the pore water pressure began to decrease (Fig. 6.22).

(2) Water inrush channel initiation stage

When the working face advanced to 88 m, the stope was pressed for the third time. The roof collapsed again, and there was still continuous communication and expansion of cracks in the horizontal direction within the failure zone. The fluctuation of the floor stress curve is relatively small. The stress of the 06, 07, 08 and 09 measuring points increases with the advancement of the working face, and the stress of the

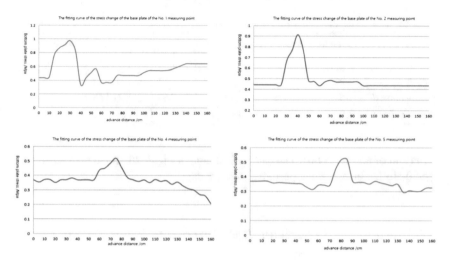

Fig. 6.21 Curves of stress change of 1, 2, 4, 5 test points in floor

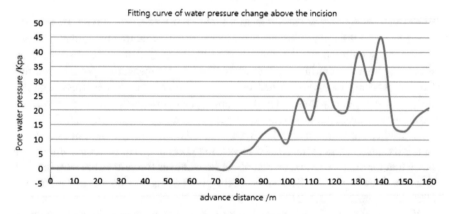

Fig. 6.22 Curve of pore water pressure over cutting-eye

01–05 measuring points reaches stability. There was no longer progress downwards. Viewed from the front of the model, the crack on the right side of the lower interface of the stope floor developed upward to below the aquifer, and had a further upward development trend, as shown in Fig. 6.23.

When the working face advanced from approximately 103–113 m, only the water pressure of the a6 and a1 sensor measuring points fluctuated up and down in the upper area of the working face (Fig. 6.24). The overlying rock was seriously damaged, and the fissures increased and expanded and communicated. The mining fissures of the overlying rock communicated with the roof aquifer to form a water inrush channel.

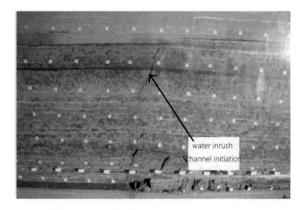

Fig. 6.23 Initiation stage of water inrush channel

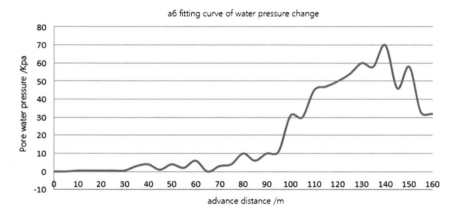

Fig. 6.24 Curve of pore water pressure of a6 test point

(3) Evolution stage of water inrush channel

When the working face advanced to 123 m, under the aquifer approximately 30 m behind the working face, water inflow first appeared and the water volume continued to increase. The fissure development of the bedrock below the aquifer continued to increase, as shown in Fig. 6.25.

As the working face continued to move forward, the floor pressure monitoring showed that the measuring point in the middle area was continuously in a decompressed state. The pore water pressures at the measuring points a1, b1, a6, and b6 were below the coal wall aquifers on both sides (Fig. 6.26). With the further increase of the water volume in the fissures, the water pressure increased. The inflow of water at the fracture channel increased, and the quartz sand and similar simulated materials in the aquifer were continuously washed out of the model.

Fig. 6.25 Initiation of water–sand inrush channel

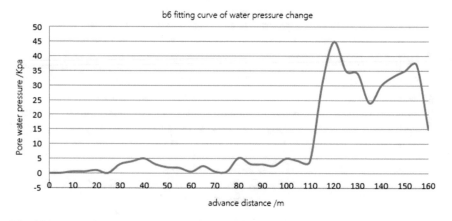

Fig. 6.26 Curve of pore water pressure of b6 test point

At 30 m away from the mining stop line, the upper overlying fissure no longer develops obliquely upward from 35 m from the bottom of the bottom plate, but develops vertically upward, forming a channel through the fissure (Fig. 6.27). The water–sand inrush occurred along the channel. In 1 min, the water and sand volume of 1 L burst out through the fracture channel. With the local outburst of water and sand and the continuous expansion and evolution of the water inrush channel, the water inrush points continued to increase. The overall water inrush volume of the stope further increased, and the water, sand and rock were washed out together, and finally evolved into roof collapse, water inundation and sand disaster.

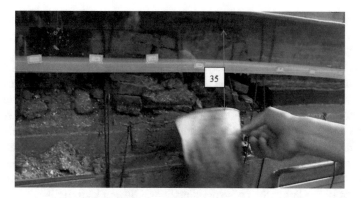

Fig. 6.27   Water–sand inrush channel near stop mining line

6.4   Analysis of Key Factors in the Process of Water–Sand Inrush

The analysis of the similar simulation experiment process indicates that the occurrence of the entire water–sand inrush disaster is divided into several stages:

(1)   The development stage of overlying rock fissures
(2)   The initiation stage of water inrush channel
(3)   The evolution stage of water inrush channel
(4)   The stage of water–sand inrush. In each stage, the stress and water pressure in the fissure area changed significantly, but after the fissure channel penetrated the aquifer, water gushing first occurred. When the fissure developed into a penetrating fissure channel and the fissure channel increased to a certain width, the water–sand outburst phenomenon occurred.

The way and scale of water sand inrush have different performances, mainly considering the following factors.

6.4.1   The Effect of Particle Size on the Permeability of Aquifers

Existing literatures show that in water–sand inrush disasters, under the interaction of water head pressure and fracture channels, the finer the particles of the water-bearing sand layer, the more prone to water–sand inrush, and the greater the amount of sand carried, the more serious the disaster consequences. But in fact, the sand particles that make up the aquifer are not composed of a single fine particle or coarse particle. Different particle gradations result in poor sorting, and the permeability of the aquifer is also different.

Regarding the research on the permeability of aquifers, scholars at home and abroad believe that the content of particle size has the greatest influence on the permeability. Many scholars have studied the effect of particle size and gradation on soil permeability, and realized that the hydraulic conductivity decreases with the increase of fine particle content; the particle size has an obvious control effect on the residual water content of the unsaturated soil, and the control effect of the silt content is the largest, followed by the clay particle, and the breccia content is relatively small. Soil particle gradation is an important evaluation index that affects soil hydraulic conductivity. The hydraulic conductivity is closely related to the fine particle content and the inhomogeneous coefficient of the soil. The hydraulic conductivity is negatively correlated with the fine particle content in general. The smaller the inhomogeneous coefficient is, the larger the corresponding hydraulic conductivity is.

The permeability of the aquifer will greatly affect the migration capacity of groundwater in the aquifer, and then affect the speed, extent and hydraulic gradient of the groundwater funnel formed near the water-conducting fracture channel and will inevitably affect the scale of water–sand inrush. Therefore, it is necessary to study the effect of different particle sizes on the permeability of aquifers.

6.4.1.1 Experimental Schematic Design

The experiment is carried on from the similar simulation experiment of the previous section. To understand the effect of different particle sizes on the permeability of aquifers, the experimental steps are divided into several steps:

- The crusher performs crushing and screening experiments on natural sandstone to obtain the particle size range and ratio of sandstone. The backbone of sandstone particles is formed according to different particle gradations, and several groups of rock samples are configured according to the material ratio of similar simulation experiments. Conduct permeability experiments on rock samples under normal pressure. Conduct permeability experiments on rock samples under pressure.

(1) Sandstone strength test

Three sets of uniaxial compressive strength experiments were performed on sandstone, and the results are shown in Table 6.5.

Table 6.5 Experimental data of uniaxial compressive strength of sandstone

Specimen number	Uniaxial compressive strength/MPa	Strain/E^{-3}	Elastic modulus/GPa	Deformation modulus/GPa	Poisson's ratio
1	62.168	9.156	9.627	5.332	0.196
2	72.306	8.551	13.322	6.556	0.411
3	70.630	7.812	12.530	7.200	0.259

Fig. 6.28 The screening results of sandstone particles

(2) Sandstone particle size range and ratio

The sandstone core (coarse, medium and fine sandstone) is crushed with a small crusher to obtain the particles that make up the sandstone. The crushed sandstone particles are screened on an electric screening machine, and the screen apertures are 5 mm, 1.25 mm, 0.5 mm, 0.25 mm, 0.15 mm, and 0.075 mm, respectively. The particle composition after screening is shown in Fig. 6.28.

According to the particle size distribution range obtained after screening and the geological histogram of the mining area, the particle size ratio schemes of 5 main sandstones are obtained according to different particle gradations, as shown in Table 6.6. The non-uniformity coefficient Cu and curvature coefficient Cc can be determined from the table. The larger the inhomogeneity coefficient Cu is, the larger the difference in particle size between the coarse and fine particles in the sand is, the more inhomogeneous the sand body is, and the better its gradation is. The curvature coefficient Cc reflects whether the slope of the cumulative particle gradation curve is continuous. In engineering, sand with Cu ≥ 5 and Cc = 1 to 3 is regarded as well-graded sand, otherwise it is poorly graded sand.

(3) Content of sand and mudstone particles

To study the change of permeability when sandstone is mixed with mudstone, the ratio of sand and mudstone is as follows: 5:5, 6:4, 7:3, 8:2, 9:1.

6.4.1.2 Experimental Process and Result Analysis

(1) Simulated sandstone aquifer

The crushed sandstone particles were made into 5 groups of samples from 1 to 5 according to the gradation ratio in Table 6.6. Then, 5 groups of sand samples with different gradations were added to the mixture (calcium carbonate, gypsum, water,

Table 6.6 Particles content of different sandstone samples

Sample	Content of each grain group %						
	> 5 mm	5–1.25 mm	1.25–0.5 mm	0.5–0.25 mm	0.25–0.15 mm	0.15–0.075 mm	< 0.075 mm
1 coarse sandstone	55.00	25.00	11.00	4.00	2.00	1.00	2.00
2 medium coarse sandstone	30.00	25.00	20.00	10.00	6.00	3.00	6.00
3 medium sandstone	18.00	19.00	19.00	12.00	10.00	7.00	15.00
4 fine sandstone	9.00	10.00	15.00	12.00	14.00	14.00	26.00
5 Siltstone	3.00	4.00	6.00	8.00	14.00	20.00	45.00

Table 6.7 Proportion content of sandstone simulation

Numbering	Sand weight/kg	Calcium carbonate weight/kg	Gypsum weight/kg	Water weight/kg	Paraffin weight /kg	Remark
1	4.25	0.29	0.31	0.53	0.21	Coarse-grained sandstone
2	4.23	0.30	0.30	0.53	0.21	Medium-coarse-grained sandstone
3	4.39	0.31	0.31	0.56	0.22	Medium sandstone
4	4.15	0.42	0.27	0.53	0.21	Fine sandstone
5	4.15	0.48	0.20	0.53	0.20	Siltstone

paraffin) in a certain proportion to simulate the sandstone aquifer. The proportion of materials for the simulated aquifer experiment is shown in Table 6.7.

(2) Penetration performance test results

The permeability of the sample is measured by a permeability tester capable of giving a certain confining pressure, given the influent flow rate, the inlet pressure and the annular pressure. The average permeability of each group of graded samples is shown in Table 6.8.

(3) Analysis of test results

(1) Influence of coarse particle content on permeability

Figure 6.29 plotted the correlation between coarse particle content and permeability. When the coarse particle content in the sample is greater than 20–30%, the permeability increases with the increase of the coarse particle content. When the content of coarse particles in the sample is less than 20–30%, the permeability increases with the decrease of the content of coarse particles. Therefore, the critical range of grain

Table 6.8 Average permeability of each sample

Group number	Average permeability k (mD)	Influent flow (ml/s)	Ring pressure (MPa)
Group 1 samples	69.50416	2	1
Group 2 samples	19.6577	1	1
Group 3 samples	6.2292	1	1
Group 4 samples	33.9094	2	1
Group 5 samples	67.34006	1	1

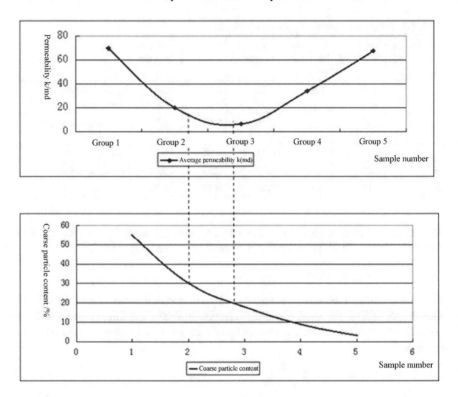

Fig. 6.29 Effect of coarse particles on permeability of sample

content in which the permeability of sandstone change significantly is between 20 and 30%.

(2)   Influence of special particle size on permeability

There is a correlation between the permeability of porous media and particle size distribution, which is exponential or logarithmic. For example, Ha Zeng's equation [2] use the characteristic particle size of d10 and d20 to represent the gradation respectively, and then deduce the equation for predicting the hydraulic conductivity with particle gradation. The content of particles with different particle sizes in each group of samples in this experiment is presented in Table 6.9, and the content of characteristic particle size uses d10, d20, d30, and d60, as summarized in Table 6.10.

The relationship between characteristic particle size and permeability is shown in Fig. 6.30. There is a certain correlation between special particle size and permeability. The fitting degree of the characteristic particle size of the sample d10 is 0.99, indicating that the permeability increases with the increase of d10. The increase of d10 means that the fine particles in the gradation are decreasing, and the particle size with the cumulative content of fine particles at 10% becomes larger, which will appear in the skeleton composed of coarse particles, and the content of fine particles

Table 6.9 The content of different particle size in samples

Particle size/mm	The percentage of particles smaller than a certain particle size%				
	Group 1	Group 2	Group 3	Group 4	Group 5
10	100	100	100	100	100
5	45	70	82	91	97
1.25	20	45	63	81	93
0.5	9	25	44	66	87
0.25	5	15	32	54	79
0.15	3	9	22	40	65
0.075	2	6	15	26	45

Table 6.10 Characteristics of particles in samples

Parameter	Group 1	Group 2	Group 3	Group 4	Group 5
d60	6.350	3.125	1.13	0.375	0.131
d50	5.450	1.644	0.740	0.222	0.094
d30	2.750	0.673	0.230	0.096	0.050
d10	0.567	0.170	0.050	0.03	0.016

is not enough to fill the gaps between particles, while the larger the pores of the coarse particles, the greater the permeability.

(3) Influence of sand particle inhomogeneity coefficient and curvature coefficient on permeability

To explore the influence of the non-uniformity coefficient and curvature coefficient of sand particles on the permeability, Fig. 6.31 is drawn according to the calculation results of the non-uniformity coefficient and curvature coefficient of each group of samples (Table 6.11). The non-uniformity coefficient is negatively correlated with

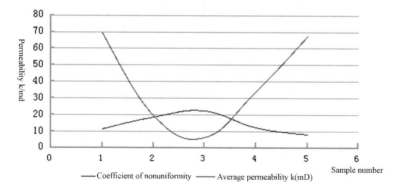

Fig. 6.30 Relation diagram of characteristic particle size and permeability

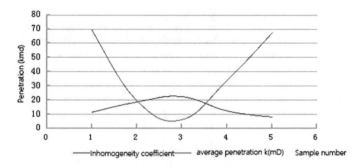

Fig. 6.31 Curve of nonuniform coefficient and curvature coefficient

Table 6.11 Nonuniform coefficient and curvature coefficient

	Sample 1	Sample 2	Sample 3	Sample 4	Sample 5
d10	0.567	0.170	0.050	0.030	0.016
d30	2.750	0.673	0.230	0.096	0.050
d60	6.350	3.125	1.130	0.375	0.131
Cu = d60/d10	11.20	18.38	22.60	12.50	8.19
Cc = (d30 * d30)/(d60 * d10)	2.100	0.852	0.936	0.819	1.193

the permeability characteristics of the sample, i.e., the larger the non-uniformity coefficient is, the lower the permeability of the sample.

(4) Experimental summary

The particle size distribution is closely related to the permeability of aquifers, among which the content of coarse particles has a particularly significant effect on the permeability, followed by the characteristic particle size, particle inhomogeneity coefficient and curvature coefficient. (1) When the content of coarse grains in sandstone reaches 20–30%, the permeability increases with the content of coarse grains. When the content of coarse grains in sandstone is not in this range, the permeability decreases with the content of coarse grains. (2) When the particles in the sandstone form a certain skeleton, the content of the special particle size increases, and the permeability also increases, and the correlation of d10 is the highest. (3) The better the gradation, the smaller the permeability; the larger the non-uniformity coefficient, the smaller the permeability.

6.4.2 Key Factors of Sand Body Movement After Fracture Formation

From the field and experimental phenomena, the sand body movement after fracture generation is not the starting and migration of a single particle analyzed in the mechanical model. Due to the existence of shale in the sand body, the movement should be the movement of tiny, certain volume blocks composed of sand and mud.

The following will analyze the causes and conditions of the induced sand collapse from the overall start-up and migration of a certain volume of sand bodies. The conditional expressions that should be satisfied when the sand body is critically collapsed. The factors that affect the overall collapse of the sand body are given below.

(1) Movement law of sand bodies

The migration of sand bodies is closely related to fractures. Under the influence of mining, when the fractures develop into loose layers, the initiation and migration of sand bodies can be roughly divided into the following processes:

(1) The development of fissures to the loose layer makes A bulk block (as shown in Fig. 6.32a) likely to fall into the fissures first due to the loss of the support of the underlying rock mass.
(2) When the A bulk block in the loose layer collapses, the B bulk block (as shown in Fig. 6.32a) may subsequently collapse because it loses the support of the A bulk block.
(3) With the collapse of A loose block and B loose block, if the upper loose layer also reaches the corresponding collapse conditions, two steps (1) and (2) will occur one by one. In severe cases, it will cause a subsidence area on the surface, which will affect the corresponding buildings or vegetation on the surface.
(4) After the bulk blocks in the loose layer collapse into the cracks, the accumulation body D in the cracks (as shown in Fig. 6.32a) may collapse into the working face due to the accumulation of mass, thus burying the working face and mechanical equipment and so on, leading to the eventual occurrence of disasters.

(2) Conditions and force analysis of sand body start-up

As the working face advances, the fracture develops to the mechanical model diagram of the loose layer, as shown in Fig. 6.32a. The critical start-up conditions of the

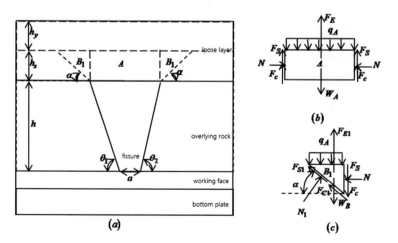

Fig. 6.32 Mechanical model and analysis of block A and B1

discrete blocks in each process and their corresponding influencing factors are analyzed below.

A. The conditions for the collapse of A bulk block:

When the fissures develop to the loose layer under the influence of mining, the loose block A may first begin to collapse due to the loss of the support of the underlying rock mass and the extrusion of the overlying loose layer. The force analysis of A bulk block is shown in Fig. 6.32b. When the bulk A does not collapse, the friction force Fs on it satisfies, i.e.

$$0 \le F_S \le N \tan \varphi_1 \tag{6.4}$$

where φ_1 − A is the internal friction angle of the bulk block.

It can be seen from the force analysis diagram of the scattered block A (as shown in Fig. 6.32b) that the condition for the collapse of the scattered block A is that the sum of its vertical downward force reaches (critical state) or exceeds (transportation state) the sum of the maximum static friction force of its shear plane and other forces in the vertical direction, i.e.

$$W_A + q_A b \left(\frac{h}{\tan \theta_1} + \frac{h}{\tan \theta_2} + a \right) \ge F_E + 2F_S + 2F_c \tag{6.5}$$

where, W_A—the gravity of A bulk block; i.e.

$$W_A = \gamma_s h_s b \left(\frac{h}{\tan \theta_1} + \frac{h}{\tan \theta_2} + a \right) \tag{6.6}$$

F_C Cohesion of the A bulk block;

F_E The environmental resistance of A loose body block (depending on the compactness of the loose body, which can be expressed by porosity, fine material content and inhomogeneity coefficient);

q_A The unit area load concentration of the extrusion force of the overlying loose layer on the loose block A, i.e.

h The thickness of the overlying rock mass on the working face;

θ_1 The angle between the left side of the crack and the horizontal direction;

θ_2 The angle between the right side of the crack and the horizontal direction;

a The horizontal crack width at the lower part of the crack;

h_s Thickness of the bulk block A;

γ_s Volumetric weight of bulk block A;

h_y Thickness of the loose layer overlying the loose block A;

γ_y The volumetric weight of the loose layer overlying the bulk A;

b Crack vertical crack width.

When the A scattered block is in a critical state, the force analysis diagram of the B1 scattered block is in the equilibrium state, as shown in Fig. 6.32c. According to the equilibrium condition:

$$N = N_1 \sin \alpha - (F_{S1} + F_{C1}) \cos \alpha \tag{6.7}$$

$$N_1 = \left(W_B + q_A b \frac{h_s}{\tan \alpha} - F_{E1} + F_S + F_C \right) \cos \alpha + N \sin \alpha \tag{6.8}$$

When the B1 bulk block is in a critical equilibrium state, the friction force on it satisfies Coulomb's law, i.e.

$$F_{S1} = N_1 \tan \varphi_1) \tag{6.9}$$

Then,

$$N = \left(\frac{\gamma_s b h_s h_s}{2 \tan \alpha} + \frac{\gamma_y b h_y h_s}{\tan \alpha} - F_{E1} + F_C \right) \frac{\sin \alpha - \cos \alpha \tan \varphi_1}{\cos \alpha \left(1 + \tan^2 \varphi_1 \right)}$$
$$- F_{C1} \frac{1}{\cos \alpha \left(1 + \tan^2 \varphi_1 \right)} \tag{6.10}$$

where, W_B—the gravity of B1 bulk block; i.e.:

$$W_B = \gamma_s \frac{1}{2} h_s b \frac{h_s}{\tan \alpha} \tag{6.11}$$

F_{C1} Cohesion of B1 bulk block;
F_{E1} Environmental resistance of B1 bulk block;
α The angle of influence of the A1 bulk block on the surrounding bulk when it is in a critical state.

Substituting Eqs. (6.1) and (6.7) into Eq. (6.2), the control conditions for the A bulk block to reach the critical state and the migration state are expressed as follows.

$$\cot \alpha \geq \frac{1}{\tan \phi_1} - \frac{\gamma_s h_s + \gamma_y h_y}{(h_s \gamma_s + 2\gamma_y h_y) h_s} \left(\frac{h}{\tan \theta_1} + \frac{h}{\tan \theta_2} + a \right) \frac{1 + \tan^2 \phi_1}{\tan^2 \phi_1} \tag{6.12}$$

When the environmental resistance and cohesion are not considered, Eq. (6.8) is simplified to

$$\cot \alpha \geq \frac{1}{\tan \phi_1} - \frac{\gamma_s h_s + \gamma_y h_y}{(\gamma_s h_s + 2\gamma_y h_y) h_s} \left(\frac{h}{\tan \theta_1} + \frac{h}{\tan \theta_2} + a \right) \frac{1 + \tan^2 \phi_1}{\tan^2 \phi_1} \tag{6.12a}$$

$$\text{Let } e = \frac{1}{\tan \phi_1} - \frac{\gamma_s h_s + \gamma_y h_y}{(\gamma_s h_s + 2\gamma_y h_y) h_s} \left(\frac{h}{\tan \theta_1} + \frac{h}{\tan \theta_2} + a \right) \frac{1 + \tan^2 \phi_1}{\tan \phi_1^2}$$

i.e. $\cot \alpha \geq e$
$$\tag{6.12b}$$

From Eq. (6.8b), it can be obtained that when the environmental resistance and cohesion are not considered, if the cotangent of the influence angle of the A scattered body on the surrounding loose body is equal to e, the A scattered body reaches the critical state. When the cotangent of the influence angle of the block on the surrounding loose body is greater than e, the A loose body block reaches the transport state. The factors affecting the migration of the A loose block are the geometrical conditions of the fracture, the material of the loose layer and the shear resistance factor.

B. The conditions for the collapse of B bulk block

Due to the collapse of the A scattered block, the B scattered block loses the support of the A scattered block and may collapse after the A scattered block. The force analysis diagram of the B scattered block in the equilibrium state when it is not collapsed, as shown in Fig. 6.33b. According to the equilibrium condition:

$$N_2 = \left(W_{B_1} - F_{E2} + q_A b \frac{h_s}{\tan \alpha_1} \right) \cos \alpha_1 \tag{6.13}$$

The frictional force on B bulk block when it is in equilibrium, then satisfies:

$$0 \le F_{S2} \le N_2 \tan \varphi_1 \tag{6.14}$$

The condition for the start and migration of the bulk B is that the sum of its downward forces along the slip surface is equal to (critical state) or exceeds (migration

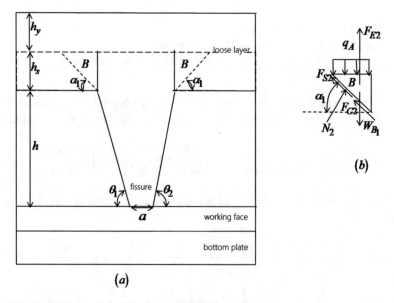

(a)

(b)

Fig. 6.33 Mechanical model and analysis of block B

state) its maximum static friction force along the slip surface and its other upward forces, as shown in Fig. 6.33b, i.e.,

$$\left(W_{B_1} - F_{E2} + q_A b \frac{h_s}{\tan \alpha_1} \right) \sin \alpha_1 \geq F_{S2} + F_{C2} \tag{6.15}$$

where, W_{B1}—the gravity of A bulk block; i.e.

$$W_{B_1} = \gamma_s \frac{1}{2} h_s b \frac{h_s}{\tan \alpha_1} \tag{6.16}$$

F_{C2} Cohesion of the B bulk block;
F_{E2} The environmental resistance of B bulk block.
α_1 The azimuth angle of the slip surface of B bulk block.

Substituting Eqs. (6.9) and (6.10) into Eq. (6.11), the control conditions for the B bulk to reach the critical state and the migration state are expressed as follows:

$$\left[\frac{(\gamma_s h_s + 2\gamma_y h_y) b h_s}{2 \tan \alpha_1} - F_{E2} \right] (\sin \alpha_1 - \cos \alpha_1 \tan \varphi_1) \geq F_{C2} \tag{6.17}$$

If environmental resistance and cohesion are not considered, Eq. (6.12) simplifies to:

$$\cos \alpha_1 (1 - \cot \alpha_1 \tan \phi_1) \geq 0 \tag{6.17a}$$

when $0 \leq \alpha_1 \leq \frac{\pi}{2}$

$$\alpha_1 \leq \phi_1 \tag{6.17b}$$

From Eq. (6.12b), when the environmental resistance and cohesion are not considered, if the azimuth angle of the slip surface of the B scattered block is equal to the internal friction angle of the A scattered block, the B scattered block reaches the critical state. When the azimuth angle of the sliding surface of the bulk block is smaller than the internal friction angle of the bulk block A, the bulk block B reaches the migration state. The factor affecting the migration of the B bulk block is the shear resistance factor of the loose layer.

C. Conditions under which the accumulation D collapses

When the size of the scattered body falling into the crack is smaller than the size of the rock mass fracture, it will directly collapse into the working face. When the size of the scattered body is larger than the size of the rock mass crack or the scattered body collapses due to the surrounding rock mass, scattered body, etc. Collision, friction, bonding, and reducing the falling energy will reduce the size of the rock mass fissures,

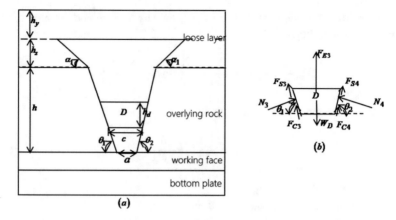

Fig. 6.34 Mechanical model and analysis of accumulation body D

resulting in the accumulation of falling loose bodies in a certain position of the rock mass fissures, as shown in Fig. 6.34a.

The force analysis diagram of the stack D in a critical equilibrium state is shown in Fig. 6.34b. The friction force F_{S3}, F_{S4}, on the stack D satisfies Coulomb's law, i.e.

$$F_{S3} = N_3 \tan \varphi_2 \tag{6.18}$$

$$F_{S4} = N_4 \tan \varphi_2 \tag{6.19}$$

where, φ_2 is the internal friction angle of the stack D.

According to the equilibrium conditions of the equilibrium state of the stack D, the normal supporting forces N_3, N_4 of the slip surface to the stack D at this time can be obtained as:

$$
\begin{aligned}
N_4 = (W_D - F_{E3}) &\frac{\cos\theta_2 - \cos\theta_1[\cos(\theta_1+\theta_2) + \tan\varphi_2\sin(\theta_1+\theta_2)]}{1 + [\cos(\theta_1+\theta_2) + \tan\varphi_2\sin(\theta_1+\theta_2)]^2} \\
&+ F_{C4}\frac{\sin(\theta_1+\theta_2)[\cos(\theta_1+\theta_2) + \tan\varphi_2\sin(\theta_1+\theta_2)]}{1 + [\cos(\theta_1+\theta_2) + \tan\varphi_2\sin(\theta_1+\theta_2)]^2} \\
&- F_{C3}\frac{\sin(\theta_1+\theta_2)}{1 + [\cos(\theta_1+\theta_2) + \tan\varphi_2\sin(\theta_1+\theta_2)]^2}
\end{aligned}
\tag{6.20}
$$

$$
\begin{aligned}
N_3 = (W_D - F_{E3}) &\frac{\cos\theta_1 - \cos\theta_2[\cos(\theta_1+\theta_2) + \tan\varphi_2\sin(\theta_1+\theta_2)]}{1 + [\cos(\theta_1+\theta_2) + \tan\varphi_2\sin(\theta_1+\theta_2)]^2} \\
&+ F_{C3}\frac{\sin(\theta_1+\theta_2)[\cos(\theta_1+\theta_2) + \tan\varphi_2\sin(\theta_1+\theta_2)]}{1 + [\cos(\theta_1+\theta_2) + \tan\varphi_2\sin(\theta_1+\theta_2)]^2} \\
&- F_{C4}\frac{\sin(\theta_1+\theta_2)}{1 + [\cos(\theta_1+\theta_2) + \tan\varphi_2\sin(\theta_1+\theta_2)]^2}
\end{aligned}
\tag{6.21}
$$

The condition for the collapse of the accumulation body D is that the sum of the vertical force is equal to (critical state) or exceeds (migration state) the maximum static friction force of the slip surface and the sum of other forces along the vertical direction, as shown in Fig. 6.34b, i.e.

$$W_D \geq F_{E_3} + \left(F_{C_3} + F_{S_3}\right) \sin \theta_1 + \left(F_{C_4} + F_{S_4}\right) \sin \theta_2$$
$$+ N_3 \cos \theta_1 + N_4 \cos \theta_2 \tag{6.22}$$

where,

f1 and f2	the cohesion of the accumulation body D;
F_{C3} and F_{C4}	the environmental resistance of the accumulation body D;
W_D	the gravity of the accumulation body D;
h_d	the thickness of the accumulation body D;
c	The width of the lower edge of the accumulation body D.

Substituting Eqs. (6.13), (6.14), (6.15), and (6.16) into Eq. (6.17), the control conditions for the accumulation D to reach the critical state and the migration state are expressed as follows:

$$
\begin{aligned}
&- \tan \phi_2 \cos \theta_2 \cos \theta_1 \sin(\theta_1 + \theta_2) \\
&\quad - [\cos \theta_2 - \cos \theta_1 \cos(\theta_1 + \theta_2)] \cos \theta_2 \\
&\quad - \tan \phi_2 \sin \theta_2 [\cos \theta_2 - \cos \theta_1 \cos(\theta_1 + \theta_2)] \\
&\frac{1 + [\cos(\theta_1 + \theta_2) + \tan \phi_2 \sin(\theta_1 + \theta_2)]^2} \\
&\geq [\cos \theta_1 - \cos \theta_2 \cos(\theta_1 + \theta_2)] \cos \theta_1 \\
&\quad + \tan \phi_2 \sin \theta_1 [\cos \theta_1 - \cos \theta_2 \cos(\theta_1 + \theta_2)] \\
&\quad + \tan \phi_2 \cos \theta_2 \cos \theta_1 \sin(\theta_1 + \theta_2) \\
&\quad + \tan^2 \phi_2 \sin(\theta_1 + \theta_2)[\cos \theta_1 \sin \theta_2 - \cos \theta_2 \sin \theta_1]
\end{aligned}
$$

$$
\geq F_{E_3} \left(1 - \frac{\cos \theta_1 - \cos \theta_2 [\cos(\theta_1+\theta_2) + \tan \varphi_2 \sin(\theta_1+\theta_2)]}{1+[\cos(\theta_1+\theta_2)+\tan \varphi_2 \sin(\theta_1+\theta_2)]^2} (\cos \theta_1 + \tan \varphi_2 \sin \theta_1) \right.
$$
$$
\left. + \frac{\cos \theta_1 - \cos \theta_2 [\cos(\theta_1+\theta_2)+\tan \varphi_2 \sin(\theta_1+\theta_2)]}{1+[\cos(\theta_1+\theta_2)+\tan \varphi_2 \sin(\theta_1+\theta_2)]^2} (\cos \theta_1 + \tan \varphi_2 \sin \theta_1) \right)
$$
$$
+ F_{C_3} \left\{ \sin \theta_1 + \frac{\sin(\theta_1+\theta_2)(\cos \theta_1 + \tan \varphi_2 \sin \theta_1)[\cos(\theta_1+\theta_2)+\tan \varphi_2 \sin(\theta_1+\theta_2)]}{1+[\cos(\theta_1+\theta_2)+\tan \varphi_2 \sin(\theta_1+\theta_2)]^2} \right.
$$
$$
\left. - \frac{\sin(\theta_1+\theta_2)(\cos \theta_2 + \tan \varphi_2 \sin \theta_2)}{1+[\cos(\theta_1+\theta_2)+\tan \varphi_2 \sin(\theta_1+\theta_2)]^2} \right\}
$$
$$
+ F_{C_4} \left\{ \sin \theta_2 - \frac{\sin(\theta_1+\theta_2)(\cos \theta_1 + \tan \varphi_2 \sin \theta_1)}{1+[\cos(\theta_1+\theta_2)+\tan \varphi_2 \sin(\theta_1+\theta_2)]^2} \right.
$$
$$
\left. + \frac{\sin(\theta_1+\theta_2)(\cos \theta_2 + \tan \varphi_2 \sin \theta_2)[\cos(\theta_1+\theta_2)+\tan \varphi_2 \sin(\theta_1+\theta_2)]}{1+[\cos(\theta_1+\theta_2)+\tan \varphi_2 \sin(\theta_1+\theta_2)]^2} \right\} \tag{6.23}
$$

If environmental resistance and cohesion are not considered, Eq. (6.18) simplifies to:

$$1 \geq \cos^2 \theta_1 - \cos^2 \theta_2 - \cos^2(\theta_1 + \theta_2)$$
$$- \tan \phi_2 \sin 2(\theta_1 + \theta_2)$$

$$- 2 \tan^2 \phi_2 \cos \theta_1 \sin \theta_2 \sin(\theta_1 + \theta_2) \qquad (6.23a)$$

$$\text{Let } g_1 = \cos^2 \theta_1 - \cos^2 \theta_2 - \cos^2 (\theta_1 + \theta_2)$$
$$g_2 = -\sin 2(\theta_1 + \theta_2)$$
$$g_3 = -2 \cos \theta_1 \sin \theta_2 \sin(\theta_1 + \theta_2)$$
$$g_3 \tan^2 \phi_2 + g_2 \tan \phi_2 + g_1 \leq 1 \qquad (6.23b)$$

From Eq. (6.18b), when the environmental resistance and cohesion are not considered, if the trigonometric function equation formed by the internal friction angle of the stack D is equal to 1, the stack D reaches a critical state; if the internal friction angle of the stack D is equal to 1 When the formed trigonometric function is less than 1, the accumulation D reaches the transport state. The factors affecting the collapse of the accumulation body D include the azimuth angle of the crack surface and the internal friction angle of the accumulation body D.

However, once the accumulation body D collapses, it represents the occurrence of a sand collapse accident. Therefore, the development of fractures and the shape and size of the accumulation body must be controlled.

(3) Analysis results

   (1) The overall migration of sand bodies is closely related to the width of the fracture and the roughness of the fracture surface;
   (2) The overall start-up and migration of sand bodies need to satisfy certain control conditional equations, namely Eqs. (6.8), (6.12) and (6.18);
   (3) When the environmental resistance and cohesion are not considered, the control conditions for the overall initiation and migration of sand bodies can be simplified to Eqs. (6.8b), (6.12b) and (6.18b);
   (4) In engineering practice, in order to prevent the occurrence of sand inrush disasters, the development of fractures during mining should be closely monitored, and when the fractures develop into loose layers, the overall start-up of the sand body and the satisfaction of the migration control conditions should be avoided.

6.5   Summary

In this chapter, on the basis of discussing the current research status of the existing mechanism and criteria of water–sand inrush, a similar simulation experiment of liquid–solid coupling was conducted. The failure pattern of roof cracks and the change regularities of water pressure in the water flowing fracture crack zone during the experiment are analyzed and summarized. The two key elements in the process of water–sand inrush are experimentally and theoretically analyzed, and their influence mechanisms are summarized. The main conclusions of the research in this chapter are as follows:

(1) The two-dimensional liquid–solid coupling test bench and the ground control-lable pressure water supply system independently developed in this work have strong applicability and reliability and can complete the predetermined experi-mental objectives. It provides an intuitive study of the phenomenon and law of the interaction between roof failure and hydraulic pressure during coal seam mining under the condition of solid–liquid coupling and reproduces the "crack generation-development-conduction-water–sand inrush" in the process of water–sand inrush".

(2) In this work, in view of the stratum conditions of the 22304 working face of Shigetai Mine, the material formula that can be compared with the on-site strata was screened out through the water resistance experiment, and the strength, deformation and water barrier properties were satisfied during the experiment.

(3) In different development stages of water–sand inrush events, the overlying rock stress and water pressure show regular changes. (1) Development stage of the overlying fissures: When the working face is advanced from 8 to 38 m, the floor stress and the monitored water pressure do not change dramatically, and there is a certain rule. As the working face advances, the stress-increasing area of the floor keeps moving forward. The stress of the floor in the goaf begins to decrease, and a pressure relief state occurs. When the working face advanced from 38 to 78 m, the pore water pressure under the aquifer began to change. After the advancing distance of 74–140 m, the water pressure shows a fluctuating upward trend. If the advancing distance exceeds 140 m, the pore water pressure drops sharply. (2) Water inrush channel initiation stage: When the working face advanced to 88 m, the stope was pressed for the third time, the roof collapsed again, and the cracks in the damage zone increased and continued to expand and communicate. The fissure on the right side of the stope floor developed upward to below the aquifer, and the water pressure monitoring value on the upper part of the working face fluctuated up and down. The mining fissure is about to be connected to the roof aquifer. Once a fissure channel through the aquifer is formed, water and sand migration can begin. (3) Evolution stage of water inrush channel: When the working face is advanced to 123 m, under the aquifer approximately 30 m behind the working face, water gushing first occurred and the water volume continued to increase, and the erosion intensified the development of fissures in the bedrock below the aquifer. With the further increase of the water volume and the water pressure in the fracture, the inflow of water at the fracture channel increased, and the carrying sand and roof material emerged together. Finally, the overlying fissures formed a vertical upward through fissure channel at 35 m below the bottom plate, and the water and sand volume burst out through the fissure channel at 1 L/min. With the interaction between the water–sand mixed flow and the fissure, the outburst point and the outburst amount continued to grow. If it increases, it will eventually lead to a water and sand inrush accident in which a large volume of water, sand and bedrock is flushed out together.

(4) The influence of particle size on the permeability of aquifers is summa-rized through laboratory simulation experiments, and the influence of coarse

particle content on permeability is particularly significant, followed by characteristic particle size, particle inhomogeneity coefficient and curvature coefficient. Aquifers with different permeability properties have different hydraulic conductivity, and this parameter will obviously affect the formation speed and scale of the falling funnel in the groundwater seepage flow field.

(5) Through the mechanical model and force analysis, the key influencing factors of sand body migration after fracture formation are studied, and it is proposed that the overall migration of sand body is closely related to the width of the fracture and the roughness of the fracture surface. Its startup and migration need to meet certain control conditions. It is very important to control the development width of the fractures in order to prevent the occurrence of sand breaking accidents.

References

1. Bai ZG, Liu QM, Liu Y (2022) Risk assessment of water inrush from coal seam roof with an AHP–CRITIC algorithm in Liuzhuang Coal Mine, China. Arab J Geosci 15(4)
2. Hazen A (1911) Discussion of "dams on sand foundation" by A. C. Koenig. Trans Am Soc Civ Eng 73:199–203
3. Yang X et al (2020) Experimental Investigation of water–sand mixed fluid initiation and migration in porous skeleton during water and sand inrush. Geofluids 2020
4. Zeng YF et al (2022) An analog model study on water–sand mixture inrush mechanisms during the mining of shallow coal seams. Mine Water Environ 41(2)
5. Zhang GB, Wang HL, Yan SL (2017) Simulated experiment of water-sand inrush across overlying strata fissures caused by mining. J Min Safety Eng 34(03):444–451
6. Zhang BY et al (2021) Experimental study on the flow behaviour of water-sand mixtures in fractured rock specimens. Int J Min Sci Technol 31(03):377–385
7. Zhang YG, Xie YL, Yang LN, Liao RP, Qiu T (2022) Intelligent prediction of coal mine water inrush based on optimized SAPSO-ELM model under the influence of multiple factors. Arab J Geosci 15(5)
8. Zhao JY, Liu WT, Shen JJ, Xu MG, Sasmito Agus P (2022) A real-time monitoring temperature-dependent risk index for predicting mine water inrush from collapse columns through a coupled thermal–hydraulic-mechanical model. J Hydrol 607

Chapter 7
Numerical Simulation Study on Mine Water–Sand Inrush Under Fluid–Solid Coupling

7.1 Overview

The role of groundwater cannot be ignored in more than 60% of coal mine accidents. After the original stress field of the overlying rock is disturbed by the excavation activities, the roof rock is ruptured. Once the overlying aquifer is connected, the groundwater infiltrates the cracks, and the rock strength, shear strength, cohesion, and internal friction are affected by the hydraulic pressure of the groundwater. The coefficient and other parameters will decrease, which will aggravate the fracture and instability of the rock, and eventually lead to the occurrence of water inrush accidents. Among them, the in-situ stress of the roof rock mass and the water pressure of the overlying aquifer are the main factors affecting the stability and deformation of the roof rock layer. The interaction between the two is the fluid–solid coupling effect of the groundwater seepage field and the in-situ stress field of the rock mass [1–3].

Numerous scholars have made remarkable progress in this area of research. Many scholars have summarized the quantitative relationship between stress–strain-permeability after rock failure under different water and perimeter pressures through full stress–strain process permeability tests. Later, based on the dual medium theory model, the porous continuous medium, the coupled field model of rock seepage field and stress field, the concentrated parameter model of rock seepage field and stress field coupling and the fracture network model were established [4–7]. To solve the model, numerical simulations such as FLAC3D and COMSOL Multiphy are mostly require. More results have been achieved in the study of the mechanism of water breakout in coal mines.

© The Author(s), under exclusive license to Springer Nature Switzerland AG 2023   211
Y. Zeng et al., *Roof Water Disaster in Coal Mining in Ecologically Fragile Mining Areas*, Professional Practice in Earth Sciences,
https://doi.org/10.1007/978-3-031-33140-4_7

7.2   Mathematical Modeling of Fluid–Solid Coupling

During the coal seam mining process, the surrounding rock is disturbed, resulting in stress redistribution, collapse zone and fracture zone. The change of stress field will change the volume strain and void rate of the rock mass, thus changing the permeability conditions, and the continuous infiltration of water from the aquifer will change its seepage field, and the action of water pressure after groundwater infiltration will further affect the stress field of the surrounding rock. To establish the calculation model, the seepage field model and the stress field model are established separately, and then the relationship between the two fields is established through the coupled intrinsic structure relationship–the relationship between stress and hydraulic conductivity, and finally the coupling calculation is carried out.

Therefore, the key to establish the mathematical model is to establish the relationship between stress–strain and rock permeability, the need for targeted laboratory experiments, usually equipped with rock mechanics electro-hydraulic servo system to complete the rock mass permeability test full stress–strain process, and for pores, cracks, pipe type rock should be established their relationship, scholars choose sandstone, argillaceous sandstone, sandy mudstone, limestone etc., were established under large deformation and micro-deformation stress–strain relationship with permeability. Exponential mathematical model for the relationship between stress and hydraulic conductivity for porous rocks, The relationship between stress and hydraulic conductivity for fractured rock descriptions has a power exponential type and cubic types mathematical models. The rock porosity loss and permeability loss change with effective stress in accordance with a quadratic parabolic mathematical model. These findings provide a useful reference for the next step in mathematical modelling [8–10].

For the process of coal seam mining disturbance leading to the development of roof fissures, which in turn interact with the overlying aquifer water, the focus is on the process of continuous destruction and development of fissures in the rock body by the action of water pressure, so the rock mass is mainly regarded as a fissured rock mass and the coupling effect of water pressure on the fissured rock mass is considered. Figure 7.1a shows a conceptual model of fractured rock masses. This simplified conceptual model considers the bedrock to be homogeneous in all values and the fractures to be uniformly distributed and orthogonal to each other, and considers fluid flow from the pores of the rock into the fractures, and considers the fluid flow pattern in the fractures to obey Darcy's law. Figure 7.1b shows force analysis model for fractured rock units. In the figure P is the expansion pressure of the fluid on the fractured rock, F is the volume force on the fractured rock and σ_m is the average stress considering only the volume change and not the shape change.

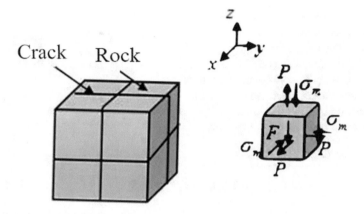

(a)Conceptual model of fractured rock masses　　(b)Force analysis of fractured rock units

Fig. 7.1 Mechanical analysis model for fractured rock coupling liquid

The controlling equations and corresponding boundary conditions for the seepage theory of fractured rock masses are expressed as follows:

Control equations:

$$\frac{\partial}{\partial x_i}\left[\frac{k_{ij}}{\zeta}\left(\frac{\partial P}{\partial x_j}+\rho_f g_j\right)\right]=cP-\gamma\varepsilon_{ii} \qquad (7.1)$$

$$Gu_{i,ij}+(\lambda+G)u_{j,ij}+\gamma P_{,i}+F_i=0 \qquad (7.2)$$

$$\gamma=1-\frac{k_{s,1}}{k_s} \qquad (7.3)$$

$$k_s=\frac{1-2\nu_s}{E_s},\ k_{s,1}=\frac{1-2\nu_{s,1}}{E_{s,1}} \qquad (7.4)$$

which,

k_{ij} permeability
P Fluid pressure against the medium
ζ Fluid viscosity
ρ_f Fluid density
g_j Gravitational acceleration
ε_{ii} Volumetric strain
c Average volume compression coefficient
γ Effective stress coefficient
F_i Volume force component
λ Lamé factor

G Modulus of shear elasticity
u_i Solid displacement vector
k_s Bulk modulus of non-fractured rock mass
$k_{s,1}$ Volume modulus of fractured rock mass
E_s Elastic modulus of non-fissure rock mass
$E_{s,1}$ Elastic modulus of fractured rock mass
v_s Poisson's ratio of non-fractured rock mass
$v_{s,1}$ Poisson's ratio of fractured rock mass

   boundary condition:

$$u_i(x_i, t) = \overline{u}_i(x_i, t)(x, y, z) \in S_1 \tag{7.5}$$

$$\sigma_{ij}(x_i, t)n_j = \overline{F}(x_i, t)(x, y, z) \in S_2 \tag{7.6}$$

$$P(x_i, t) = \overline{P}(x_i, t)(x, y, z) \in S_3 \tag{7.7}$$

$$V_i(x_i, t)n_i = \overline{V}(x_i, t)(x, y, z) \in S_4 \tag{7.8}$$

and,

$$S_1 \cup S_2 = S_3 = S_4 = S \tag{7.9}$$

initial condition:

$$P(x_i, 0) = P_0 \tag{7.10}$$

where: $\overline{u}_i, \overline{F}, \overline{P}, \overline{V}$ Indicates acknowledged displacements, exterior forces, fluid stress and fluid glide rates, respectively. S is the boundary of the computational domain.

7.3 Numerical Simulation of Stress Field and Seepage Field Coupling

Numerical solution of the fluid–solid coupling problem at the mining site using the FLAC3D seepage model, which can be used for fluid calculations alone or coupled with computational fluid mechanics to solve seepage problems that are fully saturated or have the effect of groundwater changes.

7.3.1 Field Condition

The 22,304-1 working face of Shigetai Mine is selected as the research example. This working face is mining 2–2 coal, located in the three pan area. The coal thickness ranges from 2.1 to 3.6 m, whereas the average coal thickness is 2.8 m. The dip perspective is 1–3°. The ground elevation is between 1255 and 1290 m. The backside elevation is between 1170 and 1198 m. The working face length is 2,433 m. The working face width is 235.8 m. The working face area is 573,701.40 m^2. The return wind chute is between 51 and 74 L in the upper phase of the coal seam. A layer of mudstone, siltstone and charcoal mudstone of 0.3–0.7 m is exposed.

The workface began on 23 December 2011, and water exploration and launch works have been carried out. 169 water exploration and launch holes had been built and 125 holes had water in them after the ultimate hole. During the development of the water exploration and launch gap in seventy eight link, a giant quantity of yellow sand flowed down with the water (initial extent of 32.6 m^3/h) in one hole, and no yellow soil layer was once determined when drilling via the bedrock, in the following days, three extra water exploration and launch holes in seventy seven hyperlink and seventy eight hyperlink had sand float down, the state of affairs was once comparable to the preceding one, the thickness of bedrock in these 4 holes was once 32.3–37.3 m. From 24 January to 3 February 2013, five hydrographic holes, SS52, SS51, SS50, SS49 and SS48, had been built close to the most important and auxiliary transport chute of the working face. The water degree drop values are 1.04 m, 1.28 m, 1.87 m, 1.83 m, and 1.61 m, respectively. The aquifer thickness at every gap grew to become 3.03 m, 1.49 m, 26.75 m, 35.62 m, and 34.15 m.

The upper aquifer of the overlying rocks is the pore dive (Q$_3$s) of the Salausu Formation, which can be 20–50 m thick at the ancient lake depressions and ancient washout trenches, with a maximum thickness of 53.76 m and an average thickness of 18.35 m throughout the well field. according to the pumping test data: water level lowering 1.06–15.30 m, gushing water 0.783–12.60 l/s.m, unit gushing water 0.369–1.54 l/s.m, hydraulic conductivity 3.89–10.96 m/d, water-richness moderate to strong.

7.3.2 Calculation Model Design and Selection of Calculation Parameters

In order to study mechanism of roof sand inrush water, based on Shigetai Mine 22,304 recovery geological instruction, 22,304 excavation project plan, 22,304 roadway sketch and 22,304 drilling column diagram within the working face, has been developed 3D numerical simulation model (Fig. 7.2). The model is 700 m long, 300 m wide and 109 m high, with a total of 144,000 3D units divided into a total of 152,988 nodes. The sides of the model limit horizontal movement, the bottom of the model

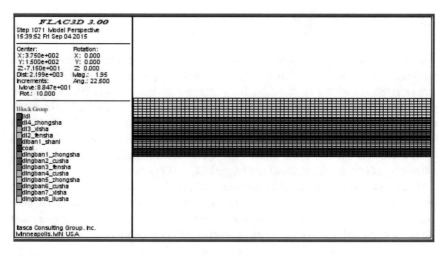

Fig. 7.2 Grid model of numerical simulation

limits vertical movement and the top of the model applies a vertical load to simulate the weight of the overlying over-lying strata.

When the load reaches the limit of strength, the rock mass is destroyed. During the post-peak plastic flow, the residual strength of rock mass decreases with the development of deformation. Therefore, the Mohr–Coulomb yield criterion was used in the calculations to determine the failure of the rock mass (Eq. 7.11), adoption of a strain softening model to reflect the nature of the progressive reduction in residual strength as deformation progresses.

$$f_s = \sigma_1 - \sigma_3 \frac{1 + \sin\phi}{1 - \sin\phi} - 2c\sqrt{\frac{1 + \sin\phi}{1 - \sin\phi}} \tag{7.11}$$

where, σ_1, σ_3 represent the maximum and minimum principal stresses respectively, c, φ stand for cohesive force and friction angle. When fs > 0, the material will undergo shear failure. Under normal stress conditions, the tensile strength of the rock mass is very low, so the tensile strength criterion ($\sigma_3 \geq \sigma T$) can be used to determine whether the rock mass is producing. According to the rock mechanics test results provided by the on-site geological survey and related research, the mechanical parameters of the main lithology in the study area are shown in Table 7.1.

Considering the scale effect of the rock, the mechanical parameters were discounted as necessary for the simulations. The uppermost layer in the model is a loose sand aquifer with a head of 30 m at the bottom, as shown in Fig. 7.3.

After the model is established, the model reaches the initial equilibrium after 1701 steps of iterative operation, as shown in Fig. 7.4. It can be seen from Fig. 7.4 that under the initial equilibrium state, the vertical stress of normal rock mass is generated by self-weight stress. With the increase of buried depth of rock mass, the

Table 7.1 Mechanical parameters of rock

Lithological description	Elastic modulus/Gpa	Tensile strength/MPa	Internal friction angle/°	Poisson's ratio	Volumetric weight/kg/m³
Loose sand layer	0.3	0.01	18	0.4	1573
Sandy shale	38.5	3.53	40	0.147	2450
Renewal Table 7.1					
Medium grained sandstone	25.5	2.5	40	0.25	2590
Silty sandstone	33.4	3.5	42	0.235	1450
Coal	3.5	0.76	32	0.32	2598
Medium coarse sandstone	20.5	2.2	40	0.27	2627

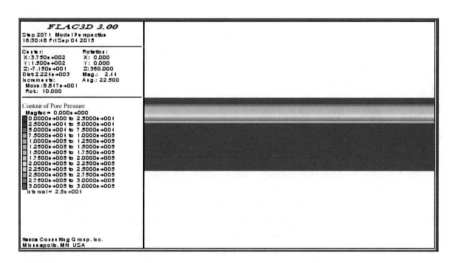

Fig. 7.3 Distribution of pore water pressure under initial conditions

vertical stress increases linearly. The minimum value of SZZ is 0.432 MPa at the upper boundary of the model, and the maximum value is 3.146 MPa at the bottom boundary of the model.

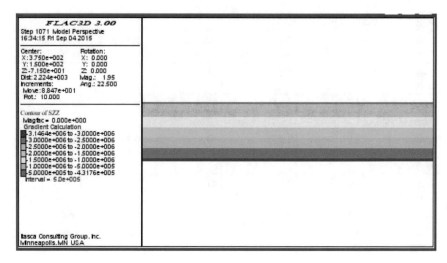

Fig. 7.4 Distribution of SZZ value under initial equilibrium state

7.3.3 Fluid–Solid Coupling Numerical Simulation Analysis of Roof Water–Sand Inrush Mechanism Under Mining Influence

7.3.3.1 Analysis of Vertical Stress Field Distribution Patterns

With the mining of the coal seam, the equilibrium state of the coal seam surrounding rock stress is broken, Figs. 7.5, 7.6 and 7.7 shows the contour of the distribution of coal seam/rock seam support pressure when the working face is advanced 10 m, 30 m and 50 m, respectively.

As can be seen in Fig. 7.5, there are no tensile stresses in the profile, it is all compressive stresses, the maximum compressive stress is located at the bottom of the model, in addition there is a compressive stress concentration 10 m before and after the mining area, the compressive stress is 3 Mpa, the stress concentration coefficient is 1.33, at this time the mining of the coal seam has basically no effect on the top loose sand seam.

When the mining face advanced 30 m, the tensile stress already appeared on the top surface of the mining area, the maximum tensile stress was 0.133 Mpa, while the stress near the bottom plate of the mining area decreased sharply. In addition, the stress concentration was more obvious at 10 m before and after the mining area, the maximum compressive stress reached 4.929 Mpa, the stress concentration coefficient reached 2.19, at this time the coal mining had some influence on the top loose sand layer, in the mining area. The stress field curve of the loose sand layer directly above the mining area was slightly bent and the stress was reduced (Fig. 7.6).

When the working face advanced 50 m, the tensile stress at the top of the mining area continued to increase, the maximum tensile stress was 0.171 Mpa, the maximum

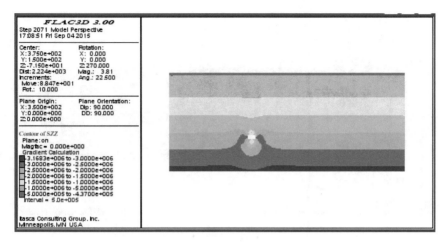

Fig. 7.5 Distribution section of coal/rock support pressure at mining 10 m

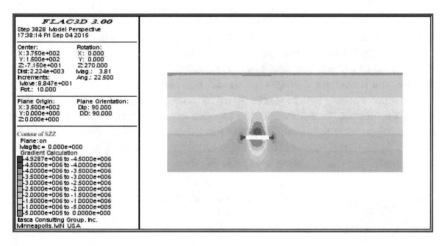

Fig. 7.6 Distribution section of coal/rock support pressure at mining 30 m

compressive stress increased to 5.12 Mpa, the stress concentration coefficient reached 2.28, coal mining has obviously affected the stress distribution of the loose sand layer at the top, which may trigger a sudden water collapse at the top (Fig. 7.7).

7.3.3.2 Analysis of Failure Field Development Patterns

After the equilibrium state of the coal seam surrounding rock stress is broken, plastic failure area appears in the top and bottom plate of the coal seam. Figures 7.8, 7.9,

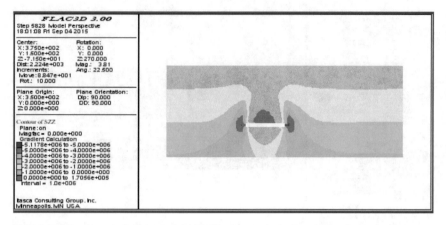

Fig. 7.7 Distribution section of coal/rock support pressure at mining 50 m

7.10 and 7.11 shows the distribution of plastic failure area around the working face when the working face is advanced 10 m, 30 m, 40 m, and 50 m respectively.

As can be seen from Fig. 7.8, when the working face is advanced by 10, the top and bottom plates of the working face have already produced plastic damage, as the top plate of the working face has not produced tensile stress at this time, so the damage type is shearing damage, and the damage height is 6 m.

When the working face advances 30 m, the damage range of the top and bottom slab of the working face increases, the top slab appears tension damage, while the damage height increases, and the damage form shows a saddle shape, as can be seen in Fig. 7.9, the damage height is approximately 15 m, the development height of the water-conducting fracture zone is 15 m at this time.

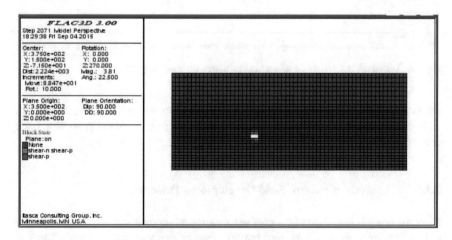

Fig. 7.8 Distribution map of plastic zone at mining 10 m

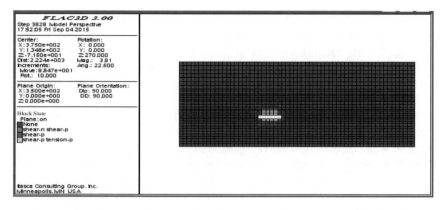

Fig. 7.9 Distribution map of plastic zone at mining 30 m

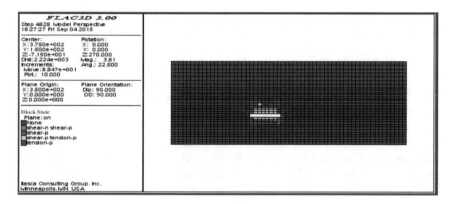

Fig. 7.10 Distribution map of plastic zone at mining 40 m

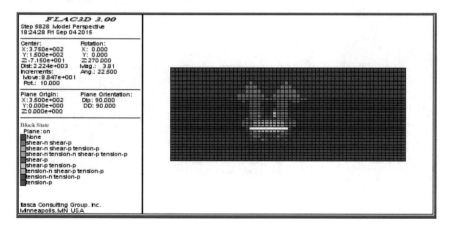

Fig. 7.11 Distribution map of plastic zone at mining 50 m

When the working face advances 40 m, the damage form of the roof plate shows an obvious saddle shape, and the development height of the roof plate fissure zone increases further, reaching a maximum of 23 m. As can be seen from Fig. 7.10, the roof plate water-conducting fissure zone has not yet developed to the loose sand layer of the roof plate, so it is not yet possible to trigger a sand collapse and sudden water accident.

Figure 7.11 shows the distribution of the plastic damage area of the top and bottom plate of the working face when the working face is advanced by 50 m. From the figure, it can be seen that the whole damage pattern shows a saddle shape, and the damage range of the top and bottom plate further increases, in which the damage height of the top plate has reached the loose sand layer of the top plate, thus communicating with the working face and the loose sand layer, and triggering a sudden water collapse accident under the joint action of groundwater and the stress of stope.

7.3.3.3 Analysis of Shear Strain Field Distribution Patterns

The shear strain increment will be triggered when the stress equilibrium state of the coal seam surrounding rock is broken. Figures 7.12 and 7.13 show the distribution of shear strain increment when the working face is advanced 40 m and 50 m, respectively.

From Fig. 7.12, it can be seen that the shear strain increment is concentrated at the two ends of the mined-out area, when the working face advances 40 m, the shear strain maximum increment is 1.588e-003, where shear strain increment in the plastic damage area minimum is 8.0e-004, and there is obvious diffidence between the loose sand layer of the roof slab and the lower bedrock, when the working face advances

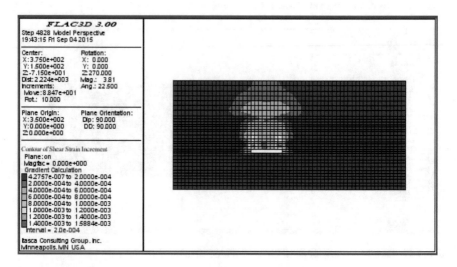

Fig. 7.12 Distribution map of shear strain increment zone at mining 40 m

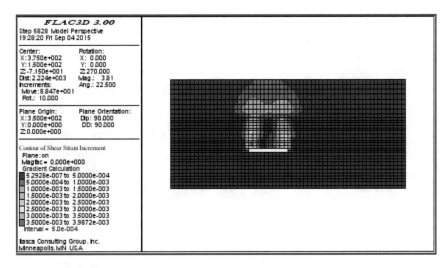

Fig. 7.13 Distribution map of shear strain increment zone at mining 50 m

50 m, the shear stress increment is still mainly distributed at the two ends of the mined area. At the same time, the value of shear stress increment increases significantly, with the maximum value of 3.987e-003 and the contour of strain increment of 1.0e-003 communicating with the loose sand layer of the roof plate, which is also the location where the fracture zone of the roof plate communicates with the loose sand layer of the roof plate, which will trigger a sudden water and sand collapse accident in this area.

7.3.3.4 Mechanism Analysis of Water–Sand Inrush

From the previous analysis, when the mining face advances 50 m, there will be a flooding and sand collapse accident. Therefore, to study the mechanism of flooding and sand collapse under the influence of mining and fluid–solid coupling, the process of fracture expansion, seepage field and stress field change when the mining face advances 50 m is analyzed in detail.

The first stage is the initial fissure extension stage. Figure 7.14 shows the distribution of the plastic zone of initial fissure expansion when the mining face is advanced 50 m on the basis of the plastic zone of 40 m excavation. From the figure compared with Fig. 7.10, the plastic zone of the roof of the mining area has increased in both scope and height, and in addition, some shear damage areas appear in the contact area between the bedrock and the loose sand layer of the roof.

The pore water pressure distribution is shown in Fig. 7.15 when the fracture initially expands when the mining face advances 50 m. It can be seen from the figure that the pore water pressure distribution at this time is the same as the water pressure

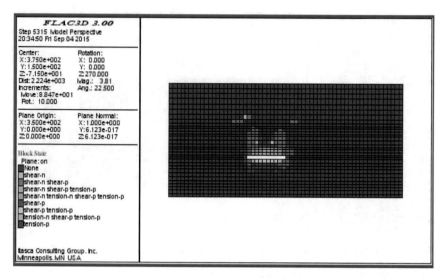

Fig. 7.14 Distribution map of plastic zone as initial crack extension at mining 50 m

distribution in the initial state, and the mining movement at this time does not affect the water pressure distribution.

The maximum principal stress distribution at the initial expansion of the fissure when the face is advanced 50 m is shown in Fig. 7.16, from which in the plastic distribution region, the maximum principal stress transforms from compressive stress

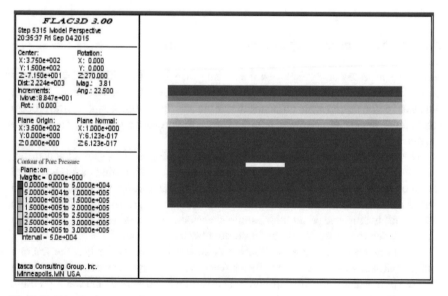

Fig. 7.15 Distribution map of pore water pressure as initial crack extension at mining 50 m

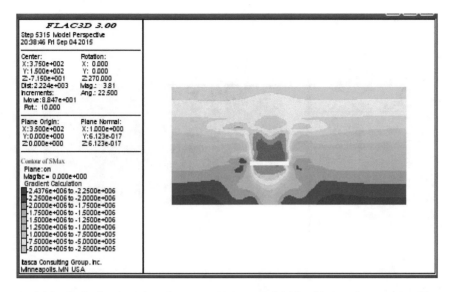

Fig. 7.16 Distribution map of maximum principal stresses initial crack extension at mining 50 m

to tensile stress and the compressive stress decreases significantly in the plastic region where the bedrock and the loose sand layer of the roof are in contact.

With the further expansion of the fissure, under the combined influence of mining and groundwater, the fissure conducts to the loose sand layer of the top slab and makes the loose sand layer of the top slab further loose (Fig. 7.17), which created a good condition for the collapse of sand and sudden water.

Figure 7.18 shows the distribution of pore water pressure after the fracture of the top slab conducts through the loose sand layer. It can be seen from the figure that at this time, groundwater and sand begin to penetrate through the fracture zone to the working face, opening the stage of collapsing sand and water.

Figure 7.19 shows the distribution of the maximum principal stress when the groundwater starts to infiltrate into the working face after the top slab fissure conducts through the loose sand layer. It can be seen from the figure that as the groundwater infiltrates, the distribution and magnitude of the maximum principal stress also changes to a certain extent, and according to the effective stress principle, this change will create conditions for the further expansion of the fissure and the further loosening and destruction of the loose sand layer.

As groundwater and sand infiltrate into the working face, a precipitation funnel is gradually formed on the roof of the goaf, as shown in Fig. 7.20. After the precipitation funnel is formed, the effective compressive stress here increases relatively due to the decrease or disappearance of the water pressure in the central area of the funnel. The effective compressive stress around the funnel decreases due to the presence of water pressure, as shown in Fig. 7.21. The distribution of this differential force will cause uneven settlement and further loosening of the loose sand layer, see Fig. 7.22.

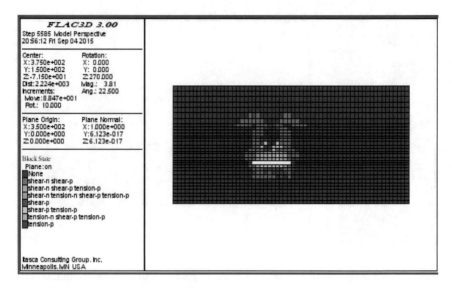

Fig. 7.17 Map of roof fissure conducting upon loose sand aquifer at mining 50 m

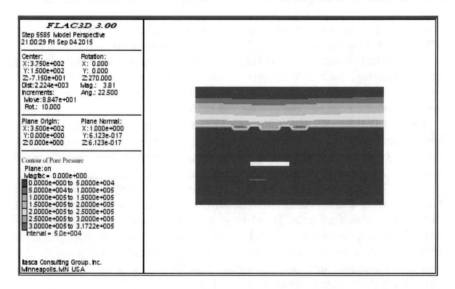

Fig. 7.18 Map of pore water stress after roof fissure conducting upon loose sand aquifer

In the process of groundwater and sand collapse into the work surface flow, the formation of precipitation funnel, so that the loose sand layer loose, but also erosion of the hydraulic fracture zone, so that the fracture zone water-conducting ability to strengthen, thus causing the collapse of sand and sudden water accident, Figs. 7.23

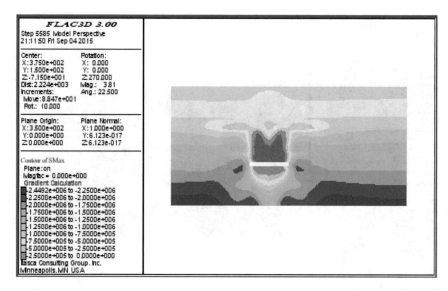

Fig. 7.19 Distribution map of maximum principal stress as groundwater permeating

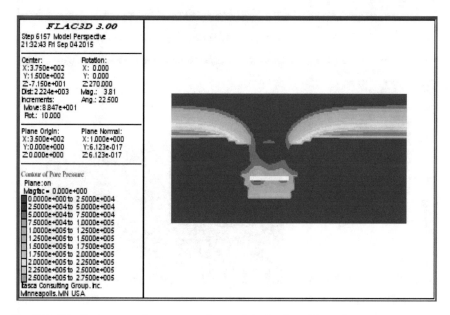

Fig. 7.20 Distribution map of water stress after precipitation funnel formation

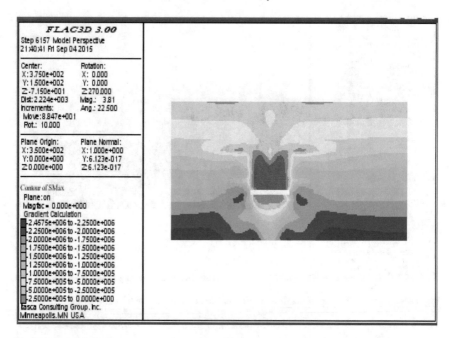

Fig. 7.21 Distribution map of maximum principal stresses after precipitation funnel formation

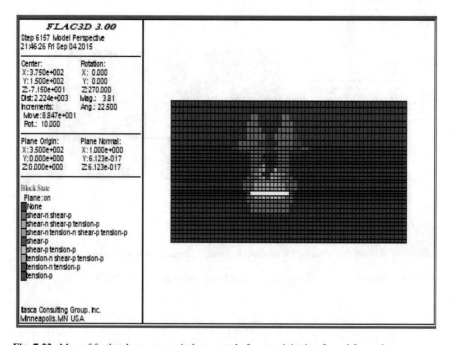

Fig. 7.22 Map of further loosen zone in loose sand after precipitation funnel formation

and 7.24, respectively for the sudden water collapse of sand after a certain period of time, water pressure and loose sand layer loose graph.

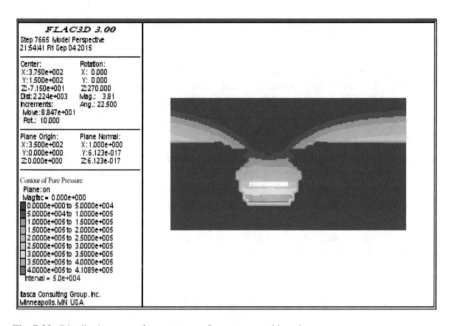

Fig. 7.23 Distribution map of water stress after water–sand inrush

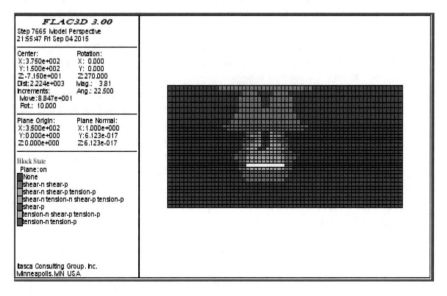

Fig. 7.24 Map of further loosen zone in loose sand after water–sand inrush

7.4 Summary

In this chapter, through the intrinsic relationship between stress and rock perme-ability, a flow-solid coupling mechanical and mathematical model based on fractured rock is established, including control equations and definite solution conditions, and the FLAC3D seepage module is used to carry out the actual settlement of flow-solid coupling at the working face 22,304 in Shigetai mine. The vertical stress field, damage field development law, shear strain distribution law at different advancing distances of the working face and the analysis of the mechanism of the sand breakout was carried out. The main conclusions obtained through the research in this section are as follows:

(1) In the process of sand inrush water development, the development of fracture generation cannot ignore the interaction with water, the established fracture medium as the main body of the flow-solid coupling mechanical model and mathematical model, a better reflection of the mechanical mechanism of water-pressure interaction in the fracture system, for the subsequent analysis of the mechanism of suddenly fractured sand to lay the foundation.

(2) The numerical simulation results show that the overall water–sand inrush process can be divided into three stages:

1. The initial stage of roof plate fissure development. The working face has just started to advance, and compressive stresses are concentrated at 10 m before and after the hollow area, the top and bottom plates of the working face have produced plastic damage, but the top plate has not yet produced tensile stress, the damage type is shear damage, the damage height is 6 m, there is no effect on the upper aquifer at all, and there is no change in the pore water pressure at this stage.

2. The roof plate fissure development and expansion stage. When the working face continues to advance to 30 m, the damage range of the top and bottom plate increases, tensile stress and tensile damage appear on the top surface of the mining area, while the stress near the bottom plate of the mining area decreases sharply, and obvious stress concentration appears 10 m before and after the mining area, the damage height and the height of the water-conducting fissure zone both increase, and coal mining begins to have an impact on the top loose sand layer. When the working face advanced to 40 m, the roof damage form showed obvious saddle shape, the shear strain increment was concentrated at both ends of the mining area, there was obvious zoning between the loose sand layer of the roof and the lower bedrock, the water-conducting fracture zone of the roof was further developed but not yet to the upper loose sand layer, so it could not trigger the sand collapse and sudden water accident yet.

3. The fissure leads to the upper aquifer stage. When the working face advances 50 m and beyond, the tensile stress at the top of the mining area continues to increase, the plastic damage form of the top and bottom plate of the working face shows a saddle shape, the damage height of the top plate has reached the loose sand layer, and the accident of sand inrush water will occur at this stage.

(3) When the working face advances to 50 m and beyond, the process of fracture expansion, seepage field and stress field changes reveal the mechanism of sudden water collapse under the effect of fluid–solid coupling. (1) Before the fracture zone leads to the upper aquifer, although tensional and plastic damage occurs at the top and bottom of the mined-out area and before and after, the deformation and stress field of the top and bottom plate changes more obviously, but as long as the fracture does not lead to the aquifer, the seepage field of the groundwater is not disturbed i.e., it will not collapse, and the role of water can be disregarded at this stage. (2) After the fissure zone leads to the upper aquifer, the water–rock interaction is not negligible, and the compressive stress is transformed to tensile stress in the plastic distribution area, especially the compressive stress in the contact area between the bedrock and the loose sand layer of the roof plate is obviously reduced, under the joint action of coal mining disturbance and groundwater, the fissure leading to the loose sand layer of the roof plate continues to develop, so that the loose sand layer begins to loosen, and at this stage, groundwater and sand begin to transport to the working face through the fissure zone. With the extension of time, the infiltration of groundwater intensified the expansion of the fissure and the loosening of the loose sand layer, and the water and sand continuously gushed into the mouth of the fissure, and the stage of sudden water collapse of sand began. (3) After the start of the sand inrush water, groundwater gradually forms a precipitation funnel in the roof of the mined out area, and the effective compressive stress in the central area of the funnel increases relatively due to the reduction or disappearance of water pressure, while the effective compressive stress around the funnel decreases due to the presence of water pressure, this distribution of differential forces will produce uneven settlement and cause further loosening of the loose sand layer. At the same time, in the process of groundwater and sand collapsing into the working face, the water–sand mixture can also erode the water-conducting fracture zone, which strengthens the water-conducting capacity of the fracture zone and eventually leads to sudden water collapse accidents.

References

1. Di ST (2020) Study on the macroscopic evolution mechanism and trend prediction of ground subsidence caused by groundwater mining [D]. Shandong University
2. Li Q (2019) Study on numerical simulation method of gas-liquid-solid coupling for hydraulic fracture of rocks [D]. Tianjin University
3. Niu L (2015) Mechanism and risk evaluation of karst trap column burst water in North China type coalfield [D]. China University of Mining and Technology (Beijing)
4. Fan GW, Lin BQ, Zhi XW (2006) Numerical simulation of outburst in Majiagou mine based on RFPA-flow [J]. J Xi'an Univ Sci Technol (02):0167–03
5. Gong QS (2016) Determination of the spacing of gas extraction boreholes in this coal seam based on COMSOL Multiphy [J]. Inner Mongolia Coal Econ (16):134–135+153

6. Luan HJ, Cao YW, Jiang YJ, Guan YT, Li CP, Zhang SH, Liu JR (2022) Implementation and application of anchor rod tension-shear coupled failure mode in FLAC3D. J Min Rock Control Eng 06:1–11
7. Xia LL, Zhang, SB, Chen, NH Shu, H, Liu. Shang G (2022) Development and validation of a fluid-solid weakly coupled program based on TOUGH2 and FLAC3D [J]. J Zhejiang Univ (Eng Ed) 56(08):1485–1494
8. Liu ZB (2014) Study on the coupling effect of seepage and stress in rock mass and the water inrush effect of coal seam floor [D]. Xi'an Research Institute, General Academy of Coal Sciences
9. Yang J, Wang JW (2010) Study on fractal characteristics of sandy soil seepage [J] Coal Field Geol Explor (2):42–45
10. Zhang YJ (2013) Experimental study on permeability characteristics of sandy soil with different particle sizes [D]. Xi'an University of Architecture and Technology

Chapter 8
Water Inrush Mode and Prevention Technology Under Bedrock Fissure Aquifers

8.1 Analysis of Main Factors Affecting Water Yield of Aquifer

Groundwater mainly occurs in rock voids, and the development of karst and fissures in rock mass is often heterogeneous and discontinuous, and their water conductivity and water storage capacity are also very different, which also makes the distribution of aquifer water abundance with great heterogeneity [1–4]. By means of theoretical analysis, experience summary and field tests, combined with the analysis of a large number of water inrush accidents, a variety of main control factors affecting the water yield of the aquifer are determined: the thickness of the aquifer, the brittle plastic ratio of the aquifer, the unit water inflow, the hydraulic conductivity and the rock core recovery rate. The indicators of each main control factor are further studied to lay a foundation for the quantitative study of the space–time distribution law of each main control factor. It also provides information support for quantitative study of aquifer water yield under the influence of multiple factors.

(1) Aquifer thickness. Aquifer thickness here generally refers to the total thickness of the aquifer group, the greater the total thickness of the aquifer group, the stronger its water storage capacity, and the water abundance of the corresponding aquifer is generally larger.

(2) The ratio of brittleness to plasticity of the aquifer. Because of the effect of stress, the brittle rock with high mechanical strength in the aquifer generally develops fissures and joints, which are easy to form the water channel of confined water, enrich a certain amount of water, and increase its permeability. However, the plastic rock with lower mechanical strength in the aquifer is generally more compact due to plastic deformation, and its water abundance and transmissivity are weak. In addition, the greater the thickness of brittle rock is, the larger the aquifer storage space is, and the hydraulic connection between aquifers is

Y. Zeng et al., *Roof Water Disaster in Coal Mining in Ecologically Fragile Mining Areas*, Professional Practice in Earth Sciences, https://doi.org/10.1007/978-3-031-33140-4_8

correspondingly enhanced, and the ability of aquifer to release water is correspondingly increased. Comprehensively, the greater the thickness ratio of brittle-plastic rock of aquifer, the stronger the water storage capacity of aquifer, and the stronger the hydraulic connection of each aquifer group, which generally corresponds to the stronger water abundance of aquifer.

(3) Structural distribution (Faults, collapse columns and fold axes, etc.). The fracture in the aquifer is a good channel for groundwater, and the fracture zone of the fault and the influence zone with a certain width of water flowing fracture in both sides are often the enrichment area of groundwater. Especially for the faults connecting multiple aquifers, once the faults have the function of water conduction, the water abundance of aquifers will be greatly enhanced. In addition, the water conductivity and water richness of fracture zone and affected zone are often different with different fault properties, generally for the normal fault, which is mostly formed under low confining pressure conditions, the degree of fracture is large, the fault fracture zone is relatively loose, and its water conductivity and water yield are often better. The reverse fault is mostly formed under the condition of high confining pressure, the fracture zone is relatively compacted and dense, and its water conductivity and water abundance are relatively weak, but the fracture influence zone of the two sides often has a certain water conductivity.

(4) Unit water inflow: The unit water inflow is the most direct parameter reflecting the water abundance of the aquifer. Generally, the larger the unit water inflow is, the better the permeability of the aquifer is, and the more sufficient the water supply is, and the stronger the water abundance of the corresponding aquifer is. In the case of less data, the unit inflow of boreholes can also be directly used to divide the water abundance of the aquifer.

(5) Hydraulic conductivity: Hydraulic conductivity mainly reflects the permeability of water in aquifer, which depends on the properties of rock on the one hand, and the physical properties of fluid on the other hand. For groundwater, the greater the permeability, the better the permeability of rock, and the stronger the water abundance of aquifer.

(6) Collecting rate of drill core rate: The rock core recovery rate mainly reflects the integrity of the rock mass. Generally, the lower the rock core recovery rate is, the worse the integrity of the rock mass is. The more developed the fractures or karst of the corresponding rock mass is, the better the water storage capacity and permeability of the corresponding aquifer are, and the stronger the water yield of the corresponding aquifer is. In addition, for soluble rock, if the rock core recovery rate is low, the corresponding karst is relatively developed, which also indicates that the aquifer may be located in the strong runoff area of groundwater, and the water yield of the corresponding aquifer is relatively strong.

(7) Consumption of flushing fluid: The consumption of flushing fluid can reflect various information characteristics of the formation. On the one hand, it reflects the integrity of rock mass. On the other hand, it reflects the hydraulic properties of the aquifer. Generally, the greater the consumption of flushing fluid after drilling into the aquifer, the worse the integrity of the aquifer, and the stronger

the water storage capacity of the rock mass with more karst and fissures. It also reflects that the water pressure in the aquifer is relatively small, but the permeability of the aquifer is relatively good. On the whole, the consumption of flushing fluid reflects the water abundance of the aquifer to some extent; The larger the value is, the stronger the water abundance of the aquifer is.

8.2 Basic Principle of Water Abundance Index Method

In order to further reveal the influence of the main controlling factors on the water yield of the aquifer, by analyzing and summarizing the characteristics of the main controlling factors, it is found that there are two common points in the influence of various main controlling factors on the water yield of the aquifer: one is that the influence degree of each main controlling factor on the water yielding of the aquifer is different at different spatial coordinate points in the spatial domain. Second, under different hydrogeological conditions, the degree of influence of the main controlling factors on the water yield of the aquifer is not the same. The weight of each main controlling factor affecting aquifer water abundance is different. By using modern information and mathematical technology, the information fusion model based on GIS is established, and the comprehensive analysis and research are carried out on the various main controlling factors and the complex interaction relationship of multiple factors affecting the aquifer water abundance, and finally the aquifer water abundance division is obtained [5–9]. Based on GIS information fusion technology, a new aquifer water abundance evaluation technology is put forward.

Firstly, GIS, which has powerful functions of spatial information analysis and processing, is used to make quantitative analysis of various main controlling factors affecting the water abundance of aquifers, study the spatial distribution law of the main controlling factors, and generate thematic maps of the main controlling factors. Secondly, using modern information fusion technology, according to the known sample information training learning or inversion identification, the weight of each main control factor affecting the water richness of aquifer is quantitatively determined, and the information fusion technology based on GIS is further applied to establish the evaluation and prediction model of aquifer water richness index under the influence of multiple factors; Then, through the data processing and spatial analysis function of GIS, the water abundance index values of each section in the study area are statistically analyzed, and the water abundance zoning threshold is determined according to its frequency histogram, and the aquifer water abundance evaluation zoning is carried out.

According to the different mathematical methods of information fusion, the information fusion technology based on GIS can be divided into two categories: linear and nonlinear. Linear information fusion technology includes AHP information fusion technology based on GIS and so on. The nonlinear information fusion technology includes the ANN information fusion technology based on GIS, the evidence weight information fusion technology based on GIS and the Logistic regression information fusion based on GIS.

The GIS-based aquifer water abundance evaluation method is a prediction and evaluation method which couples GIS with powerful spatial information analysis and processing functions and the determination of the weight of the main controlling factors affecting the aquifer water abundance. On the one hand, it reflects a variety of main controlling factors affecting the water yield of the aquifer. On the other hand, it reflects the weight of the main factors affecting the water abundance of the aquifer, and realizes the water abundance division of the aquifer, which truly reflects the water abundance of the water-filled aquifer under the influence of multiple factors. The evaluation method of water abundance based on GIS, water abundance index method, has a set of systematic theoretical systems and research technical route (Fig. 8.1).

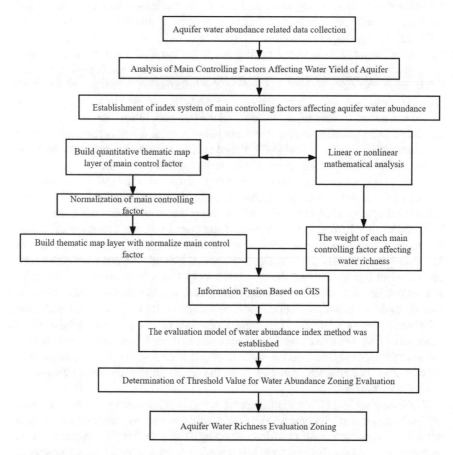

Fig. 8.1 Roadmap of information fusion evaluation technology based on GIS

(1) Data collection and analysis of the study area

Mainly on the geological conditions of the study area, hydrogeological conditions and other related drilling, geophysical exploration, hydrogeological tests, water chemical analysis results, as well as research reports related to the study of aquifer data collection, and data collation and analysis.

(2) Analysis of Main Controlling Factors Affecting Water Yield of Aquifer

Based on the collection of data analysis, combined with the actual study area, the main control factors affecting the water abundance of the aquifer are determined from the aspects of the lithology field, groundwater dynamic field, karst fissure development degree, hydrochemical field and geological structure field of the aquifer.

(3) Establishment of index system of main controlling factors affecting aquifer water abundance

Based on the analysis of the main control factors affecting the water yield of aquifer, the index system of main control factors affecting the water yield of aquifer is established. Further combined with the actual data of the study area, the indicators of the main controlling factors were determined, and the information of each indicator was collected.

(4) Quantification of main controlling factors and establishment of thematic map

Through the information collection of the quantitative indicators of the main controlling factors, the relevant information is quantified, the spatial database and the attribute database of each geological entity are established, and the GIS is further used to display the graphs of the main controlling factors, and the thematic map layer of each main controlling factor is established.

In order to eliminate the influence of dimension of different factors on the evaluation results of water richness, the data of main controlling factors should be normalized to make the data more comparable and statistically significant according to Eq. (8.1):

$$A_i = a + \frac{(b - a) \times (x_i - \min(x_i))}{\max(x_i) - \min(x_i)} \tag{8.1}$$

where, Ai is the data after normalization, and b are respectively the lower limit and upper limit of the normalization range, and 0 and 1 are taken as the original data before normalization. $\min(x_i)$ and $\max(x_i)$ are the minimum and maximum values of the quantized values of the main control factors, respectively.

Equation (8.1) is a normalized treatment method for evaluating the water abundance of a single aquifer. When evaluating multiple aquifers in the same region, the unified normalization treatment can also be carried out, so that the water abundance evaluation results of each aquifer are comparable. By normalizing the data of main controlling factors, the normalized thematic map layer of main controlling factors is established by GIS.

(5) Determination of Weight of Main Controlling Factors Affecting Water Yield of Aquifer

Application of information fusion different mathematical methods, to determine the weight of each main control factor affecting the water abundance of aquifer.

(6) Information Fusion Based on GIS

The spatial information processing and analysis functions of GIS are used to couple the main controlling factors and the weight of aquifer water yield.

(7) Establishment of Water Abundance Evaluation Model

On the basis of information fusion, the evaluation model of aquifer water abundance is established. The initial model (Eq. 8.2) of water-richness index is introduced to evaluate the water-richness of aquifer。

$$WI = \sum_{k=1}^{n} W_k \cdot f_k(x, y) \tag{8.2}$$

where, WI is the vulnerability index; W_k is that weight of the influence factor; $f_k(x, y)$ is a single factor influence value function; x, y is a geographical coordinate; n is the number of influencing factor.

(8) Determination of Water Richness Zoning Threshold

After the initial model of water abundance index method is established, the information of each thematic layer map is processed and analyzed, the water abundance index value of each block is statistically analyzed by using frequency histogram, and the water abundance evaluation zoning threshold is finally determined by further fitting analysis and inversion identification of known point.

(9) Water Richness Evaluation Zoning

According to the zoning threshold, the aquifer is zoned for water abundance evaluation, and the aquifer water abundance evaluation zoning map is generated. The water abundance of coal seam roof aquifer is comprehensively analyzed from many aspects and angles. The fractured aquifer of weathered bedrock and fractured aquifer of burnt bedrock in coal seam roof were evaluated by water abundance index method based on the composite superposition of multi-source geoscience information. The unit water inflow directly reflects the water abundance of the aquifer, and it is the final index to test the water abundance of the aquifer, so the unit water inflow is not taken as the main control factor in this aquifer water abundance evaluation but is taken as the measured index to correct the aquifer water abundance evaluation results. For the fractured aquifer in weathered bedrock, four main controlling factors are determined, which are aquifer thickness, hydraulic conductivity, collecting rate of drill core and brittle-plastic lithology ratio; For 2^{-2} coal, 3^{-1}coal and 4^{-2} coal, the thickness of aquifer, hydraulic conductivity and collecting rate of drill core rate are determined as the main control factors.

8.3　Water Abundance Evaluation and Division of Coal Roof Aquifer Based on Water Abundance Index Method

There are many types of water-filled aquifers in coal measures strata in China, and the conditions are complex, especially the water-filled aquifers such as semi-cemented pores and fissures, sandstone fissures, thin carbonate rock and karst fissures with huge thickness, because of the characteristics of high heterogeneity, anisotropy and discontinuity of permeability, the water yield of aquifers is extremely uneven, with discontinuous block distribution characteristics. Therefore, it is of great significance for mine safety production to study the law of aquifer water abundance and make a reasonable prediction and evaluation of the extremely uneven water abundance distribution [10–12].

　　The water abundance of coal seam roof aquifer is comprehensively analyzed from many aspects and angles. The fractured aquifer of weathered bedrock and fractured aquifer of burnt bedrock in coal seam roof were evaluated by water abundance index method based on the composite superposition of multi-source geoscience information. The unit water inflow directly reflects the water abundance of the aquifer, and it is the final index to test the water abundance of the aquifer, so the unit water inflow is not taken as the main control factor in this aquifer water abundance evaluation but is taken as the measured index to correct the aquifer water abundance evaluation results. For the fractured aquifer in weathered bedrock, four main controlling factors are determined, which are aquifer thickness, hydraulic conductivity, collecting rate of drill core and brittle-plastic lithology ratio; The thickness of aquifer, hydraulic conductivity and collecting rate of drill core rate are three main control factors for fractured aquifer of burnt rock.

8.3.1　Water Abundance Evaluation and Zoning of Fractured Aquifers in Weathered Bedrock

1. Multi-geoscience information reflecting the water yield of water-filled aquifer

(1) Aquifer thickness

As one of the multi-source geoscience information to be excavated, the thickness of aquifer is important in the study of water yield zoning, because the water yield of aquifer is proportional to the thickness of aquifer when other factors affecting the water yield of aquifer are constant, which is consistent with our consistent experience.

Based on the collation and statistics of borehole data (Table 8.1), this work uses the powerful interpolation function of Surfer to draw the thematic map of weathered bedrock fissure aquifer (Fig. 8.2). As shown in Fig. 8.2, the sandstone thickness in the middle, northwest, and northeast of the weathered bedrock fissure aquifers in the study area is relatively thick, and the sandstone thickness in other areas is relatively thin.

(2)   Hydraulic conductivity

Hydraulic conductivity is an important hydrogeological parameter which can and indicate the permeability of rock. Generally, the greater the hydraulic conductivity is, the stronger the permeability of rock is. The hydraulic conductivity depends

Table 8.1   Statistics of thickness of fissure aquifer of weathered bedrock

Name of borehole	Aquifer thickness (m)	Name of borehole	Aquifer thickness (m)	Name of borehole	Aquifer thickness (m)	Name of borehole	Aquifer thickness (m)
221-1	7.80	H14	4.10	ZK3	10.70	H4	0.00
221	11.30	H15	14.49	ZK7	30.94	168	4.10
222	6.42	H16	26.00	ZK8	15.00	198	0.00
223	18.60	H18	2.95	ZK9	4.90	316	0.00
224	26.35	H19	6.71	ZK10	0.00	323	0.00
225	17.20	H22	5.50	ZK11	0.00	324	0.00
226	13.00	H23	0.00	ZK12	3.15	N366	13.80
227	9.00	H24	0.00	ZK14	7.95	N370	2.86
228-1	7.15	H25	4.40	B1	3.04	N391	17.11
228-2	5.75	12–1	1.90	BS1	0.00	N455	4.34
228	7.55	12–2	8.54	FJ3	0.00	N481	9.31
3–1	10.00	H5-2	10.3	ZJ4	0.00	N482	18.98
3–2	4.00	H6-2	7.80	6–2	9.7	N493	0.00
H02	3.95	B1	7.55	6–3	4.50	N505	10.18
H03	10.00	B2	3.66	7–1	3.00	N507	13.58
H04	10.00	B3	11.00	7–3	9.35	N509	5.13
H04-2	5.40	B4	5.72	7–4	0.00	N513	0.00
H05	11.63	B5	0.00	8–1	0.00	N587	12.51
H06	10.50	B6	20.20	8–5	2.30	N594	8.84
H07	42.20	B7	4.60	9-1S	17.45	N596	5.18
H08	0.00	B8	5.80	9–6	3.00	N597	9.01
H09	28.08	B9	0.00	10–2	24.06	N610	12.89
H10	11.36	B10	1.10	10–4	15.00	N611	21.08
H12	2.38	ZK1	11.82	10–5	13.85	N612	14.46
H13	0.00	ZK2	12.15	H3	23.48	J2	13.00

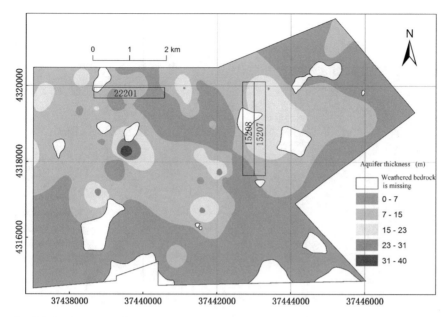

Fig. 8.2 Thematic map of fractured aquifer thickness of weathered bedrock

not only on the grain size, composition, grain arrangement, fracture properties and development degree of rock, but also on the physical properties of seepage fluid, such as unit weight and viscosity.

On the basis of collation and statistics of hydrological drilling data (Table 8.2), the hydraulic conductivity thematic map of weathered bedrock fissure aquifer (Fig. 8.3) is established by using the powerful interpolation function of Surfer. Among them, the hydraulic conductivity of BK2 hole 4.06700 m/d is quite different from other hydrological hole data, and the authenticity of the data is in doubt. Out of respect for the actual situation and drawing on previous experience, the data of 222 boreholes 0.16680 m/d is used to replace this hole. It can be seen from Fig. 8.2 that the hydraulic conductivity of the weathered bedrock fissure aquifer in the northwest and southwest of the study area is relatively high, and that of other areas is relatively low.

Table 8.2 Statistical table of hydraulic conductivity of weathered bedrock fissure aquifer

Name of borehole	Permeability coefficient (m/d)	Name of borehole	Hydraulic conductivity (m/d)	Name of borehole	Hydraulic conductivity (m/d)	Name of borehole	Hydraulic conductivity (m/d)
BK3	0.08880	222	0.16680	B8	0.00710	BK2	4.06700
BK4	0.00460	B1	0.06460	BK1	0.04553	SK5	0.00497
SK4	0.08500	B2	0.02510				

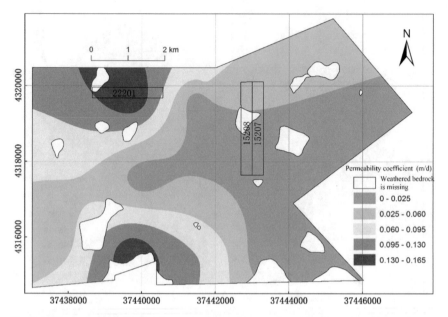

Fig. 8.3 Thematic map of hydraulic conductivity of fractured aquifer of weathered bedrock

(3) Collecting rate of drill core

From the point of view of water content or permeability, the cracks in rock mass can be divided into open cracks or connected cracks and non-open cracks or non-connected cracks. The so-called open fracture or connected fracture refers to the fracture zone which is connected with the water-bearing body or the surface or intersected by different fracture combinations. If these fissures are connected with the surface or shallow water-bearing body, they must contain water or become the channel of groundwater flow because of the water storage body as a recharge source. Non-opened or non-connected fracture is a kind of fracture which is not connected with the surface or groundwater and terminates in the rock mass, and this kind of fracture cannot form a water inflow channel. In this study, we use the coring rate of aquifer core to study the lithological field, where the coring rate refers to the ratio of the length of aquifer core to the thickness of aquifer, expressed as a percentage. This index reflects the integrity of rock mass, i.e., the index reflects the intersection degree of rock fracture.

Based on the collation and statistics of borehole data (Table 8.3), the collecting rate of drill core rate thematic map of weathered bedrock fissure aquifer (Fig. 8.4) is established by using the powerful interpolation function of Surfer. It can be seen from Fig. 8.4 that the core sampling rate of weathered bedrock fissure aquifers in the study area is low as a whole, and it is higher only in the northeast and local areas in the north.

Table 8.3 Statistics of collecting rate of drill core rate of weathered bedrock fissure aquifer

Name of borehole	Collecting rate of drill core (%)	Name of borehole	Collecting rate of drill core (%)	Name of borehole	Collecting rate of drill core (%)	Name of borehole	Collecting rate of drill core (%)
221-1	76	H13	42	ZK2	69	168	57
221	88	H14	42	ZK3	75	198	33
222	82	H15	46	ZK7	90	316	14
223	87	H16	52	ZK8	64	323	70
224	81	H18	41	ZK9	73	324	12
225	85	H19	58	ZK10	64	N366	6
227	75	H23	42	ZK12	72	N391	38
228-1	76	H25	42	ZK14	77	N455	60
228-2	72	12-1	100	B1	76	N481	57
228	80	12-2	90	BS1	80	N482	78
3–1	52	H5-2	63	6–2	86	N493	72
3–2	50	H6-2	78	6–3	96	N505	73
H02	61	B1	63	7–1	65	N507	80
H03	54	B2	80	7–3	93	N509	74
H04	42	B3	71	7–4	94	N513	25
H04-2	96	B4	63	8–1	20	N587	69
H05	45	B5	52	8–5	88	N594	81
H06	47	B6	68	9–1	85	N596	83
H07	33	B7	70	10–3	83	N597	77
H08	41	B8	68	10–4	86	N610	46
H09	43	B9	81	10–5	88	N611	39
H10	47	B10	56	H3	84	N612	35
H12	63	ZK1	79	H4	85	J2	65

(4)   Brittle-ductile lithologic ratio

The coal seam roof aquifer is mainly composed of medium, coarse and fine sandstone, siltstone, mudstone and sandy mudstone. From the vertical view, the mudstone or sandy mudstone and other plastic strata with different thickness and weak permeability are deposited between some sandstone layers of this layer, and the mudstone and sandy mudstone are mainly concentrated in the adjacent layers of coal seam and coal line. Under the condition of tectonic destruction, the fracture characteristics reflected by lithology with different mechanical properties are quite different. The brittle sandstone layer releases stress in the form of fracture after being stressed, so the fissures and joints in the sandstone layer are relatively developed, and the permeability is greatly enhanced. Plastic argillaceous rock releases stress in the form of plastic deformation under the action of stress load, which makes the permeability of plastic rock change little after stress. Therefore, the ratio of brittle rock thickness

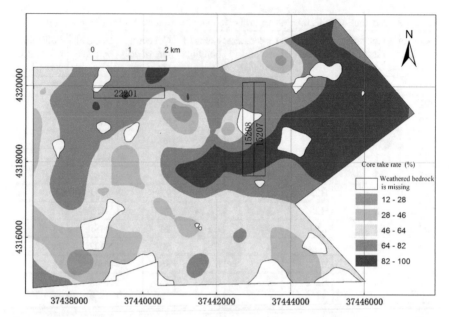

Fig. 8.4 Collecting rate of drill core rate thematic diagram of weathered bedrock fissure aquifer

to plastic rock thickness can be used as a qualitative index to judge the permeability of sandstone fractured aquifer. In general, the larger the brittle-plastic lithology ratio is, the better the permeability of the aquifer is, and vice versa.

Based on the collation and statistics of hydrological borehole data (Table 8.4), the brittle-plastic lithology ratio thematic map of weathered bedrock fissure aquifer is established by using the powerful interpolation function of Surfer (Fig. 8.5). From Fig. 8.5, the brittle-plastic lithology ratio of weathered bedrock fissure aquifers in the study area is low as a whole, and it is high only in the central and northwest local areas.

2. Evaluation of Aquifer Water Richness by Water Richness Index Method Based on Coupling of GIS and AHP

(1) Establishment of AHP Structure Model

Based on the analysis of multiple geoscientific information reflecting the water abundance of the aquifer, it is divided into three levels. The goal of this problem is to evaluate the water abundance of the fractured aquifer in weathered bedrock, which is taken as the target layer of the model. The water field, aquifer and lithologic field reflect the water abundance of aquifer, but the way of their influence is reflected by the specific factors related to them, which is the middle link to solve the problem, i.e., the criterion layer of the model. Each specific multivariate geoscience information constitutes the decision level of this model, and through the decision of this level, the goal of the required solution can be finally achieved (Fig. 8.6).

Table 8.4 Statistics of brittle-plastic lithology ratio of weathered bedrock fissure aquifers

Name of borehole	Brittle-ductile lithologic ratio	Name of borehole	Brittle-ductile lithologic ratio	Name of borehole	Brittle-ductile lithologic ratio	Name of borehole	Brittle-ductile lithologic ratio
221-1	2.03	H14	0.46	B1	0.77	323	0.00
221	0.69	H23	0.00	BS1	0.00	324	0.00
222	0.51	H25	1.10	FJ3	0.00	N370	0.20
223	6.41	12–2	1.22	ZJ4	0.00	N455	0.55
224	1.68	B1	0.38	6–2	4.85	N481	1.04
225	2.12	B2	0.34	6–3	0.56	N493	0.00
226	2.71	B3	2.84	7–1	0.48	N505	7.77
227	1.42	B4	0.56	7–4	0.00	N509	0.90
228-1	1.11	B5	0.00	8–1	0.00	N513	0.00
228-2	0.60	B6	3.88	8–5	0.15	N587	1.81
228	2.29	B7	0.46	9–1S	1.67	N596	1.03
3–2	0.36	B8	0.55	9–6	0.30	N597	3.83
H02	0.25	B9	0.00	10–4	2.10	N610	21.48
H03	10.00	B10	0.08	H4	0.00	N611	12.40
H08	0.00	ZK10	0.00	168	0.76	N612	0.47
H10	0.42	ZK11	0.00	198	0.00	J2	0.97
H13	0.00	ZK12	0.20	316	0.00	J3	0.97

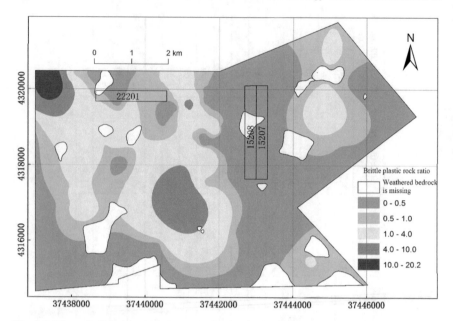

Fig. 8.5 Thematic diagram of brittle plastic lithology ratio of weathered bedrock fissure aquifer

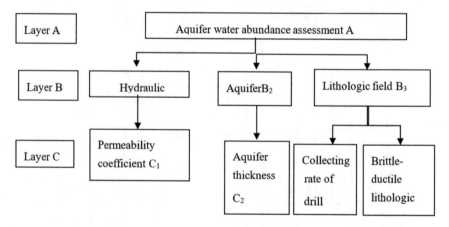

Fig. 8.6 Analytic hierarchy process structure model for water yield evaluation of weathered bedrock fissure filling aquifer

(2)   Construct Judgement Matrix

According to the analysis of the multivariate information reflecting the water abundance of the aquifer and the 9/9–9/1 scaling method based on the expert opinion, the relative importance of each factor is evaluated, the quantitative score of each information is given, and the judgment matrix of AHP evaluation on the water abundance of the coal seam roof water-filled aquifer is constructed. Refer to Tables 8.5, 8.6 and

8.7 and Column W in Table 8.8 for the weight value of each single layer sorting calculated according to the judgment matrix.

(3) Evaluation of Water Abundance in Water-filled Aquifer

The aquifer water-richness evaluation model based on analytic hierarchy process has been established. On this basis, the powerful spatial data analysis function of GIS is used to calculate the water-richness index value of each unit after the superposition of each information thematic map. The size of the water-richness index value of each unit identifies the degree of water-richness of the water-filled aquifer in the unit. The value of the water-richness index is between 0 and 1, and the larger the value is, the better the water-richness of the area is. It is based on this that the zoning study of the water-richness of the water-filled aquifer becomes possible. To facilitate

Table 8.5 Judgment matrix a–Bi (I = 1 − 3)

A_1	B_1	B_2	B_3	W (A/B)
B_1	1	9/6	1	0.375
B_2	6/9	1	6/9	0.25
B_3	1	9/6	1	0.375

λmax = 3, CI_1 = 0, CR_1 = 0 < 0.1

Table 8.6 Judgment matrix B3–Ci (I = 3 − 4)

B_3	C_3	C_4	W
C_3	1	9/5	0.64
C_4	5/9	1	0.36

λmax = 2, CI_{21} = 0, CR_{21} = 0 < 0.1

Table 8.7 Weight of multi-source information in water abundance classification of weathered bedrock fissure water-filled aquifer

A/C_i	B_1/0.50	B_2/0.25	B_3/0.25	W_A/C_i
C_1	1			0.375
C_2		1		0.25
C_3			0.64	0.25
C_4			0.36	0.125

Table 8.8 Weight of geoscience information in water yield analysis of weathered bedrock aquifer

Multi-source information	Hydraulic conductivity (W_1)	Aquifer thickness (W_2)	Coring rate (W_3)	Brittle-ductile lithologic ratio (W_4)
Weighting Wi	0.375	0.25	0.24	0.135

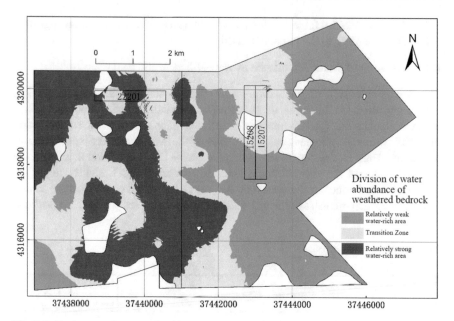

Fig. 8.7 Zoning map for water abundance evaluation of fractured aquifer in weathered bedrock

our use in production, we use the cluster analysis method to divide the whole area into three areas with different water abundance degrees by means of SPSS software according to the research on water abundance index, as shown in Fig. 8.7. The order of water-bearing capacity of water-filled aquifer from small to large is relatively weak water-bearing area, transition area and relatively strong water-bearing area.

(4) Correction of aquifer water abundance zoning result

In Fig. 8.8, the water-rich evaluation zoning is drawn based on the weight calculated by analytic hierarchy process completely according to the expert opinion. The expert's experience is reasonable, so the water-rich zoning result is relatively reliable. However, to more accurately divide the study area into water-rich zones, it is necessary to judge it according to the actual situation. This time, the division result of water abundance evaluation is tested by taking the actually measured unit water inflow as the standard. If the test result is not ideal, the weight of each main control factor is modified until the evaluation result is consistent with the actual situation and meets the production requirements.

In order to judge the zoning results more intuitively, the ten collected measured unit water inflow data are also divided into three categories (Table 8.9) by cluster analysis method to correspond to the three zones with strong or weak water abundance. The cluster analysis result is compared with the water abundance evaluation zoning result, and the matching degree of the two is calculated (Fig. 8.8, Table 8.10). It is known that the predicted zoning position of 7 boreholes in 10 hydrological boreholes matches the actual position, and the matching degree is 70%.

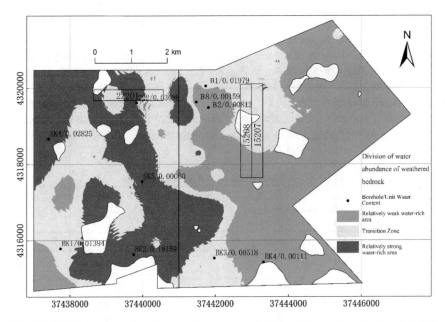

Fig. 8.8 Divisional inspection of water abundance evaluation of fractured aquifer in weathered bedrock Fig. 8.1

Table 8.9 Cluster analysis results of measured unit water inflow

Name of borehole	SK5	BK4	B8	BK3	B2	BK1	B1	SK4	222	BK2
Unit water inflow (10^{-4}L/s.m)	6.0	11.1	15.9	51.8	81.2	139.4	197.9	282.5	308.8	1915.9
Corresponding partition	Relatively weak water-rich area					Transition zone		Relatively strong water-rich area		

Table 8.10 Table of test conditions of water richness evaluation result

Name of borehole	SK5	BK4	B8	BK3	B2	BK1	B1	SK4	222	BK2
Unit water inflow (10^{-4}L/s.m)	6.0	11.1	15.9	51.8	81.2	139.4	197.9	282.5	308.8	1915.9
Actual partition	Relatively weak water-rich area (1)					Transition zone (2)		Relatively strong water-rich area (3)		
Forecast partition	3	1	1	2	1	2	2	3	2	3

The matching degree of 70% is far from meeting the needs of actual production. In order to improve the accuracy of water abundance evaluation results, the weights calculated by the analytic hierarchy process are modified (Table 8.11). The modified evaluation zoning effect is shown in Fig. 8.9 and Table 8.12. It can be seen that the matching degree has increased to 80%.

By continuously modifying the weights of the main control factors, when the weights of the main control factors are shown in Table 8.13, the evaluation zoning effect is as shown in Fig. 8.10 and Table 8.14, only borehole BK3 cannot match the

Table 8.11 Weights of main controlling factors of water richness

Multi-source information	Hydraulic conductivity (W$_1$)	Aquifer thickness (W$_2$)	Coring rate (W$_3$)	Brittle-ductile lithologic ratio (W$_4$)
Weighting Wi	0.46	0.22	0.21	0.11

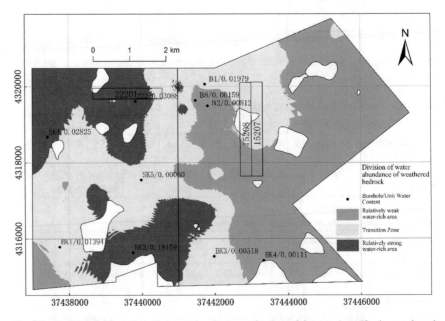

Fig. 8.9 Divisional inspection of water abundance evaluation of fractured aquifer in weathered bedrock Fig. 8.2

Table 8.12 Table of test conditions of water richness evaluation result

Name of borehole	SK5	BK4	B8	BK3	B2	BK1	B1	SK4	222	BK2
Unit water inflow (10^{-4}L/s.m)	6.0	11.1	15.9	51.8	81.2	139.4	197.9	282.5	308.8	1915.9
Forecast partition	2	1	1	2	1	2	2	3	3	3

actual result, and the matching degree reaches 90%. The matching degree cannot reach 100% even if the weight is modified, which is related to the accuracy of drilling data. If the matching degree is blindly pursued to reach 100%, the evaluation results may deviate from reality, but the loss outweighs the gain. Considering that the matching degree of 90% can fully meet the actual production needs, the evaluation results are taken as the final division results of water abundance evaluation of weathered bedrock fissure aquifer.

It can be seen from Fig. 8.10 that the fractured aquifer of weathered bedrock in Zhangjiamao Coal Mine has strong water abundance in the northwest and south of the mining area, weak water abundance in the east and southwest of the mining area, and other areas are transitional areas. The overall water abundance of the second panel is strong, most of the areas are relatively strong water abundance areas, and the overall water abundance of the first panel is weak. The fractured aquifer of weathered bedrock in 22,201 working face is rich in water and water content, which is consistent

Table 8.13 Weights of main controlling factors of water richness

Multi-source information	Hydraulic conductivity (W_1)	Aquifer thickness (W_2)	Coring rate (W_3)	Brittle-ductile lithologic ratio (W_4)
Weighting Wi	0.5	0.20	0.20	0.10

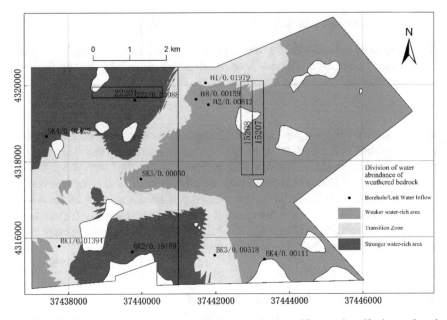

Fig. 8.10 Divisional inspection for water abundance evaluation of fractured aquifer in weathered bedrock Fig. 8.3

Table 8.14 Table of test conditions of water richness evaluation result

Name of borehole	SK5	BK4	B8	BK3	B2	BK1	B1	SK4	222	BK2
Unit water inflow (10^{-4}L/s.m)	6.0	11.1	15.9	51.8	81.2	139.4	197.9	282.5	308.8	1915.9
Actual partition	Relatively weak water-rich area (1)					Transition zone (2)		Relatively strong water-rich area (3)		
Forecast partition	1	1	1	2	1	2	2	3	3	3

with the water inflow event of weathered bedrock in the roadway of this working face, and it is a strong water-rich area.

8.3.2 Water Richness Evaluation and Zoning of Pore-Fissure Aquifer in Burnt Bedrock of 2^{-2} Coal, 3^{-1} Coal and 4^{-2} Coal

1. Multi-geoscience information reflecting the water yield of water-filled aquifer

(1) Thickness of pore-fissure aquifer of each burnt bedrock

According to the collected drilling data, the thickness of pore-fissure aquifer in burnt bedrock of 2^{-2} coal, 3^{-1} coal and 4^{-2} coal is calculated (Table 8.15), and the thematic map of aquifer thickness is drawn (Fig. 8.11).

(2) Hydraulic conductivity

According to the collected pumping test drilling data, the hydraulic conductivity data of burnt bedrock of 2^{-2} coal, 3^{-1} coal and 4^{-2} coal are obtained (Table 8.16). Among them, the hydraulic conductivity of ZK8 borehole is 197.7087 m/d, the hydraulic conductivity of B6 borehole is 56.25 m/d, which is quite different from the data of other hydrological boreholes in the same layer, and the authenticity of the data is doubtful. Respecting the actual situation and using the past experience for reference, 11.97 m/d for borehole H17 and 14.1738 m/d for water B5 are used as substitutes, and the corresponding thematic maps are drawn (Fig. 8.12).

(3) Collecting rate of drill core

According to the collected drilling data, the data of core recovery rate of burnt bedrock of 2^{-2} coal, 3^{-1} coal and 4^{-2} coal are obtained (Table 8.17), and the corresponding thematic map is drawn (Fig. 8.13).

Table 8.15 Statistical table of the thickness of pore-fissure aquifer of burned bedrock of 2^{-2}, 3^{-1} and 4^{-2} coal the thickness of pore-fissure aquifer of burned bedrock of 2^{-2}

Name of borehole	Aquifer thickness (m)	Name of borehole	Aquifer thickness (m)	Name of borehole	Aquifer thickness (m)	Name of borehole	Aquifer thickness (m)
H07	4.4	H09	18.14	H10	11.94	H11	2.13
H13	1.4	B2	12.16	B1	1.67	B4	18.58
B5	17.77	B6	27.85	B20	43.57	B24	18.73
BS1	3.14	BH5	45.91	6–1	35	350	5.47
The thickness of pore-fissure aquifer of burned bedrock of 3^{-1}							
3–2	9.32	H08	12.27	B15	6.68	B16	11.47
B30	12.44	B6	2	B20S	4.95	B3	13.6
The thickness of pore-fissure aquifer of burned bedrock of 4^{-2}							
H22	11.5	H24	10.05	ZK4	6.25	ZK5	11.5
ZK6	19.5	ZK8	13.5	ZK9	10	ZK10	27.4
ZK11	22.5	ZK12	21.75	B15	7.2	202	7.4

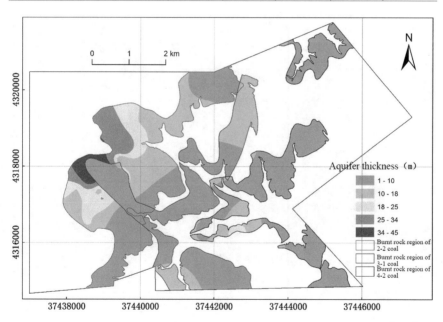

Fig. 8.11 Thickness thematic map of pore-fissure aquifer of burned bedrock of coal 2^{-2}, 3^{-1} and 4^{-2}

Table 8.16 Statistical table of hydraulic conductivity of pore-fissure aquifer of burned bedrock of coal 2^{-2}, 3^{-1} and 4^{-2}

Name of borehole	Hydraulic conductivity (m/d)	Name of borehole	Hydraulic conductivity (m/d)	Name of borehole	Hydraulic conductivity (m/d)	Name of borehole	Hydraulic conductivity (m/d)
Hydraulic conductivity of pore-fissure aquifer of burned bedrock of 2^{-2}							
B5S	14.1738	B6S	56.25	B21S	2.5952	BS1	0.1026
Hydraulic conductivity of pore-fissure aquifer of burned bedrock of 3^{-1}							
Name of borehole	Hydraulic conductivity (m/d)	Name of borehole		Hydraulic conductivity (m/d)		Name of borehole	Hydraulic conductivity (m/d)
B20S	0.4242	BH8S		0.0408		B30	0.2995
Hydraulic conductivity of pore-fissure aquifer of burned bedrock of 4^{-2}							
Name of borehole	Hydraulic conductivity (m/d)	Name of borehole		Hydraulic conductivity (m/d)		Name of borehole	Hydraulic conductivity (m/d)
ZK8	197.7087	H17		11.97		ZK2	0.0016

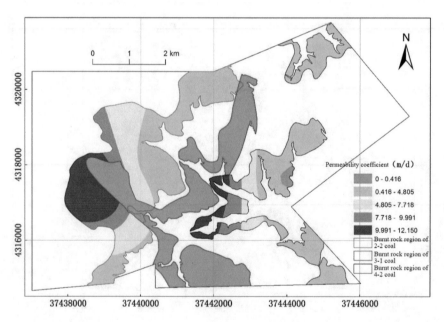

Fig. 8.12 Thematic diagram of hydraulic conductivity of bedrock pore-fracture aquifer of coal 2^{-2}, 3^{-1} and 4^{-2}

Table 8.17 Statistics of collecting rate of drill core rate of pore-fissure aquifers of burnt bedrock coal 2^{-2}, 3^{-1} and 4^{-2}

Name of borehole	Collecting rate of drill core (%)	Name of borehole	Collecting rate of drill core (%)	Name of borehole	Collecting rate of drill core (%)	Name of borehole	Collecting rate of drill core (%)
2^{-2} Collecting rate of drill core rate of pore-fissure aquifers in burnt bedrock							
H07	18	H09	45	H10	42	H11	38
H13	57	B2	86	B1	77.8	B4	40.2
B5	79.3	B6	60.8	B20	35.6	B24	41.6
BS1	79.6	BH5	39.4	B1	61		
3^{-1} Collecting rate of drill core rate of pore-fissure aquifers in burnt bedrock							
3–2	77	H08	19	B15	70.4	B16	72.4
B30	72.5	B6	67				
4^{-2} Collecting rate of drill core rate of pore-fissure aquifers in burnt bedrock							
H22	41	H24	38	ZK4	70	ZK5	57
ZK6	71	ZK8	57	ZK9	56	ZK10	55
ZK11	64	ZK12	62	B15	66.7	202	17

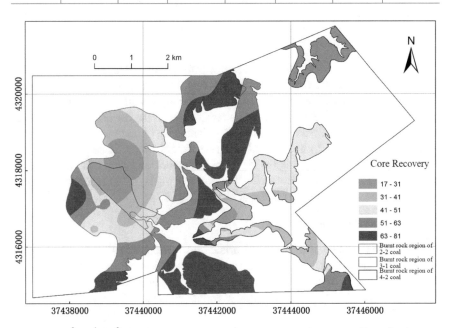

Fig. 8.13 2^{-2}, 3^{-1}, 4^{-2} Coal burning metamorphic bedrock pore-fracture aquifer collecting rate of drill core rate thematic map

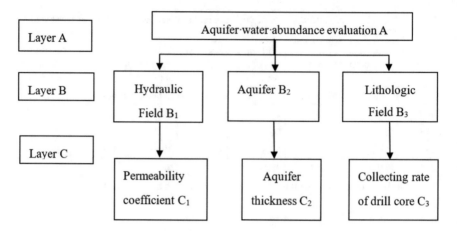

Fig. 8.14 Analytic hierarchy process structure model for evaluation of water yield of pore-fracture water-filled aquifer of burned and transformed bedrock

2. Evaluation of Aquifer Water Richness by Water Richness Index Method Based on Coupling of GIS and AHP

(1) Establishment of AHP Structure Model

According to the analysis of the multi-source information reflecting the water abundance of the aquifer, it is divided into three levels. The ultimate goal of this problem is to evaluate the water abundance of pore-fissure aquifer in burnt bedrock of 2^{-2} coal, 3^{-1} coal and 4^{-2} coal, which is the target layer of the model. The water field, aquifer and lithologic field reflect the water abundance of aquifer, but the influence mode is also reflected by the related concrete factors, which is the middle link to solve the problem, i.e., the criterion layer of the model, and each concrete multiple geoscience information constitutes the decision-making layer of the model. The goal of the required solution is eventually achieved (Fig. 8.14).

(2) Construct Judgement Matrix

According to the analysis of the multivariate information reflecting the water abundance of the aquifer, the relative importance of each factor is evaluated according to the 9/9–9/1 scale method, and the quantitative score of each information is given. The judgment matrix of AHP evaluation of water abundance of coal seam roof water-filled aquifer is constructed. Refer to Table 8.18 and Column W in Table 8.19 for the weight value of each single layer sorting calculated according to the judgment matrix.

(3) Evaluation of Water Abundance in Water-filled Aquifer

The aquifer water abundance evaluation model based on analytic hierarchy process has been established, and on this basis, the water abundance index value of each

Table 8.18 Judgment matrix A-Bi (i = 1 − 3)

A_1	B_1	B_2	B_3	W (A/B)
B_1	1	5/9	9/7	0.27
B_2	9/5	1	9/3	0.53
B_3	7/9	3/9	1	0.20

λmax = 3, CI_1 = 0, CR_1 = 0 < 0.1

Table 8.19 Weight of geoscience information in water abundance analysis of pore-fissure water-filled aquifer of burnt bedrock

Multiple information	Hydraulic conductivity (W_2)	Aquifer thickness (W_3)	Collecting rate of drill core (W_4)
Weighting W_i	0.27	0.53	0.20

unit after the superposition of each information thematic map is calculated by means of the powerful spatial data analysis function of GIS. The water abundance index value of each unit indicates the water abundance degree of the water-filled aquifer in the unit, the value of the water abundance index is between 0 and 1, and the larger the value is, the better the water abundance of the area is, which makes it possible to divide the water-filled aquifer into different areas. The units with the same water abundance index or its value in a certain interval can be merged to divide the areas with different water abundance degrees of the water-filled aquifer. In order to facilitate our use in production, according to the study of water abundance index, we use cluster analysis method and SPSS software to divide the whole area into three areas with different water abundance degrees. As shown in Fig. 8.15. 2^{-2} coal, 3^{-1} coal, 4^{-2} coal burning bedrock pore-fissure aquifer water from small to large order is relatively weak water-rich area, transition area, relatively strong water-rich area.

(4) Correction of aquifer water abundance zoning result

In Fig. 8.15, 2^{-2} coal, 3^{-1} coal and 4^{-2} coal are burned into bedrock water abundance evaluation zoning, which is drawn according to the weight calculated by analytic hierarchy process based on the expert opinion. The expert's experience is reasonable, so the water abundance zoning result is relatively reliable. However, in order to more accurately divide the study area into water-rich zones, it is necessary to judge it according to the actual situation. This time, the division result of water abundance evaluation is tested by taking the actually measured unit water inflow as the standard. If the test result is not ideal, the weight of each main control factor is modified until the evaluation result is consistent with the actual situation and meets the production requirements.

In order to judge the zoning results more intuitively, the nine measured unit water inflow data collected are also divided into three categories (Table 8.20) by cluster analysis method to correspond to the three zones with strong or weak water abundance. The cluster analysis result is compared with the water abundance evaluation

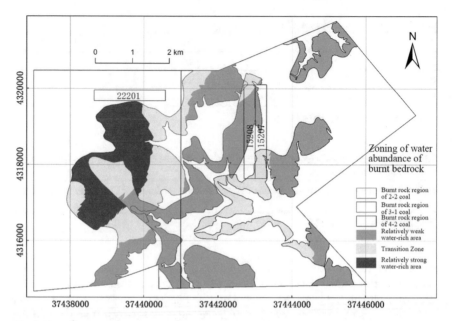

Fig. 8.15 2^{-2} Coal, 3^{-1} Coal and 4^{-2} Coal burning transformation bedrock porous-fracture water rich evaluation zoning diagram

zoning result, and the matching degree of the two is calculated (Fig. 8.16, Table 8.21). It is known that the predicted zoning position of 7 boreholes in 9 hydrological boreholes matches the actual position, and the matching degree is 78%.

The matching degree of 78% is far from meeting the needs of actual production. In order to improve the accuracy of the water richness evaluation result, the weights of the main control factors are continuously modified. Finally, when the weights of the main control factors are shown in Table 8.22, the evaluation zoning effect is shown in Fig. 8.17 and Table 8.23. Only borehole H17 could not match the actual results, and the matching degree reached 89%. The matching degree cannot reach 100% even if the weight is modified, which is related to the accuracy of drilling data. If the matching degree is blindly pursued to reach 100%, the evaluation results may deviate from reality, but the loss outweighs the gain. Taking into account the

Table 8.20 Cluster analysis results of measured unit water inflow

Name of borehole	ZK2	BH8S	B30	B21S	B20S	H17	B6S	B5S	ZK8
Unit water inflow	0.0008	0.0043	0.0487	0.1685	0.2004	0.5063	2.1041	2.1547	9.417
Ideal partition	Relatively weak water-rich area (1)			Transition zone (2)			Relatively strong water-rich area (3)		

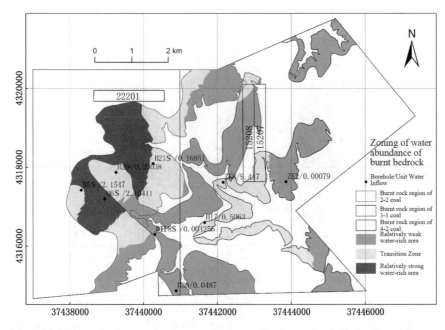

Fig. 8.16 Diagram 1 of water abundance evaluation zoning inspection for bedrock pore-fracture aquifer of combustion transformation of coal 2^{-2}, 3^{-1} and 4^{-2}

Table 8.21   Table of test conditions of water richness evaluation result

Name of borehole	ZK2	BH8	B30	B21S	B20S	H17	B6S	B5S	ZK8
Unit water inflow	0.0008	0.0043	0.0487	0.1685	0.2004	0.5063	2.1041	2.1547	9.417
Ideal partition	Relatively weak water-rich area (1)			Transition zone (2)			Relatively strong water-rich (3)		
Forecast partition	1	1	1	2	2	2	3	2	2

matching degree of 89% to meet the actual production needs, so the evaluation results as 2^{-2} coal, 3^{-1} coal, 4^{-2} coal burnt bedrock pore-fissure aquifer final water-rich evaluation zoning results.

It can be seen from Fig. 8.18 that the fissure aquifer of burnt bedrock in Zhangjiamao Coal Mine has strong water richness in 2^{-2} coal burnt area in the middle of second panel and 4^{-2} coal burnt area in the middle of first panel, and the water richness in 3^{-1} coal burnt area in the middle of mine field is weak. The water

Table 8.22   Weights of main controlling factors of water richness

Multiple information	Hydraulic conductivity (W_1)	Aquifer thickness (W_2)	Collecting rate of drill core (W_3)
Weighting Wi	0.4	0.25	0.35

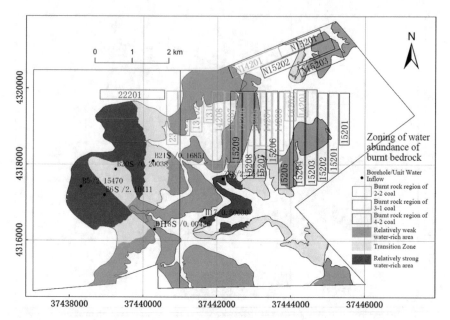

Fig. 8.17 Coal 2^{-2}, 3^{-1} and 4^{-2} burning bedrock pore-fissure aquifer water abundance evaluation zoning inspection diagram-2

Table 8.23 Table of test conditions of water richness evaluation result

Name of borehole	ZK2	BH8S	B30	B21S	B20S	H17	B6S	B5S	ZK8
Unit water inflow	0.0008	0.0043	0.0487	0.1685	0.2004	0.5063	2.1041	2.1547	9.417
Ideal partition	Relatively weak water-rich area (1)			Transition zone (2)			Relatively strong water-rich area (3)		
Forecast partition	1	1	1	2	2	3	3	3	3

richness of burnt rock in the south of 15,207 and 15,208 working faces is medium. Although burnt rock does not exist in the upper part of 22,201 working face, its south boundary is close to burnt rock, and the fracture zone formed by coal mining will also develop to the strong water richness area of 2^{-2} coal burnt rock, which poses a certain threat to mining.

8.4　Safety Evaluation and Zoning of Roof Caving in Coal Seam

Through the analysis of the water flowing fractured zone in the roof of Zhangjiamao Coal Mine, using the measured data and comparing the calculated results of two equations with the measured values, a equation for calculating the development height

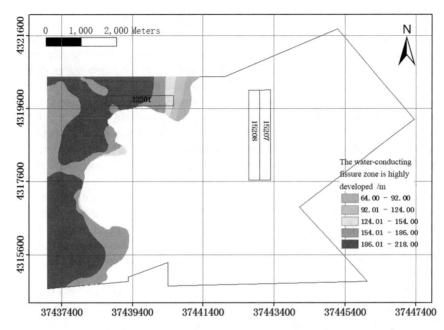

Fig. 8.18 Thematic map of development height of water flowing fractured zone in 2^{-2} coal seam roof

of the water flowing fractured zone suitable for coal seam mining in Zhangjiamao Coal Mine is preliminarily obtained, and the development height of the water running fractured zone in the roof is determined by using the equation. By comparing the development height of water flowing fractured zone with the thickness of overlying strata, the safety zoning of roof caving is obtained.

8.4.1   Calculation of Height of Water Conduct Fractured Zone in 2^{-2} Coal Seam Roof

The thickness of 2^{-2} coal seam in Zhangjiamao Coal Mine is 1.85–10.60 m, with an average thickness of 7.81 m. The structure of the coal seam is complex, generally containing 2–3 layers of parting, the parting thickness is approximately 0.30 m, the lithology is mainly mudstone and siltstone, followed by carbonaceous mudstone. According to the test of rock physical and mechanical properties, the saturated compressive strength of 2^{-2} coal seam roof rock is generally 20–40 MPa, therefore, the coal seam roof is classified as medium-hard rock roof.

Zhangjiamao Coal Mine is located in the Jurassic coalfield of northern Shaanxi Province, where the coal seam is shallow, the overlying bedrock is thin, and the coal seam mining thickness is large. The empirical equation for the development height of water flowing fractured zone in the previous regulations and specifications such as Regulations for Coal Pillar Reservation and Pressure Coal Mining of Buildings, Water Bodies, Railways and Main Shafts and Roadways is not applicable to the calculation of the development height of water flowing fractured zone in fully mechanized mining of thick coal seam under the condition of shallow and thin bedrock in this area.

Therefore, this study refers to the actual measurement results (Table 8.24) of the development height of water flowing fractured zone during the mining of 2^{-2} coal seam in Ningtiaota Coal Mine with adjacent geographical location and similar geological conditions, and adopts the following scheme to study and analyze the development height of water flowing fractured zone of 2^{-2} coal seam roof in Zhangjiamao Coal Mine.

Equation (8.2) is the empirical calculation equation for the development height of water flowing fractured zone proposed in the Research Report on Water Resources Dynamics and Water Conservation Technology for Coal Mining in Shennan Large Mining Area (December 2010, Shennan Mining Company of Shaanxi Coal Industry Chemical Group, 185 Team of Shaanxi Coal Geology Bureau and China University of Mining and Technology): $H_{li} = 9.59\,M + 13.55$.

In this scheme, the development height of water flowing fractured zone is calculated by using the empirical equation of Coal Mine Water Prevention and Control Manual, and compared with the measured height, the proportional relationship between them is obtained, and then the development height calculation equation of water flowing fractured zone is obtained. The specific methods are as follow:

The empirical calculation equation for the development height of water flowing fractured zone proposed by China University of Mining and Technology, Beijing in the Manual of Water Prevention and Control in Coal Mine is applicable to the conditions of medium-hard overlying strata and fully mechanized caving mining (Eq. 8.3):

$$H_{1i} = \frac{100 \sum M}{0.26 \sum M + 6.88} \pm 11.49 \tag{8.3}$$

Table 8.24 List of comparison between calculation and actual measurement of development height of water flowing fractured zone in 2^{-2} coal seam roof

Measured borehole	Coal seam mining thickness (m)	Buried depth of coal seam floor (m)	Equation calculation height (m)		Measured height (m)
			Equation 1	Equation 2	
Borehole 1	4.8	150.91	70.55	59.58	125.8
Borehole 3	5.8	171	80.64	69.17	147.66
Borehole 4	5.46	186.1	77.28	65.91	143.82
Borehole 6	5.46	188.91	77.28	65.91	139.77

Note Eq. (8.1) is the empirical calculation equation for the development height of water flowing fractured zone proposed by China University of Mining and Technology, Beijing in the Manual of Water Prevention and Control in Coal Mine, which is applicable to the conditions of medium-hard overburden and fully mechanized caving mining: $H_{li} = 100\,M/(0.26\,M + 6.88) \pm 11.49$

where:

H_{li} Water flowing fractured zone, meter
M   Coal Seam Thickness, meter.

Calculate the development height of the water flowing fractured zone in the roof of Coal Seam 2^{-2} according to the above equation and compare it with the measured height to obtain the proportional relationship between the measured height and the height calculated by the equation (Table 8.25).

It can be seen from Table 8.25 that the development height of the water flowing fractured zone in the roof of Coal Seam 2^{-2} is 1.822 times of the height calculated by Eq. (8.1), and then the equation for calculating the development height of the water flowing fractured zone in the roof of Coal Seam 2^{-2} in Zhangjiamao Coal Mine is determined:

$$H_{li} = \frac{182.2 \sum M}{(0.26 \sum M + 6.88)} \pm 20.93 \qquad (8.4)$$

where:

H_{li} Water flowing fractured zone, meter
M   Coal Seam Thickness, meter.

The above Eq. (8.4) is used to calculate the development height of the water flowing fractured zone generated by the mining of Coal Seam 2^{-2} in Zhangjiamao Coal Mine (Table 8.26). It can be seen from the Table that the development height of the water flowing fractured zone is 49.61–219.08 m, with an average height of 173.47 m. By using the interpolation function of Surfer and the spatial processing function of GIS, the corresponding thematic map is drawn (Fig. 8.18).

Table 8.25 Comparison between measured height and calculated height of water flowing fractured zone development in 2^{-2} coal seam roof

Measured borehole .	Coal seam mining thickness (m)	Buried depth of coal seam floor (m)	Equation calculation height (m)	Measured height (m)	Coefficient of proportionality	Average value of proportional coefficient
Borehole 1	4.8	150.91	70.55	125.8	1.7833	1.822
Borehole 3	5.8	171	80.64	147.66	1.8312	
Borehole 4	5.46	186.1	77.28	143.82	1.8611	
Borehole 6	5.46	188.91	77.28	139.77	1.8087	

Table 8.26 Development height of water flowing fractured zone in 2–2 coal seam roof

Borehole	Coal seam thickness (m)	Height of water flowing fractured zone (m)	Borehole	Coal seam thickness (m)	Height of water flowing fractured zone (m)	Bore-hole	Coal seam thickness (m)	Height of water flowing fractured zone (m)
222-1	6.61	160	9-1SWK	1.85	66.69	N455	8.53	191.67
221	9.335	203.58	150	8.71	194.38	N481	8.23	187.09
222	7.3	172.37	154	8.53	191.67	N482	8	183.52
224	2.1	72.42	157	9.87	211.20	N483	7.69	178.64
225	9.89	211.49	161	8.45	190.46	N493	9.69	208.66
226	9.39	204.37	168	7.08	168.77	N507	7.54	176.25
227	7.32	172.70	169	9.78	209.94	N508	10.15	215.11
228-2	8.89	197.06	176	7.72	179.12	N509	8.2	186.63
228	1.13	49.61	192R	9.92	211.91	N513	2	70.14
H08	3.6	104.80	198	8.57	192.27	N516	9.27	202.64
B3	7.46	174.96	324	8.47	190.76	N587	2.62	84.03
B25	7.27	171.88	N353	8.5	191.21	N590	5.78	146.49
B26	7.95	182.74	N354	8.37	189.24	N591	5.1	134.11
B27	9.95	212.33	N357	4.73	127.14	N594	9.19	201.47
BS1	8.08	184.77	N358	9.91	211.77	N596	9.44	205.09

8.4.2   Safety Zoning of Roof Caving in 2^{-2} Coal Seam

According to the above analysis and research on the development height of the water flowing fractured zone in the roof of 2^{-2} Coal Seam, for the bedrock fractured aquifer in the roof of 2^{-2} Coal Seam, the water flowing fractured zone generated by coal seam mining all develops and leads to this aquifer, so the whole area is a dangerous area of caving, and water inrush accidents are prone to occur. Based on this, the roof caving safety zoning of 2^{-2} coal seam is obtained (Fig. 8.19).

8.4.3   Calculation of Height of Water Conduct Fractured Zone in 5^{-2} Coal Seam Roof

The thickness of Coal Seam 5^{-2} in Zhangjiamao Coal Mine varies from 4.4 to − 7.87 m with an average thickness of 6.09 m. The thickness of the coal seam does not change much. The 5^{-2} coal seam structure is simple, basically without parting, and there is a layer of parting at the bottom of some coal points, the thickness of parting is 0.10 m, and most of the parting is mudstone and siltstone. The coal seam roof is mainly composed of siltstone and sandy mudstone, followed by medium-grain sandstone and

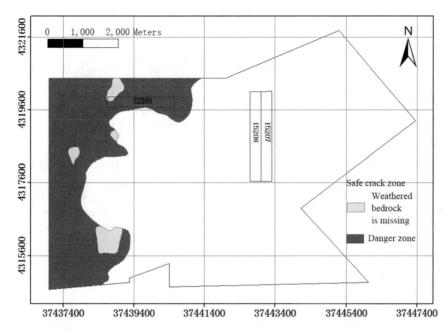

Fig. 8.19 Safety zoning map for caving of fractured aquifer in weathered bedrock of 2^{-2} coal seam roof

fine-grain sandstone. The coal seam floor is mainly composed of siltstone, followed by mudstone. According to the test of rock physical and mechanical properties, the saturated compressive strength of 5^{-2} coal seam roof rock is generally 20–40 MPa, therefore, the coal seam roof is classified as medium-hard rock roof.

Zhangjiamao Coal Mine is located in the Jurassic coalfield of northern Shaanxi Province, where the coal seam is shallow, the overlying bedrock is thin, and the coal seam mining thickness is large. The empirical equation for the development height of water flowing fractured zone in the previous regulations and specifications such as Regulations for Coal Pillar Reservation and Pressure Coal Mining of Buildings, Water Bodies, Railways and Main Shafts and Roadways is not applicable to the calculation of the development height of water flowing fractured zone in fully mechanized mining of thick coal seam under the condition of shallow and thin bedrock in this area.

Therefore, based on the collected actual measurement results (Table 8.27) of the development height of the water flowing fractured zone during the mining of Coal Seam 5^{-2} in Zhangjiamao Coal Mine, the development height of the water flowing fractured zone in the roof of Coal Seam 5^{-2} in Zhangjiamao Coal Mine is studied and analyzed with the following schema.

Equation (8.2) is the empirical calculation equation for the development height of water flowing fractured zone proposed in the Research Report on Water Resources Dynamics and Water Conservation Technology for Coal Mining in Shennan Large Mining Area (December 2010, Shennan Mining Company of Shaanxi Coal Industry

Table 8.27 List of calculation and actual measurement comparison of development height of water flowing fractured zone in 5^{-2} coal seam roof

Measured borehole	Coal seam mining thickness (m)	Buried depth of coal seam floor (m)	Equation calculation height (m)		Measured height (m)
			Equation (8.1)	Equation (8.2)	
Borehole 7	5.6	157.88	78.67	67.254	152.28
Borehole 8	5.6	165.11	78.67	67.254	159.51
Borehole 9	5.6	165.9	78.67	67.254	160.3

Note Eq. (8.1) is the empirical calculation equation for the development height of water flowing fractured zone proposed by China University of Mining and Technology (Beijing) in the Manual of Water Prevention and Control in Coal Mine, which is applicable to the conditions of medium-hard overburden and fully mechanized caving mining: $H_{li} = 100 \, M/(0.26 \, M + 6.88) \pm 11.49$

Chemical Group, 185 Team of Shaanxi Coal Geology Bureau and China University of Mining and Technology): $H_{li} = 9.59 \, M + 13.55$.

In this scheme, the development height of water flowing fractured zone is calculated by using the empirical equation of Coal Mine Water Prevention and Control Manual, and compared with the measured height, the proportional relationship between them is obtained, and then the development height calculation equation of water flowing fractured zone is obtained. The specific methods are as follow:

The empirical calculation Eq. (8.4) for the development height of water flowing fractured zone proposed by China University of Mining and Technology (Beijing) in the Manual of Water Prevention and Control in Coal Mine is applicable to the conditions of medium-hard overburden and fully mechanized caving mining:

$$H_{li} = \frac{100 \sum M}{0.26 \sum M + 6.88} \pm 11.49 \qquad (8.5)$$

where:

H_{li} Water flowing fractured zone, m
M   Coal Seam Thickness, m

Calculate the development height of the water flowing fractured zone in the roof of Coal Seam 5^{-2} according to the above equation and compare it with the measured height to obtain the proportional relationship between the measured height and the height calculated by the equation (Table 8.28).

It can be seen from Table 8.28 that the development height of the water flowing fractured zone in the roof of Coal Seam 5^{-2} is 2.000 times of the height calculated by Eq. (6.1), and then the equation for calculating the development height of the water flowing fractured zone in the roof of Coal Seam 5^{-2} in Zhangjiamao Coal Mine is determined by:

Table 8.28 Comparison of measured height and calculated height of water flowing fractured zone in 2^{-2} coal seam roof

Measured borehole	Coal seam mining thickness (m)	Buried depth of coal seam floor (m)	Equation calculation height (m)	Measured height (m)	Coefficient of proportionality	Average value of proportional coefficient
Borehole 7	5.6	157.88	78.67	152.28	2.0376	2.000
Borehole 8	5.6	165.11	78.67	159.51	2.0276	
Borehole 9	5.6	165.9	78.67	160.3	1.9357	

$$H_{li} = \frac{200 \sum M}{\left(0.26 \sum M + 6.88\right)} \pm 22.98 \tag{8.6}$$

where:

H_{li} Water flowing fractured zone, m
M   Coal Seam Thickness, m

   Equation (8.5) is used to calculate the development height of the water flowing fractured zone generated by the mining of Coal Seam 5^{-2} in Zhangjiamao Coal Mine (Table 8.29). It can be seen from the Table that the development height of the water flowing fractured zone is 132.67–199.35 m, with an average height of 167.17 m. By using the interpolation function of Surfer and the spatial processing function of GIS, the corresponding thematic map is drawn (Fig. 8.20).

8.4.4 Safety Zoning of Roof Caving in 5^{-2} Coal Seam

According to the above analysis and research on the development height of the water flowing fractured zone in the roof of Coal Seam 5^{-2}, for the bedrock fractured aquifer and burnt rock aquifer in the roof of Coal Seam 5^{-2}, most of the water flowing fractured zones generated by coal seam mining are developed and conducted to this aquifer, and water inrush accidents are prone to occur. Based on this, the safety zoning of roof caving in Coal Bed 5^{-2} is obtained (Figs. 8.21 and 8.22).

Table 8.29 Development height of water flowing fractured zone in 5^{-2} coal seam roof

Borehole	Coal seam thickness-s (m)	Height of water flowing fractured zone	Borehole	Coal seam thickness-s (m)	Height of water flowing fracture-d zone	Borehole	Coal seam thickness-s (m)	Height of water flowing fracture-d zone
3–2	5.8	161.30	B16	5.7	159.33	9–4	6.1	167.11
H13	5.8	161.30	B17	5.65	158.35	9–5	6.25	169.98
H16	5.8	161.30	B18	5.82	161.69	9–6	6.25	169.98
12–1	6.61	176.75	B19	7.4	191.11	10–1	6.1	167.11
12–2	4.88	142.77	B20	7.4	191.11	10–2	6.2	169.03
H4-2	4.96	144.43	B21	6.05	166.15	10–3	6.35	171.87
H5-2	6.49	174.51	B22	6.04	165.96	10–4	6.05	166.15
H6-2	5.1	147.30	B23	4.4	132.67	10–5	6.1	167.11
ZK1	6.9	182.10	B24	6.2	169.03	11–1	6.2	169.03
ZK2	6.83	180.82	B25	5.9	163.25	11-2SWK	6.2	169.03
ZK3	6.06	166.34	B26	5.65	158.35	H1	6.35	171.87
ZK4	5.98	164.80	B27	5.95	164.22	150	6.15	168.07
ZK5	6.05	166.15	B29	5.85	162.27	151	6.05	166.15
ZK6	5.75	160.32	B30	5.3	151.36	154	5.34	152.17
ZK7	6.28	170.55	B2Y	5.81	161.49	157	6.41	173.01
ZK8	6.21	169.22	B3Y	5.5	155.37	168	6.48	174.32
ZK9	6.08	166.73	FJ1	6.04	165.96	169	6.04	165.96
ZK10	6.02	165.57	FJ3	6.1	167.11	176	6.43	173.38
ZK11	6.16	168.26	ZJ4	6.55	175.63	198	5.65	158.35
ZK12	7.14	186.46	6–1	7.35	190.23	201	6.13	167.69

(continued)

Table 8.29 (continued)

Borehole	Coal seam thickness-s (m)	Height of water flowing fractured zone	Borehole	Coal seam thickness-s (m)	Height of water flowing fracture-d zone	Borehole	Coal seam thickness-s (m)	Height of water flowing fracture-d zone
ZK14	5.94	164.02	6-2	6.1	167.11	202	6.28	170.55
B1	5.78	160.91	6-3	5.9	163.25	203	6.27	170.36
B2	5.7	159.33	7-1	6.05	166.15	263	6.07	166.53
B3	6.15	168.07	7-2	6.35	171.87	316	6.18	168.64
B4	6.5	174.70	7-3	6	165.18	318	5.93	163.83
B5	6.1	167.11	7-4	5.55	156.37	324	6.16	168.26
B6	6.35	171.87	7-5	6	165.18	349	5.99	164.99
B7	5.65	158.35	8-1	5.9	163.25	N353	7.87	199.34
B8	6.01	165.38	8-2	6.1	167.11	N354	6.59	176.38
B9	5.95	164.22	8-3	6.1	167.11	N357	7.51	193.06
B10	6.84	181.00	8-4	6	165.18	N358	5.85	162.27
B11	5.8	161.30	8-5	6	165.18	N359	6.59	176.38
B12	5.75	160.32	9-1SWK	6.25	169.98	N370	5.83	161.88
B13	5.85	162.27	9-2	6.1	167.11	N371	5.77	160.71
B14	7.3	189.33	9-3	5.85	162.27	N462	6.26	170.17

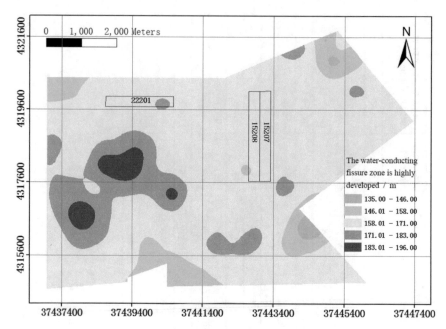

Fig. 8.20 Thematic map of development height of water flowing fractured zone in 5^{-2} coal seam roof

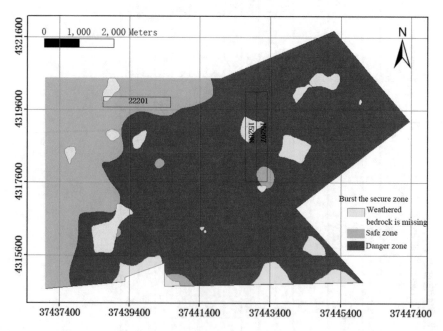

Fig. 8.21 Safety zoning diagram for caving of fractured aquifer in weathered bedrock of 5^{-2} coal seam roof

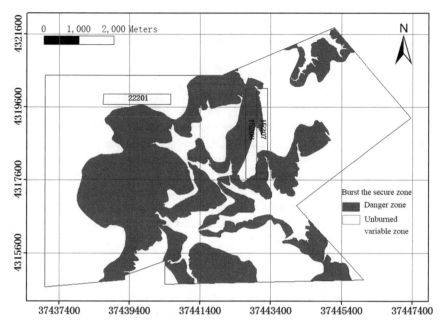

Fig. 8.22 Safety zoning diagram for caving of 5^{-2} coal seam roof burnt bedrock porous-fractured aquifer

8.5   Risk Assessment and Zoning of Water Inrush from Water-Filled Aquifer in Coal Seam Roof

According to the safety analysis of roof caving of 2^{-2} coal seam and 5^{-2} coal seam, the roof water flowing fractured zone of 2^{-2} coal seam mining is all connected with the roof weathered bedrock aquifer, the roof water flowing fractured zone of 5^{-2} coal seam mining is all connected with the burnt bedrock aquifer, and the weathered bedrock aquifer is not connected in the west and other local areas. Therefore, in the conduction area, the water inrush condition of coal seam roof mainly depends on the water abundance of the water-filled aquifer, and the non-conduction area can be considered as a relatively safe area. According to the water abundance evaluation result of the roof aquifer by the water abundance index method and the caving safety zoning result, the water inrush risk evaluation and zoning of the 2^{-2} coal and 5^{-2} coal roof water-filled aquifer of Zhangjiamao Coal Mine are finally generated.

8.5.1 Risk Assessment and Zoning of Water Inflow in Fractured Aquifer of Roof Weathered Bedrock of 2^{-2} Coal Seam

Based on water abundance zoning and caving safety zoning of 2^{-2} coal seam roof weathered bedrock fissure aquifer, the water inrush risk zoning map of 2^{-2} coal seam roof weathered bedrock fissure water-filled aquifer in Zhangjiamao Coal Mine is obtained (Fig. 8.23). It can be seen from the figure that for the weathered bedrock fissure aquifer, the risk of water inrush in the northwest of 2^{-2} coal seam of Zhangjiamao Coal Mine is higher than that in the south, and the risk of water inrush increases gradually from the southeast to the northwest, especially the northwest should be taken as the key area for water prevention and control. Among them, most of the 22,201 working face is in the more dangerous area, and the risk of water inrush increases gradually from east to west.

8.5.2 Water Inrush Risk Assessment and Zoning of 5^{-2} Coal Seam Roof Weathered Bedrock and Burnt Bedrock Aquifer

On the basis of water abundance zoning and caving safety zoning of roof weathered bedrock fissure aquifer of 5^{-2} coal seam and burned bedrock pore-fissure aquifer of 2^{-2} coal seam, 3^{-1} coal seam and 4^{-2} coal seam, The water inrush risk zoning map (Fig. 8.24) of the fractured aquifer filled with water in the weathered bedrock of the roof of Coal Seam 5^{-2} of Zhangjiamao Coal Mine and the water inrush risk zoning map (Fig. 8.25) of the pore-fractured aquifer filled with water in the burned bedrock of Coal Seam 5^{-2} are obtained.

It can be seen from Fig. 8.24 that for the weathered bedrock fissure aquifer, the water inrush risk is higher in the west, south and northeast, the water inrush risk is weaker in the middle and east, and the water irruption risk increases gradually from east to west. The risk of water inrush in 15,208 and 15,207 working faces increases gradually from south to north. The large area of the second panel is a dangerous area, so it is necessary to focus on preventing the occurrence of water inrush accidents. The most western relative safety zone is the area where the calculated water flowing fractured zone is not conducted. In the actual production, attention should be paid to the area where the water flowing fractured zone is developed. This area is rich in water, and once conducted, it will produce a larger water inflow.

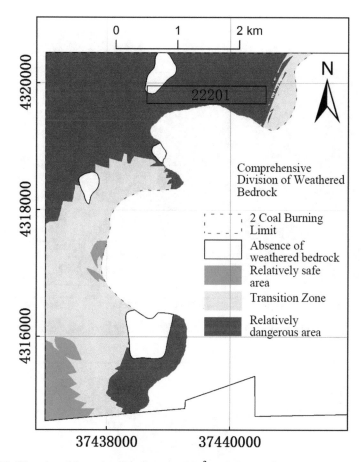

Fig. 8.23 Water inrush hazard zoning diagram of 2^{-2} coal seam roof weathered bedrock fracture aquifer

It can be seen from Fig. 8.25 that for 2^{-2} coal seam, 3^{-1} coal seam and 4^{-2} coal seam, the risk of water inrush in the middle of the second panel and the first panel is relatively high, and the risk of water inrush in the middle of the mine field is relatively low. The risk of water inrush in 15,208 and 15,207 working faces gradually increases from north to south, especially in the southernmost part of the working face near the area of 4^{-2} coal burnt rock, which is exposed in reservoirs and other rivers with strong water content, so attention should be paid to water prevention and control. The 2^{-2} coal burning area in the middle of the second panel is also the key area to prevent water inrush because of its thick aquifer and large unit water inflow.

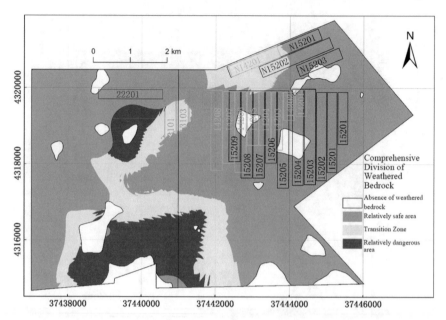

Fig. 8.24 Zoning diagram of water inflow hazard of fractured aquifer in weathered bedrock of 5^{-2} coal seam roof

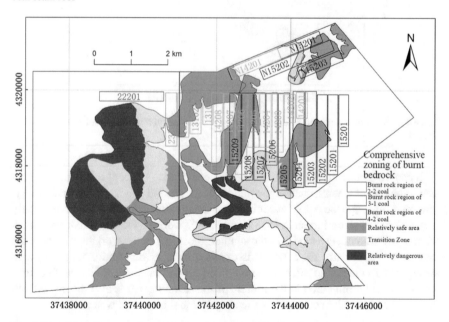

Fig. 8.25 Zoning diagram of water inflow (inrush) hazard of 5^{-2} coal seam roof burnt bedrock pore-fracture aquifer

References

1. Hamraz BS, Akbarpour A, Pourreza Bilondi M, Sadeghi Tabas S (2015) On the assessment of ground water parameter uncertainty over an arid aquifer [J]. Arab J Geosci 8(12)
2. Chen K, Xue GQ, Chen WY, Zhou NN, Li H (2019) Fine and quantitative evaluations of the water volumes in an aquifer above the coal seam roof, based on TEM [J]. Mine Water Environ 38(1)
3. Hu YB, Li WP, Liu SL et al (2019) Risk assessment of water inrush from aquifers underlying the Qiuji coal mine in China [J]. Arab J Geosci 12(3)
4. Li B, Chen LX, Chen YL (2019) Assessment technique of karst aquifer water abundance in shale gas exploitation area based on multi-source information fusion technology—FengGang shale gas area case [J]. Arab J Geosci 12(12)
5. Gao R, An H, Ju F, Mei XC, Wang XL (2018) Influential factors and control of water inrush in a coal seam as the main aquifer [J]. Int J Min Sci Technol 28(2)
6. Li B (2016) Water abundance analysis of aquifer in coal roof based on grey relational analysis and analytic hierarchy process [J]. Water Pract Technol 11(2)
7. Sun J, Hu Y, Zhao GM (2017) Relationship between water inrush from coal seam floors and main roof weighting [J]. Int J Min Sci Technol 27(5)
8. Wu Q, Liu YZ, Zhou WF et al (2015) Evaluation of water inrush vulnerability from aquifers overlying coal seams in the Menkeqing coal mine, China [J]. Mine Water Environ 34(3)
9. Zeng YF, Wu Q, Liu SQ et al (2018) Evaluation of a coal seam roof water inrush: case study in the Wangjialing coal mine, China [J]. Mine Water Environ 37(1)
10. Wu Q, Guo XM, Shen JJ et al (2017) Relevamiento del riesgo de irrupción de agua desde los acuíferos subyacentes de la mina de carbón Gushuyuan, China [J]. Mine Water Environ 36(1)
11. Yao BH, Bai HB, Zhang BY (2012) Numerical simulation on the risk of roof water inrush in Wuyang coal mine [J]. Int J Min Sci Technol 22(2)
12. Zhang B, Cao SG (2015) Study on first caving fracture mechanism of overlying roof rock in steep thick coal seam [J]. Int J Min Sci Technol 25(1)

Chapter 9
Water–Sand Inrush Mode and Prevention Technology Under Bedrock Fissure and Loose Pore Aquifer

9.1 The Main Controlling Factors Affecting the Occurrence of Water–Sand Inrush Break Disasters

According to the four elements (i.e., provenance, channel, power source, and space) of mine water–sand inrush accident, when the overlying rock movement caused by coal seam mining touches the water-rich water body, it will give mine safety production has a very serious impact. When the fracture zone touches the loose layer aquifer with flow characteristics, under the influence of high hydraulic gradient and mining intensity, the flowing water body can easily drive the sand layer through the fracture and enter the goaf, thereby causing water–sand inrush accidents occurred in mines [3, 9, 10]. Therefore, this chapter mainly analyzes the main factors affecting the water–sand inrush accidents of the working face caused by the total fracture of the overlying rock under the conditions of strong mining of thick coal seams in the Yushenfu mining area based on the existing theoretical basis.

9.1.1 Thickness and Physical Characteristics of Water-Bearing Sand Layer

According to the four factors of water inrush and water break accident in mine working face, the occurrence of water–sand inrush accident is closely related to water-rich characteristics of water-bearing sand layer, water–sand combination characteristics and other factors. The sand layer is the largest source of mine water–sand inrush accident. The storage of sand layer in water-bearing sand layer mainly depends on the spatial combination relationship between aquifer and sand layer, and its thickness plays an important decisive role in the degree and grade of disaster. At the same time, the physical characteristics of sand layer is also an important factor to determine the occurrence of sand collapse accident. When the diameter of the sand particles is large, the pores of the particles between the sand layers are large, and the water passage is larger, the aquifer is easy to drain, and it will not be replaced to the goaf by the gravity potential generated by the water flow. According to the existing statistical data analysis of water–sand inrush accidents, the diameter of the sand particles in the accident is 0.005–0.074 mm, and there is no bond between the particles. Therefore, the thickness of water-bearing sand layer, the thickness of sand layer particles and the bond degree between sand layer particles play a prerequisite role in the occurrence of mine water–sand inrush accident.

9.1.2 Influence of Aquifer Abundance and Recharge Quantity

According to the existing rock mass hydraulics theory and relevant studies on the occurrence mechanism of mine water disaster [5, 6], the hydrostatic pressure value and water potential energy of the aquifer are directly determined by the water-rich of the aquifer and the supply of the accessory water bodies. Under the continuous action of this hydrostatic pressure, fracture scour phenomenon occurs in the primary fracture of rock mass, resulting in the continuous increase of crack gap. At the same time, during the mining process of the coal seam, the overlying rock continuously generates water flow channels, and the original hydrostatic pressure becomes the dynamic water pressure under the action of the change of the hydraulic gradient; The water body moves instantaneously with the newly generated mining fissures. When the mining-induced fissures cause the hydraulic connection between the phreatic aquifer and the bedrock aquifer, the water-bearing sand layer gushes to the working face under the carrying effect of the hydrodynamic force, causing a water disaster accident [1, 2, 8].

Before the coal seam mining, with the drainage of the weak water-rich aquifer, the When the water flowing fractured zone triggers the aquifer, the water body does not produce a large hydraulic slope, which cannot drive the sand layer to move zone triggers the aquifer, the water body will not produce a large hydraulic gradient, and thus cannot drive the sand. However, for aquifers with strong water abundance, after being affected by mining, the water body contained in the aquifer quickly converts water potential energy into kinetic energy and forms a large hydraulic pressure difference. When the value exceeds the water pressure value that drives the movement of the sand layer, the sand body enters the goaf together with the water body, which eventually leads to an accident. However, according to the current research scholars ' research on the critical hydraulic gradient of water and sediment, the critical hydraulic gradient is related to the initial water level of the aquifer, the physical characteristics of the sand layer and the crack size of the water channel. Relevant experimental demonstrations show that when the physical characteristics of the sand layer particle size are determined, the amount of sand inrush at the working face has a linear relationship with the fracture size of the water inflow channel and the initial water head. Therefore, the strength of water-rich aquifers and the development of fissures play an important role in the occurrence of mine flood accidents.

9.1.3 Effective Clay Layer Thickness and Physical Properties

According to the geological situation of the Yushenfu mining area in the second chapter, there is a laterite clay aquifuge of Baode Formation at the bottom of the Quaternary loose layer. According to the research content of Chap. 4, when the overlying rock fissure develops to the bottom of the soil layer, whether the overlying Salawusu Formation aquifer water body will enter the goaf through the mining-induced bedrock fissure, the thickness and distribution characteristics of laterite clay layer play a decisive role. If there is a clay layer with a certain effective aquifuge thickness after mining, the probability of water–sand inrush is low. If the clay layer has no effective aquifuge thickness after mining, and the overlying water-bearing sand layer is highly water-rich and the diameter of the sand layer particles is small, the accident probability of water–sand inrush at the working face is high. However, the effective water-resisting thickness of clay layer after mining is directly related to the thickness and mining strength of clay layer before mining. Through the analysis and summary of the roof water–sand inrush accidents in the working face of the Yushenfu mining area in recent years, the scientific researchers know that, after the clay layer is affected by mining, it will not completely lose the water isolation performance due to the up and down cracks generated by the clay layer and presents the critical failure value of the clay layer in a certain range. When the failure strength is greater than the critical failure value, the clay layer is broken.

At the same time, according to the laboratory study of the clay layer, such as composition diffraction and expansion test, it can be seen that the clay layer contains a large number of montmorillonite, illite, kaolinite and other viscous minerals, but the content of the composition is controlled by the formation period of the clay layer. However, the content of these minerals plays an important role in the water-resisting strength of the clay layer. Therefore, the effective clay layer thickness and its physical characteristics are considered as one of the important factors in the qualitative evaluation of the main factors affecting the occurrence of water–sand inrush accidents.

9.1.4   The Ratio of the Thickness of Loose Bed to the Thickness of Bedrock

According to the fourth chapter, it is discussed that the bedrock layer acts as the first waterproof barrier and the soil layer acts as the second water barrier, which has a direct effect on preventing the occurrence of mine water damage. However, according to the distribution map of bedrock thickness shown in Fig. 9.1, there are great differences in the stratigraphic combination of the roof overlying bedrock and the loose bed at different positions above the same coal seam. However, the ratio of bedrock thickness to loose layer thickness has an important influence on the occurrence of mine flood accidents. Under the same mining intensity, if the thickness of the bedrock is greater than the thickness of the loose layer, according to the key stratum theory, when the overlying fissures develop to the position of the main key layer, the development of the fissures can be prevented, as shown in Fig. 9.1c and d. When the thickness of the bedrock layer is less than the thickness of the loose layer, due to the influence of mining intensity, the mining caving cracks can easily enter into the overlying loose layer, as shown in Fig. 9.1a, cutting through the overlying aquifer causes the water body to carry sand and enter to the goaf. Therefore, qualitative and quantitative analysis of the ratio of loose layer to bedrock thickness is of great significance for the evaluation of water–sand inrush.

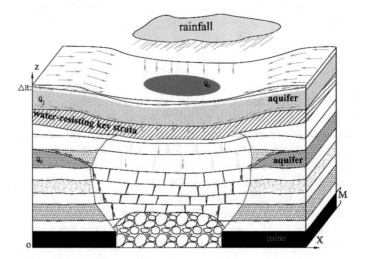

Fig. 9.1 Geological conceptual model of water–sand inrush under strong mining conditions

9.1.5 Development Range and Height of Mining Fissures

Water inrush channel is one of the indispensable factors for mine water–sand inrush accident. Under the condition that the material source, power source and flow space are determined, the affected range and development height of the water–sand inrush fissure channel caused by coal seam mining determine the amount of water inflow in the working face, and then determine the change of the hydraulic gradient of the aquifer and the magnitude of the hydrodynamic pressure. According to the existing classification of mine water gushing channels, it can be divided into natural channels and man-made channels; and the current mine water hazards are all caused by man-made and mining influence conditions. Therefore, it is an important guaranteed basis to ensure the safe mining of mines and the evaluation of water–sand inrush to carry out relevant research work on the development range and height of mining fissures [4, 7]. According to the theoretical research on the movement and failure of overburden rock, under the premise of the determination of lithology combination, the thickness of coal seam mining directly determines the development degree of overburden rock fracture, and the change value of water pressure increases with the increase of fracture development height, and then the initial water potential energy is converted into larger water kinetic energy. When the failure range and development height of overburden fracture reach the maximum, the water inflow area and water discharge capacity of water body increase, and the water inflow of working face increases instantaneously. Especially in the fully mechanized caving mining currently used in the west, this mining method has a large degree of damage to the roof, which is extremely unfavorable for safe mining under water bodies. However, filling mining and height-limited mining, as the key mining methods to limit the movement of overlying strata, play an important role in the application of water-preserved mining in the mining area, especially in the process of building a 'coal-water' dual-resource mine, which provides technical guarantee for the safe mining of the mining area. The influence of mining intensity on aquifers is carefully studied in the fourth chapter in the fluid–solid coupling simulation and analysis of overburden deformation and seepage characteristics.

9.2 Mechanism Analysis of Water–Sand Inrush Disaster Induced by Mining

The water–sand inrush accident occurred in the process of coal seam mining in the coal mine is under the coupling action of strong mining pressure and dynamic water pressure, and the sand body is broken into the coal mining face through the mining fissure. In view of the above research contents in Chap. 4 and the research results in Sect. 9.1, after the bedrock is completely broken under the influence of strong mining, a certain thickness of N_2 aquiclude will show the role of key strata, supporting the overlying Quaternary strata. At the same time, according to the analysis results of the main factors affecting the occurrence of water–sand inrush accidents in the above Sect. 9.1, whether water–sand inrush accidents occur in the working face during the mining process of thin bedrock coal seams mainly depends on the second water-proof soil layer barrier whether the effective water-insulating soil layer damaged by mining still has water-insulating effect. Therefore, the water–sand inrush accidents induced by strong mining in thin bedrock coal seams mainly depend on whether the hydraulic connection between the weathered bedrock aquifer and the upper Salawusu Formation aquifer can be blocked under the coupling action of strong mining pressure and hydrodynamic pressure. In this section, the mechanism of water–sand inrush disaster induced by mining is analyzed based on the key strata theory.

9.2.1 Key Layer Definition

Strata with different properties are deposited in the roof of coal measure strata. In the process of coal seam mining, strata in the middle-stratified position will bear the upper strata and hinder the movement of upper lithology. The strata that can be regarded as key strata in the strata shall have the following characteristics:

(1) Characteristics of strata thickness: the strata that can be regarded as key strata should still have a certain effective thickness to support the overlying strata after being affected by mining.
(2) Mechanical characteristics of rock strata: relative to the underlying rock strata, the compressive and shear strength of the key rock strata, which can be regarded as the bearing layer, should be greater than the force exerted during the mining process.
(3) Cooperative deformation characteristics: when the key rock strata break and sink after the load is greater than its ultimate bearing capacity, the upper rock strata should synchronously sink with it, and cause a wide range of changes in the overlying rock strata; As shown in Fig. 4.11, when the red soil layer is unstable and sinks, the overlying strata sink instantaneously and cracks appear on the surface. The laterite representing a certain thickness in the field can be regarded as the key stratum supporting the overlying strata.
(4) Supporting structure characteristics: the key stratum changes from the original beam structure to the masonry beam structure to support the upper rock layer of the key stratum during the process of fracture.

According to the key stratum theory, the position of the key stratum of coal seam roof is determined mainly by adopting the calculation equation shown in Eq. 9.1 below; Firstly, the overburden load carried by the rock beam is calculated layer by layer by Eq. 9.1. Secondly, the load size of the upper and lower positions of the rock beam is compared. Finally, if the calculated load of layer n is greater than that of the lower layer, then this layer is the key stratum.

The load formed from the first layer to the layer n is shown in the following equation:

$$(q_n)_1 = \frac{E_1 h_1^3 (\gamma_1 h_1 + \gamma_2 h_2 + \cdots + \gamma_n h_n)}{E_1 h_1^3 + E_2 h_2^3 + \cdots + E_n h_n^3} \qquad (9.1)$$

where: E_n is the elastic modulus of rock in the layer n, GPa; n is the number of layers of rock; h_n is the deposition thickness of the layer n, m; γ_n is the density of the NTH layer, g/cm3.

According to the beam structure theory in mechanics of materials, the calculation equations are different when the boundary conditions of the beam are different. The boundary conditions of the beam structure can be divided into the fixedly supported beam structure and the simply supported beam structure. The calculation equations of the two different structures are Eqs. 9.2 and 9.3:

The beam structure is fixed at both ends:

$$L_s = h \sqrt{\frac{2R_\tau}{nq}} \qquad (9.2)$$

The beam structure is simply supported at both ends:

$$L_s = h \sqrt{\frac{2R_\tau}{3nq}} \qquad (9.3)$$

where: R_τ is the tensile strength of the key stratum, GPa; n is the safety factor when the key stratum is broken. According to the characteristics of the rock beam in this study area, n = 6.

9.2.2 Establishment of Mechanical Catastrophe Model of Water–Sand Inrush

Based on the above key layer definition analysis and the fourth chapter research content, it can be seen that the first bedrock water barrier after high-intensity mining in the Yushenfu mining area is destroyed by strong mining; after coal seam mining, the stability of the second aquifuge layer barrier under the combined influence of the strong dynamic water pressure above and the strong mine pressure caused by the instability and fracture of the first bedrock aquifuge barrier determines whether there is a strong hydraulic connection between the aquifer of the overlying Salawusu Formation and the weathered bedrock aquifer, and also determines whether there is a water–sand inrush accident in the mine. Therefore, analyzing the stability of the aquifuge layer has a great effect on the safe mining of the mine. Based on the theory of plane fixed beam and simple support, this book simplifies the water-resisting key stratum into two fixed beams and establishes the geological conceptual model of water–sand inrush under strong mining conditions (Fig. 9.1).

According to the above analysis, when the water-resisting soil layer is under the coupling effect of the mine pressure caused by the fracture of the bedrock layer below and the dynamic water pressure above, the water-resisting key soil layer will bend. When the bending deformation deflection generated under the coupling force exceeds its inherent limit deflection, the key stratum will break. According to the calculation model of beam theory in material mechanics, the key stratum of aquiclude is simplified as a beam under uniform load:

When the beam is broken, the maximum bending moment will be generated at both ends of the beam: $M_{max} = -\frac{1}{12}ql^2$. Therefore, the ultimate tensile stress at both ends of the beam is:

$$\sigma_{max} = \frac{ql^2}{2h^2} \tag{9.4}$$

where: q is the tensile strength of rock, kN/m; l is rock length, m; h is for rock length, m.

Whether the aquiclude is broken is determined by the maximum tensile stress borne by the aquiclude. If the ultimate tensile strength of the key stratum of the aquiclude is R_T, $\sigma_{max} > R_T$, then the key stratum of the aquiclude is broken and the thickness of the aquiclude is $h = l\sqrt{\frac{q}{2\sigma_{max}}}$.

The position where the shear fracture occurs is at both ends of the beam. $F\frac{ql}{2\sigma_{max}}$, therefore, the ultimate tensile stress at both ends of the beam is:

$$\tau_{max} = \frac{3ql}{4h} \tag{9.5}$$

The maximum tensile stress of the aquiclude is used to determine whether the aquiclude is broken. The ultimate shear strength of the key stratum of the aquiclude is set as R_S, if τS_{max}, the key stratum of the aquiclude is broken.

When the water-resisting key soil layer breaks, the overlying rock layer bends and sinks with the soil layer at the same time. The fracture phenomenon of the soil layer in Chap. 4.2 demonstrates the correctness of the theory. Therefore, the limit fracture step and limit subsidence value that the aquiclude can bear can be calculated by the following Eqs. 9.6 and 9.7:

Limit fracture step of aquiclude is:

$$L_D = \frac{\pi}{\sqrt{2}}\left[\frac{EI}{q(1-\mu^2)}\right]^{\frac{1}{6}} \tag{9.6}$$

The limit subsidence value of aquiclude is:

$$\omega = 2\left[\frac{L_D}{\pi}\right]\left[\frac{q(1-\mu^2)}{EI}\right]^{\frac{1}{3}} \tag{9.7}$$

where: E is the elastic modulus of the aquiclude, GPa; I is the moment of inertia of the aquiclude, $I = \frac{bh^3}{12}$, where b is the span of the aquiclude, m, h is the thickness of the aquiclude, m.

9.2.3 Mechanical Catastrophe Criterion of Water–Sand Inrush

9.2.3.1 Horizontal Tensile Limit Span of Aquiclude

According to the analysis of the above research content, when the bedrock is completely broken, the damage degree of the water-resistant soil layer directly determines whether water–sand inrush occurs in the mine. For the fully fractured mines with thick coal seams and strong mining thin bedrocks discussed in this study area, the core of ensuring that water hazards do not occur in the working face is to reduce the damage height of the water-resistant soil layer to ensure that it can still be maintained under strong mining conditions. It can realize the function of the key stratum of water insulation. According to the simulation study of similar materials in Chap. 4, the main factors affecting the stability of the water-resistant soil layer are: the distance between the water-resistant soil layer and the underlying rock mass and its own limit span. According to the relevant content, the horizontal tensile capacity of the water-resistant soil layer is greater than that of the rock layer. At present, the horizontal tensile limit of water-resistant soil layer is 1–2 mm/m.

According to the simulation of the soil layer instability and fracture process during the excavation process with similar materials, the cementation between the soil layers is strong, and the phenomenon of layer separation will not occur during the subsidence process, and the underlying bedrock layer will sink synchronously. Before the water-resisting soil layer is judged, the separation height of the interface with the lower rock layer determines whether it breaks during the sinking process. When the separation height is large, the bending deflection of the soil layer will exceed its limit value in the process of sinking, and the soil layer will be stretched and broken. When the separation distance between the lower strata is small, the settlement process of the water-resisting soil layer will not reach its limit deflection value. Therefore, according to the water–sand inrush model of strong mining overburden strata and the related research contents proposed in Sect. 9.2.2, the aquiclude is simplified as a fixed beam structure under uniform load (Fig. 9.2). The rotation angle θ and deflection curve ω at any section in the beam are calculated as:

$$\theta = \frac{1}{EI}\left(-\frac{1}{6}qx^3 + \frac{1}{4}qlx^2 - \frac{1}{24}ql^3\right) \tag{9.8}$$

$$\omega = \frac{1}{EI}\left(\frac{1}{12}qlx^3 - \frac{1}{24}qx^4 - \frac{1}{24}ql^3x\right) \tag{9.9}$$

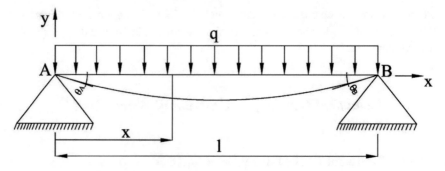

Fig. 9.2 Simplified model of fixed beam

Note: The unit symbols used in this equation have the same meaning as the above symbols.

When $x = \frac{l}{2}$, the deflection of the water-proof soil layer is the largest. Combined with Eq. 9.9, it can be obtained:

$$\omega = -\frac{5ql^4}{384EI} \tag{9.10}$$

Then the curvature equation is:

$$\frac{1}{\rho} = \frac{d\theta}{dx} = -\sum_{i=1}^{n} \frac{6ql}{(2n-1)^2\pi^2 Eh^3} \cos\frac{(2n-1)2\pi x}{l} \tag{9.11}$$

The horizontal tensile deformation when the aquiclude breaks is:

$$\varepsilon = -\sum_{i=1}^{n} \frac{6qly}{(2n-1)^2\pi^2 Eh^3} \cos\frac{(2n-1)2\pi x}{l} \tag{9.12}$$

When $\cos\frac{(2n-1)2\pi x}{l} = -1, n \to \infty$, the horizontal deformation value at the position of $y = \frac{h}{2}$ is maximum, the maximum value is:

$$\varepsilon = -\frac{3ql}{8Eh^2} \tag{9.13}$$

The critical horizontal tensile deformation value of the soil layer is 1.0 mm/m. At this time, the span of the maximum horizontal tensile deformation value generated by the force bending of the soil layer is:

$$l_b = \frac{2Eh^2}{375q} \tag{9.14}$$

According to the similar material simulation in the fourth chapter, if the distance between the water-resisting soil layer and the lower rock layer is small, the soil layer will not break.

9.2.3.2 Water-Resisting Performance Criterion of Water-Resisting Soil Layer

The ultimate deflection and ultimate span that the soil layer can bear is derived from Eqs. (9.10) and (9.14), it can be seen that when the load carried by the water-resistant soil layer is greater than its limit value, the water-resistant soil layer will bend and break, thus This leads to the hydraulic connection between the weathered bedrock aquifer and the aquifer of the Salawusu Formation. Therefore, the height of the separation space between the water-resistant soil layer and the underlying rock layer and its ultimate tensile capacity are the main factors that affect its tensile fracture. However, through the above research, it is found that the sinking motion of the aquiclude after the bending and breaking of the aquiclude is synchronous with the sinking of the lower rock layer. If the space height between the aquiclude and the lower rock layer is greater than its own limit deflection during the sinking process, the aquiclude will bend and break during the sinking process. In the process of subsidence, if its own limit deflection value is greater than the height of its separation space, bending fracture will not occur, which will not lead to weathered bedrock aquifer and Salawusu aquifer hydraulic connection. At the same time, the horizontal tensile value of the water-resistant soil layer during the subsidence process determines whether it will break or not, and if the maximum horizontal tensile strain exceeds its maximum value, cracks will occur. Therefore, in the fully broken water-retaining mining of the overlying rock under strong mining, the criteria for the water-resistant soil layer to maintain its complete water-resistant performance are as follows:

$$\omega_{max} \tag{9.15}$$

$$l_p < l_b \tag{9.16}$$

where, ω_{max} is the maximum bending subsidence of the water-resistant soil layer, m; Δ is the height of the separation layer between the water-resistant soil layer and the underlying rock layer, m; l_b is the span when the water-resistant soil layer breaks, m; l_p is the fracture span of the lower rock stratum of the aquiclude, m.

The equation for calculating the free space height under each rock layer is:

$$\Delta_i = M - \sum_{j=1}^{i-1} h_j(k_j - 1) \tag{9.17}$$

where, Δ_i is the maximum bending subsidence of the i-th layer in the stratum; M is the thickness of the mined coal seam; h_j is the thickness of the j-th layer in the stratum; k_j is the residual bulking factor of rock, 1.04 for hard rock and 1.02 for soft rock.

9.3   Weight Coefficient Determination of Main Controlling Factors of Water–Sand Inrush in Overburden Rock Based on AHP

According to the above analysis of the main influencing factors affecting the occurrence of water–sand inrush and the analysis of the mechanism of water–sand inrush disasters induced by mining, the occurrence of water–sand inrush accidents in mines is a geological disaster phenomenon that occurs in the mining process controlled by various influencing factors. This phenomenon has very complex nonlinear dynamic characteristics. At present, scientific research scholars mainly carry out feasibility analysis and risk assessment of coal seam mining based on the rules of *Coal Mine Water Prevention and Control Rules, Buildings, Railways, Water Body* and *Main Roadway Coal Pillar Setting and Coal Mining Specification* and the research results of related contents before coal seam mining. However, with the change of strata characteristics of coal seam mining, especially the change of hydrological conditions and coal seam mining conditions, and the increasing requirements of relevant departments on ecological and environmental protection issues. Therefore, based on traditional research, we should continue to study the work, so as to realize the accurate evaluation of water–sand inrush in the working face.

However, according to the research content of Chap. 5, it is found that different mining intensities have different influence characteristics on overburden and surface ecological damage. Therefore, it can be seen that the formation of water–sand inrush risk of mining overburden is a multi-factor-controlled disaster process with spatial and temporal evolution characteristics. At the same time, according to the statistical data in Sect. 5.1.3, the bedrock of the mine in this area is basically broken and water-conducting fractures are formed after high-intensity mining. However, due to the existence of clay layers with different thicknesses and good plasticity in the upper part, the fissures in some areas do not directly cut through the strata. Therefore, the mine water–sand inrush accident in this area is different from the water inrush accident in the shallow coal seam mine of the Shendong mining area in the early stage. The main difference is that the water–sand inrush accident in the Shendong mining area is mainly caused by the collapse zone directly cutting through the quaternary loose layer under the action of gravity and water pressure. The accident in this study area is mainly caused by the strong mine pressure appearance and strong dynamic water pressure under the action of the whole fracture of the bedrock, which promotes the water and sand to flow into the goaf. Therefore, it is necessary to comprehensively consider and analyze the various controlling factors affecting the occurrence of water–sand inrush accidents in Sect. 9.2 and establish a risk assessment method for water–sand inrush under the condition of strong mining overburden strata in thick coal seam, to realize the safe mining of working face.

9.3.1 Main Controlling Factors System of Water–Sand Inrush Based on AHP

The Analytic Hierarchy Process (AHP) selected in this book is a decision analysis method that solves the complex problems of multiple main control factors through qualitative and quantitative analysis. This method can decompose the main control factors layer by layer; the method used in the decomposition is mainly the analysis method proposed by Saaty in 1970. It mainly divides the complex problem into three levels around the decision-making goal, namely the target layer, the criterion layer and the factor layer. Each layer has its own influencing factors; therefore, the method has the characteristics of systematization and hierarchy. According to the main control factors affecting the occurrence of water–sand inrush discussed in Sect. 9.2, the occurrence is controlled by many complex factors. Therefore, this book uses AHP to construct the main control factor system of water–sand inrush risk. This chapter constructs the main control factor system of water–sand inrush risk, which is mainly based on the content of Sect. 9.2 as the theoretical guidance. At the same time, according to the hydrogeological characteristics of the Yushenfu mining area and the ecological environment requirements of the mining area, the hierarchical structure diagram (Fig. 9.3) is finally established to meet the conditions of water–sand inrush in the fully broken working face of the thick coal seam in the Yushenfu mining area.

The water–sand inrush accident occurred in the mining process of the Yushenfu mining area is mainly because the mining-induced fracture cuts through the clay layer and enters the aquifer of Salawusu Formation. In the process of rapid decline of water level in Salawusu Formation, the overlying aeolian sand layer surges together to the

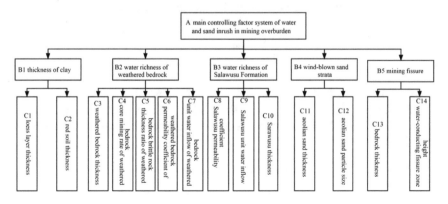

Fig. 9.3 Main control factor system of water–sand inrush in mining overlying strata

working face. Therefore, under the premise of determining the hydrological conditions of the strata, the mining intensity of the coal seam is the main factor leading to the occurrence of water disasters in the working face. The weathered bedrock aquifer is a direct water-filling aquifer in the Yushenfu mining area. The coal seam is drained in advance in accordance with the regulations during the mining process. However, when the water-richness of the weathered bedrock aquifer is strong, it is constantly recharged by the overlying Salawusu Formation aquifer in areas where the clay layer is think. The weathered bedrock aquifer water body is difficult to drain to the specified value. The Baode Formation red soil and Lishi Formation loess deposited between the weathered bedrock aquifer and the Salawusu Formation aquifer are in the mining area. According to the research content of Chap. 2, the Baode Formation red soil layer is the main aquiclude in the area, and the Lishi Formation loess has a certain weak permeability. But the combination layer can be used as the relative aquiclude in the area. Therefore, it is necessary to comprehensively consider the thickness of clay layer, water abundance of weathered bedrock, water abundance of Salawusu Formation, aeolian sand layer, mining fissure and other factors to evaluate the risk of water–sand inrush in overlying strata.

9.3.2 Normalization of Main Control Factors

For the comprehensive evaluation of water–sand inrush because the unit dimension of each factor is inconsistent, the evaluation criteria are different. For example, the thickness of bedrock, the thickness of red soil layer and the height of water flowing fractured zone are negatively correlated with water–sand inrush, i.e., the greater the above values, the smaller the risk of water–sand inrush; such as weathered bedrock aquifer thickness, Salawusu aquifer thickness, unit water inflow and other factors are positively correlated, i.e., the greater the value, the greater the risk of water–sand inrush. Therefore, before comprehensive evaluation, it is necessary to eliminate the wrong judgment of the evaluation results caused by the inconsistency of dimensions. The following equation is used to normalize the index data:

$$\begin{cases} x' = -x_i \\ y = \frac{(x - MinValue)}{(MaxValue - MinValue)} \end{cases} \tag{9.19}$$

where, x' is the positive quantification value of the negative correlation factor; x_i is the original value, x is the normalized initial value, y is the normalized use value, $MaxValue$ is the maximum value in the data, and $MinValue$ is the minimum value in the data.

(1) Weathered bedrock water-rich foundation layer

As a direct water-filled aquifer in the Yushenfu mining area, the water-rich characteristics of the weathered bedrock aquifer play an important role in the safe mining of the mine. The factors affecting the water-rich characteristics of the weathered bedrock aquifer include aquifer thickness, core mining rate, brittle-plastic bedrock ratio, hydraulic conductivity and unit water inflow. To ensure the consistency of the evaluation, this section uses Eq. (9.19) to normalize the five elements determined above. Among them, aquifer thickness, brittle-plastic bedrock ratio, hydraulic conductivity and unit water inflow are positively correlated with increasing the risk of water–sand inrush, and core recovery rate is negatively correlated. Through the powerful data processing function of ArcGIS software, this chapter establishes the topic of each main control factor, as shown in Fig. 9.4. The weight determination table of each main control factor is shown in Table 9.1, and the water-rich zoning map of weathered bedrock aquifer is shown in Fig. 9.5.

(2) Salawusu group water-rich basic layer

The Salawusu aquifer is not only the key protective layer of ecological protection in the Yushenfu mining area, but also the main aquifer of residents and industrial water. At the same time, it is an indirect water-filled aquifer for coal seam mining in the mining area. Its water-rich characteristics determine the power source of water–sand inrush accidents in the mine. I.e., the water-richness of the aquifer is strong. When the mining fissure cuts through the soil layer, the hydraulic slope of the Salawusu aquifer is larger, so the water disaster accident level is larger. According to the field hydrological test, the aquifer thickness, hydraulic conductivity and unit water inflow of Salawusu Formation were selected as the main control factors to evaluate the water-richness zoning. The above three factors were positively correlated with the risk of water–sand inrush. Thematic map of each main controlling factor (Fig. 9.6); weight determination table (Table 9.2), water abundance zoning map of aquifer in Salawusu Formation (Fig. 9.7).

(3) Clay layer thickness basic layer

The most important aquiclude in the Yushenfu mining area is composed of the Baode group laterite and Lishi group loess. Its thickness distribution range plays an important role in slowing down water–sand inrush accidents in mines. The main reason is that the laterite of Baode Formation contains a large amount of clay minerals. According to the similar physical simulation research content in Chap. 4, the soil layer has certain shear resistance. Therefore, the thickness of clay layer has an important influence on the safe mining of mine. The selection of main controlling factors of clay layer thickness mainly selects the thickness of the Lishi group loess and Baode group laterite as influencing factors, and its thickness value is negatively correlated with mine water–sand inrush. The thematic map of each main control factor (Fig. 9.8), the weight determination table of each main control factor (Table 9.3), and the partition evaluation map of clay layer thickness (Fig. 9.9).

(4) Aeolian sand layer foundation layer

As the source of mine water–sand inrush, the distribution characteristics of the aeolian sand layer is the key problem affecting mine water–sand inrush. The occurrence of

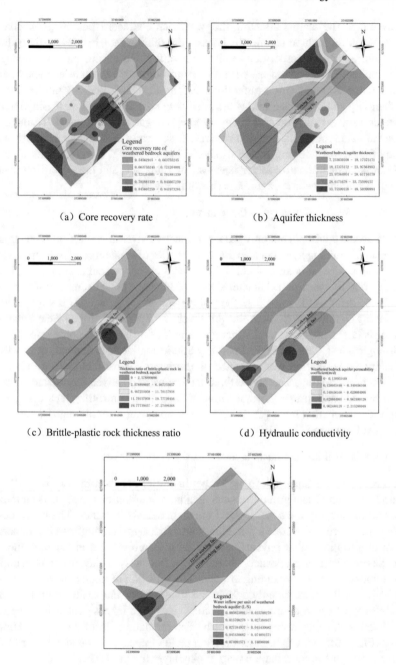

(a) Core recovery rate

(b) Aquifer thickness

(c) Brittle-plastic rock thickness ratio

(d) Hydraulic conductivity

(e) Unit inflow

Fig. 9.4 Thematic map of main controlling factors of weathered bedrock

Table 9.1   Weight determination table

Main controlling factor	Core recovery rate	Aquifer thickness	Brittle-plastic rock thickness ratio	Hydraulic conductivity	Unit inflow
Weight	0.0615	0.3999	0.0615	0.3073	0.2857

Fig. 9.5   Water abundance map of weathered bedrock aquifer

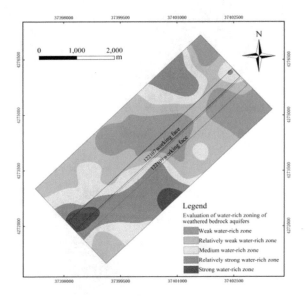

water–sand inrush mainly depends on the physical characteristics of aeolian sand layer under the premise that the water abundance of aquifer providing power source, the free surface condition providing sand inrush space and the mining fissure width providing sand inrush channel are determined. According to the description of the thickness and physical characteristics of the water-bearing sand layer in Sect. 9.2, the accident is that the particle diameter of the sand layer is 0.005–0.074 mm. Therefore, this section mainly selects the thickness of the aeolian sand layer and the proportion of the particle size of the aeolian sand layer less than 0.074 mm as the main controlling factors for evaluating the aeolian sand layer. Both are positively correlated with the mine water–sand inrush disaster. Thematic map of main controlling factors (Fig. 9.10), weight determining table of main controlling factors (Table 9.4), evaluation zoning map of aeolian sand layer (Fig. 9.11).

(5)   Mining fissure foundation layer

Working face water–sand inrush accident or not, in the source, water source and hydrodynamic determination of the premise, the water inrush channel is the only factor to determine whether the working face water–sand inrush accident will occur. Therefore, analyzing and judging whether the water inrush channel exists and judging the development degree is one of the indispensable links to accurately evaluate the

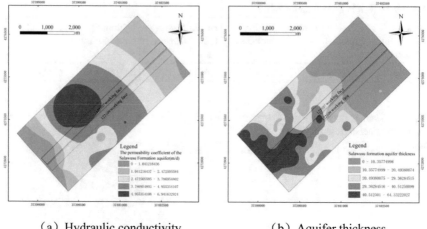

(a) Hydraulic conductivity   (b) Aquifer thickness

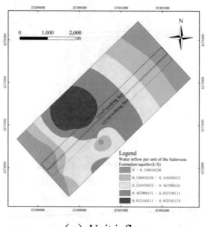

(c) Unit inflow

Fig. 9.6 Thematic map of main control factors of Salawusu formation

Table 9.2 Weight determination table

Main controlling factor	Aquifer thickness	Hydraulic conductivity	Unit inflow
Weight	0.5714	0.2857	0.1429

risk of water–sand inrush in the mine. According to the theory of water-resisting key strata and the research results of Chap. 4.4, the development height of water-conducting fracture zone in overburden rock is linearly related to the thickness of mining coal seam. When the mining fracture is connected to the Salawusu Formation,

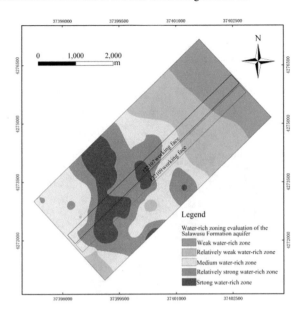

Fig. 9.7   Water abundance map of Salawusu formation aquifer

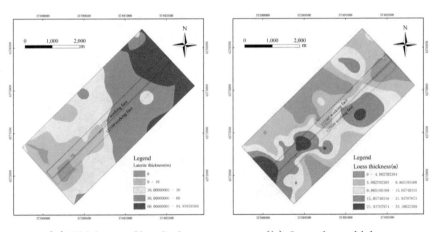

（a）Thickness of laterite layer   （b）Loess layer thickness

Fig. 9.8   Thematic map of main control factors of clay layer thickness

Table 9.3 Weight determination table

Main controlling factor	Loess layer thickness	Thickness of laterite layer
Weight	0.2	0.8

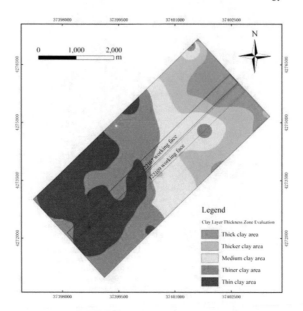

Fig. 9.9 Clay layer thickness zoning evaluation map

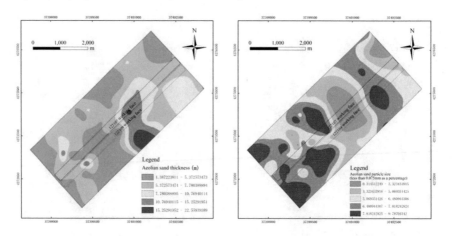

（a）Aeolian sand thickness   （b）Aeolian sand particle size

Fig. 9.10 Thematic map of main control factors for aeolian sand layer evaluation

Table 9.4 Weight determination table

Main controlling factor	Aeolian sand particle size	Aeolian sand thickness
Weight	0.6	0.4

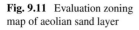

Fig. 9.11 Evaluation zoning map of aeolian sand layer

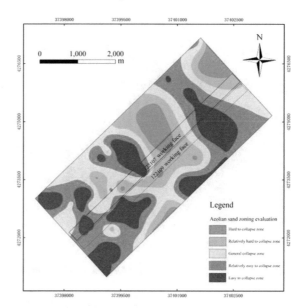

the channel of water–sand inrush in the working face is formed. When the thickness of coal seam mining is determined, the thickness of roof bedrock can also hinder the development of fracture height. Therefore, the thickness of overlying strata and the height of mining failure control the occurrence of sand break and water inrush in overlying strata, which is an important control criterion. The thickness of bedrock layer is negatively correlated with water–sand inrush in working face, and the height of water flowing fractured zone is positively correlated with water–sand inrush in working face. The thematic map of each factor (Fig. 9.12), the weight determination table of each factor (Table 9.5), and the water-rich zoning map of weathered bedrock aquifer (Fig. 9.13).

9.3.3   Weight Determination of Main Controlling Factors of Water–Sand Inrush Based on AHP

Because the occurrence of water–sand inrush accidents is a comprehensive and complex process, and the factors affecting the occurrence of disasters are different. However, the size of the weight value is mainly to reflect the relative importance of each main control factor in the analysis and evaluation of accident risk decisions. Its connotation is mainly reflected in the contribution of each main control factor to the risk of water–sand inrush in the process of risk decision evaluation and analysis. Therefore, it is of great significance to determine the weight of each factor reasonably and scientifically for disaster risk assessment.

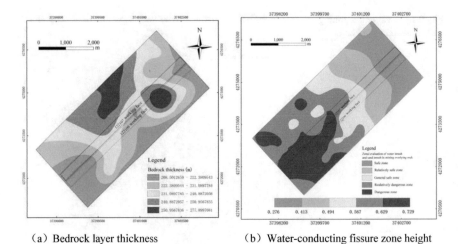

(a) Bedrock layer thickness   (b) Water-conducting fissure zone height

Fig. 9.12 Thematic map of main control factors of mining fissures

Table 9.5 Weight determination table

Main controlling factor	Bedrock layer thickness	Water-conducting fissure zone height
Weight	0.5	0.5

Fig. 9.13 Zoning map of mining fracture evaluation

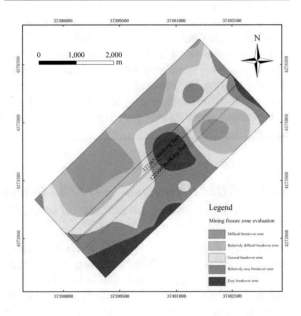

At present, researchers mainly use subjective weight and objective weight to determine the weight of the main control factors. The subjective determination of weight is mainly based on the qualitative analysis of the main control factors affecting the occurrence of disasters and accidents by experts, and the mathematical method is used to deal with the weight of the factors, such as fuzzy comprehensive evaluation, AHP and so on. Objective determination of weight is mainly for the relationship between the main control factors to determine the weight. The main methods used are gray correlation analysis, principal component analysis, etc. The method of determining weight in this book is AHP. The principle of AHP weight determination is to first evaluate and analyze the problem in a multi-factor system that affects each other and form a multi-level analysis structure model. Then the quantitative and qualitative methods are used to sort the layers, and finally the weight of each main control factor is determined to carry out the risk zoning. The specific steps are: (1) select and construct the factors of each level, (2) construct the judgment matrix, (3) establish the pairwise comparison matrix, (4) after determining the weight, check the consistency index (CI) and consistency ratio (CR) as follows:

$$CI = \frac{\lambda_{\max}}{n-1} \tag{9.20}$$

$$CR = \frac{CI}{RI} \tag{9.21}$$

where, $\lambda_{\max}$ is the maximum value in the matrix; n is the number of parameters. As a general criterion, CR should be less than or equal to 0.1, CI is a consistency index.

According to the system of main controlling factors of water–sand inrush in overburden strata determined in Fig. 9.4, based on AHP analytic hierarchy process, the judgment matrix is constructed, and the weight value of each layer is calculated (Tables 9.6, 9.7, 9.8, 9.9, 9.10 and 9.11). The weight value of each main control factor (Table 9.12).

Through the above weight analysis, it can be seen that the thickness of clay layer and mining fissure have great influence on water–sand inrush in working face, and the weight evaluation results are consistent with the existing theory.

Table 9.6 Judgment matrix A-Bi (i = 1–5)

A	B1	B2	B3	B4	B5	W(A/B)
B1	1	4	2	8	4/3	0.3766
B2	1/4	1	1/2	2	1/4	0.0889
B3	1/2	2	1	4	2/3	0.1883
B4	1/8	1/2	1/4	1	1/6	0.0471
B5	3/4	4	3/2	6	1	0.2992

$\lambda_{\max} = 5.0099$, CI = 0.0025, CR = 0.0022

Table 9.7 Judgment matrix B1-Ci (i = 1–2)

B1	C1	C2	W
C1	1	1/4	0.2
C2	4	1	0.8

$\lambda_{max} = 2$, CI = 0, CR = 0

Table 9.8 Judgment matrix B2-Ci (i = 3–7)

B2	C3	C4	C5	C6	C7	W
C3	1	6	6	6/5	3	0.3999
C4	1/6	1	1	1/5	1/3	0.0615
C5	1/6	1	1	1/5	1/3	0.0615
C6	5/6	5	5	1	5/3	0.3073
C7	1/3	3	3	3/5	1	0.17

$\lambda_{max} = 5.0198$, CI = 0.0044, CR = 0.0049

Table 9.9 Judgment matrix B3-Ci (i = 8–10)

B3	C8	C9	C10	W
C8	1	2	1/2	0.2857
C9	1/2	1	1/4	0.1429
C10	2	4	1	0.5714

$\lambda_{max} = 3$, CI = 0, CR = 0

Table 9.10 Judgment matrix B4-Ci (i = 11–12)

B4	C11	C12	W
C11	1	2/3	0.4
C12	3/2	1	0.6

$\lambda_{max} = 2$, CI = 0, CR = 0

Table 9.11 Judgment matrix B5-Ci (i = 13–14)

B5	C13	C14	W
C13	1	1	0.5
C14	1	1	0.5

$\lambda_{max} = 2$, CI = 0, CR = 0

Note The above judgment matrix *CI* and *CR* are less than 0.1, in line with the requirements, can pass the consistency test

Table 9.12 Weight table of main control factors

Main controlling factors	Thickness of clay	Water abundance of weathered bedrock	Salawusu water-rich	Aeolian sand layer	Mining fissure
Weight	0.3766	0.0889	0.1883	0.0471	0.2992

9.4 Quantitative Evaluation Criteria and Zoning Evaluation of Water–Sand Inrush Break Risk of Overlying Rock Under Mining

9.4.1 Quantitative Evaluation Criterion of Water–Sand Inrush Risk Based on AHP

To quantitatively determine the risk of water–sand inrush in the study area, and according to the evaluation results, different preventive measures are established for different dangerous areas. Therefore, based on AHP, this book proposes a risk index (risk index, RI) for quantitative discrimination of water–sand inrush in strong mining overburden rock, RI can be defined as:

$$
\begin{aligned}
RI &= \sum_{i=1}^{n} w_i' * F_i(x, y) \\
&= 0.3766 * F_1(x, y) + 0.0889 * F_1(x, y) \\
&\quad + 0.1883 * F_1(x, y) + 0.0471 * F_1(x, y) \\
&\quad + 0.2992 * F_1(x, y) + \cdots + 0.3766 * F_n(x, y) \\
&\quad + 0.0889 * F_n(x, y) + 0.1883 * F_n(x, y) \\
&\quad + 0.0471 * F_n(x, y) + 0.2992 * F_n(x, y)
\end{aligned}
\tag{9.24}
$$

where, $w_i' = (w_1, w_2, \ldots, w_n)^T$ represents the weight determined by different factors, $F_i(x, y)$ is the function applied by different factors.

To evaluate the risk of water–sand inrush more accurately in the strong mining overburden rock proposed in this book, it is necessary to partition the risk assessment results, and then provide a theoretical basis for the safe mining of the working face. The evaluation results are divided by using the natural discontinuity method in GIS software, and the threshold value of water–sand inrush risk zoning in the working face of the study area is determined (Table 9.13). The calculation principle is:

$$
SSD_{i\cdots j} = \sum_{i=1}^{n} \left(A[K] - mean_{i\cdots j} \right)^2
\tag{9.25}
$$

Table 9.13 Partition threshold of water–sand inrush risk in the study area

Risk zoning	Regional risk
$0.276 \leq RI \leq 0.413$	Safety zone
$0.413 \leq RI \leq 0.494$	Safer zone
$0.494 \leq RI \leq 0.567$	Transitional safety zone
$0.567 \leq RI \leq 0.629$	More dangerous area
$0.629 \leq RI \leq 0.729$	Danger zone

where, $SSD_{i...j}$ is the variance expression used for discontinuity, i, j represents the row and column number, $A[K]$ is the classification interval matrix, $K = i \cdots j$; $mean_{i...j}$ represents the mean of the classification.

The multi-factor criterion decision-making result of the risk of water–sand inrush in the strong mining overburden in the study area is Figs. 9.14f. (a)–(e) in Fig. 9.14 are the main controlling factors affecting the risk of water–sand inrush in the strong mining overburden in the study area.

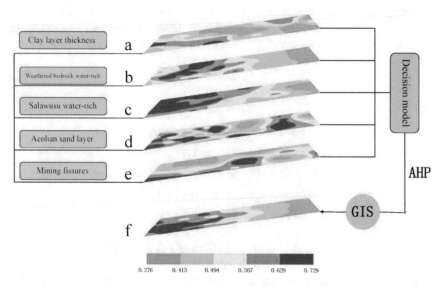

Fig. 9.14 Decision-making result diagram of multi-factor criterion for water inrush risk in mining overlying strata fissures

9.4.2 Zoning Evaluation and Parameter Uncertainty Analysis of Hazardous Results

9.4.2.1 Risk Result Zoning Evaluation

It can be seen from the zoning map of the risk of water–sand inrush in the whole mining overburden rock (Fig. 9.15) that the middle part of the study area is in the transition stage. The high-risk areas and dangerous areas are mainly concentrated in the southeastern region of the study area, and the high-risk areas are mostly concentrated in the middle of the southeastern region of the study area and overtook the southeastern boundary area of the study area. The safety zone is mainly located in the northeast of the study area near the stop line. It should be noted that due to the different stratigraphic characteristics and coal seam fluctuations at different positions, the threat of water–sand inrush is also different when the working face is mined with the same mining parameters. Therefore, the countermeasures adopted to solve different regions are also different.

(1) Type I water–sand inrush danger zone: $0.729 \leq RI < 0.629$ area, which is mainly located in the central part of the southeast area of the study area, and over to the southeast boundary area of the study area. The main reason why this location is a high-risk area is that the thickness of clay layer is relatively weak, and the Salawusu Formation is rich in water. At the same time, the mining-induced fissure in this position is channeled to the Salawusu Formation. Therefore, reasonable measures must be taken to ensure safe mining before mining at this position. The measures that can be taken before mining in this area are

Fig. 9.15 Risk zoning map of water–sand inrush in mining overlying strata

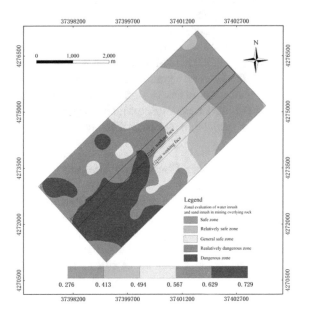

limited high mining, filling mining and other mining methods that can limit the development of water-conducting fractures. Before the coal seam is mined, the grouting reinforcement and reconstruction project can be carried out in the area where the water-resisting soil layer is missing and weak, so as to ensure the safety of the mining face.

(2) Type II water–sand inrush risk area: $0.629 \leq RI < 0.567$ area, according to Fig. 9.14, the area is more dangerous and larger, accounting for approximately 1/3 of the total study area. Therefore, under the action of strong mining, the position is easily transformed into a dangerous area. In summary, the same safety measures as Class I areas can be taken when mining coal seams in this area. When mining at this position, the advancing speed of the working face can be accelerated and the roadway support strength at this position can be strengthened, thereby reducing the roof strata behavior of the working face.

(3) Type III water–sand inrush transition safety zone: $0.567 \leq RI < 0.494$ area, which is in the transition safety zone. The thickness of the clay layer that plays a major role in the transition safety zone is medium thickness area, and the surface sand layer at this position is difficult to break into the area; at the same time, the water-richness of the Salawusu Formation aquifer is medium water-rich area, and the water-richness of the weathered bedrock aquifer is weak water-rich area. Therefore, this area is the transition safety zone of water–sand inrush, which is mainly due to the comprehensive factors such as waterpower source, sand inrush source and water-resisting soil layer, which are not conducive to the occurrence of water disaster accidents in the mine. However, it belongs to the transitional safety zone. The mine should strengthen the observation of surface water hydrological data at the position of mining and carry out real-time mine water quality analysis and the change of bearing capacity of hydraulic support in working face.

(4) Class IV Safer zone and class V safer zone: $0.494 \leq RI < 0.276$ area, which is a safe area. The main reason is that the thickness of the clay layer at this location is large, the thickness of the bedrock is large, and the Salawusu Formation aquifer and the weathered bedrock aquifer at this location are less water rich. Therefore, two locations in the area are safer. However, mining in this area still needs to be carried out in accordance with the requirements of mining to formulate relevant water prevention and control safety.

9.4.2.2 Parameter Uncertainty Analysis

The research idea of the quantitative evaluation of the risk assessment of water–sand inrush in the mining overburden fracture is based on the theoretical system of the traditional 'upper three zones' theory for the determination of the upper limit position of the mining method and the classical 'three maps-double prediction'. The related research contents of risk assessment of water–sand inrush in mining overburden fractures during mining are studied. In this study, many parameters are involved, including the hydrological parameters of the aquifer, the formation parameters, especially the thickness parameters of the clay layer and the development height of the water-conducting fracture zone. Limited by the sample selection of the above parameters, the use of interpolation method to determine the partition may cause errors with the actual formation of the mine, which will have a certain impact on the evaluation results. However, the overall idea and method of evaluation content is correct. Therefore, to further accurately divide the water–sand inrush area, more data samples need to be further analyzed and verified.

References

1. Chen B, Zhang SC, Li YY, Li JP (2020) Experimental study on water and sand inrush of mining cracks in loose layers with different clay contents. Bull Eng Geol Environ 80(prepublish)
2. Fan GW, Zhang DS, Zhang SZ, Zhang CG (2018) Assessment and prevention of water and sand inrush associated with coal mining under a water-filled buried gully: a case study. Mine Water Environ 37(3)
3. Xu YC, Luo YQ, Li HJ, et al (2018) Water and sand inrush during mining under thick unconsolidated layers and thin bedrock in the Zhaogu No. 1 coal mine, China. Mine Water Environ 37(2)
4. Yan WT, Dai HY, Chen JJ (2018) Surface crack and sand inrush disaster induced by high-strength mining: example from the Shendong coal field, China. Geosci J 22(2)
5. Yang P, Feng WL, Study on water inrush and sand-breaking disasters in shallow coal seams in Shenfu Dongsheng mining area. Coal Sci Technol 30(S1):0065–0069
6. Yang WF, Xia XH, Zhao GR, Ji YB, Shen DY (2011) Overburden failure and the prevention of water and sand inrush during coal mining under thin bedrock. Mining Sci Technol (China) 21(5)
7. Yang WF, Xia XH, Pan BL, Gu CS, Yue JG (2016) The fuzzy comprehensive evaluation of water and sand inrush risk during underground mining. J Intell Fuzzy Syst 30(4)
8. Yang WF, Jin L, Zhang XQ (2018) Simulation test on mixed water and sand inrush disaster induced by mining under the thin bedrock. J Loss Prev Process Indust 57
9. Zhang WQ, Wang ZY, Zhu XX, Li W, Gao B, Yu HL (2020) A risk assessment of a water-sand inrush during coal mining under a loose aquifer based on a factor analysis and the fisher model. J Hydrol Eng 25(8)
10. Zuo YQ, Zhang MY (1995) Introduction to rock hydraulics. Southwest Jiaotong University Press, Chengdu

Chapter 10
Feasibility Analysis of Safe Coal Mining Under Qingcaojiegou and Hezegou Water Bodies

10.1 Study Area

Two perennial ditches, Qingcaojiegou and Hezegou, are developed in Jinjie Coal Mine (Figs. 10.1 and 10.2). The ditchesaffect the rational layout of coal mine production and are associated with the current economic benefits of coal mine enterprises. Therefore, the feasibility of coal mining under Qingcaojiegou and Hezegou should be carefully studied.

10.2 Analysis of Geological Conditions of Coal Pressing Area Under Qingcaojiegou Water Body

10.2.1 Surface Water Body Conditions

The Qingcaojie gully flow is perennial flow, according to long-term observation data, the annual average daily flow is 21,417.70 m^3/d, Heze gully flow is 9832.32 m^3/d, from the southwest of the well field into the Tuwei River.

The surface watershed of gully flow turns from Gongcaowan in the southwest of Minefield to Hujialiang and Machangliang in the southeast via Hezegou Triangle, Dayingdi Beam and Baijiamiao Beam. The northwest is Hezegou watershed, and most of the rest belong to Qingcaojiegou watershed. Qingcaojiegou watershed covers an area of 92 km^2, and 90% of the watershed is in the minefield. The sand layer water is exposed along the foot of the slope in a linear seepage manner in Cuijiagou, Yangjiagou and Baijiagou branches in the head of the ditch, forming a spring collection ditch. The discharges of Spring 4, Spring 6, Spring 7, and Spring 8 exposed in Cuijiagou are 18.64 L/s, 13.99 L/s, 5.96 L/s, and 4.29 L/s, respectively. The Yangjiagou Spring 2 discharge is 16.64 L/s, Baijiagou Spring 15 discharge is 16.27 L/s, and

Y. Zeng et al., *Roof Water Disaster in Coal Mining in Ecologically Fragile Mining Areas*, Professional Practice in Earth Sciences, https://doi.org/10.1007/978-3-031-33140-4_10

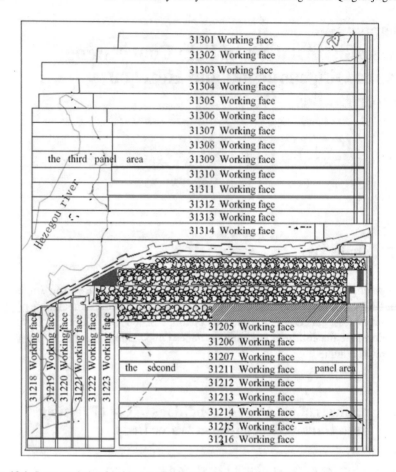

Fig. 10.1 Layout relationship between Hezegou and coal mining face

the discharges of ditch mouth of Cuijiagou, Yangjiagou and Baijiagou are 63.32 L/ s, 47.15 L/s and 28.73 L/s, respectively. The Qingcaojie gully area is an independent hydrogeological unit in the minefield, which is far away from the gully water source, with clear watershed and wide height difference, and there is no groundwater hydraulic connection.

10.2.2 Hydrogeological Conditions

The main water filling sources in the Jinjie mine field are atmospheric precipitation, surface water, loose rock phreatic water of Quaternary and weathered bedrock fissure water of Zhiluo Formation. There are two ways of mine water filling: direct and indirect. There are two main aquifers in the mine field: loose layer pore phreatic water

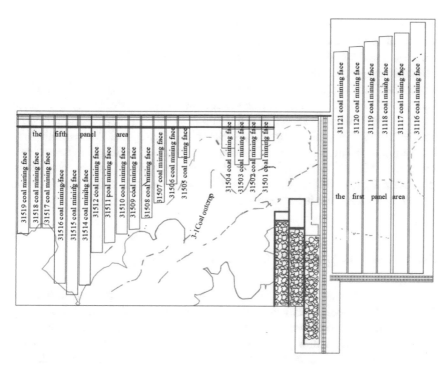

Fig. 10.2   Relationship between Qingcaojiegou and layout of coal mining face

(sand layer water) and Zhiluo Formation pore fissure confined aquifer (weathered rock water layer). The former includes valley alluvial phreatic water and Salawusu Formation phreatic water. Sand phreatic water is mainly recharged by atmospheric precipitation (infiltration coefficient is 0.1–0.6), but regional lateral recharge and condensate recharge are weak; Pore-fissure phreatic water of the Middle Jurassic Zhiluo Formation mainly receives regional lateral recharge and upper Quaternary groundwater infiltration recharge through "skylight". Sandstone (weathered bedrock) aquifer of Zhiluo Formation of Jurassic is the main direct water source, and phreatic aquifer of the Quaternary loose layer is also the main water source, which is an indirect water source. Therefore, the focus of water prevention and control work should be to prevent water inrush from weathered bedrock aquifer and Quaternary sand aquifer of Zhiluo Formation, and the possible sand collapse disaster caused by it.

10.2.3   Engineering Geological Conditions

The 3^{-1} coal seam in the area is relatively shallow, and the formation and development of caving zone and water-conducting fracture zone are usually controlled by the lithology, thickness, lithology combination and engineering geological properties of bedrock above the main roof of 3^{-1} coal seam. 3^{-1} coal seam is mainly mined in this area, and the main coal seam is stable and reliable without major faults. The inclination angle is gentle.

According to borehole data, the surface layer is the Quaternary loose layer with a thickness of 20–60 m, and the thickness of complete bedrock roof overlying 3^{-1} coal is generally less than 60 m, which is a shallow buried coal seam characterized by shallow buried depth, thin bedrock, and thick loose overburden. The roof lithology is mainly carbonaceous mudstone, followed by siltstone and fine-grained sandstone. The rock quality designation value of rock is approximately 60%, and the rock quality grade is medium, so this rock belongs to weak-semi-hard rock and the compressive strength of rock stratum is low (Fig. 10.3).

10.3 Analysis of Mining Grade of Water Body

Article 50 of "Regulations for Coal Pillar Setting and Coal Pressure Mining of Buildings, Water Bodies, Railways and Main Roadways" stipulates that when mining near water bodies, the mining influence on water bodies must be strictly controlled. According to the type, flow pattern, scale, storage conditions and allowable mining influence degree of water bodies, the water bodies affected by mining are divided into different mining grades (as summarized in Table 10.1).

According to the type, flow pattern, scale and occurrence condition of surface water body and Quaternary bottom water body in Jinjie Mine, and in comparison, with Article 50 of Regulations for Coal Pillar Setting and Coal Pressure Mining of Buildings, Water Body, Railway and Main Roadways, the mining grade of water body should be designated as Grade I. The permissible mining degree is that the water-conducting fracture zone is not allowed to spread to the water body with loose pores and weak aquifers, so it is required to establish roof waterproof safety coal pillar. In addition, considering the uneven distribution of clay water-resisting layer at the bottom of Quaternary and some blocks directly supply coal measures strata, it must keep enough waterproof coal pillars to ensure mine safety.

10.4 Height Analysis of Water-Conducting Fracture Zone in Overlying Strata

Theoretical research and production practice have proved that the water inrush accident of coal mine roof is related to the movement and failure of overlying strata and the height of water-conducting fracture zone. If the destruction of the arch connected to the upper aquifer and surface rivers, lakes and other water sources, water inrush accidents will occur. After coal seam mining, whether the overlying water body on

rock name		Lithology description	Adoption rate	core mining length	Layer thickness	Accumulation depth	Histogram 1:1000
Quaternary Holocene quicksand						39.47	39.47
Middle Jurassic Zhiluo Formation	medium grained sandstone	Purple red, purple gray, the main component is quartz, fissure development (burning)	4	0.5	1.515	0.98	
Middle Jurassic YanAn Formation	Fine grained sandstone	Purple gray, the main component is quartz, fracture development(burning)	31	0.85	2.70	53.68	
	2-2 coal seam	Macroscopic type:1.00Semi-bright type	95	0.95	1.00	54.68	
	muddy siltstone	Gray,Massive bedding	71	1.80	2.54	57.22	
	siltstone	Gray,Contains more gray matter,Massive bedding	57	0.40	0.70	57.92	
	muddy siltstone	Gray,Massive bedding	43	0.30	0.70	58.62	
	carbon mudstone	Gray,Containing plant fossil fragments	77	0.10	0.13	58.75	
	muddy siltstone	Gray,Interbeding silty mudstone	57	1.10	1.94	50.69	
	siltstone	Gray,Interbeding silty mudstone	77	0.85	0.10	51.79	
	mudstone	Gray,lumpy,interbedinbg Microwaved bedding,Locally containing plant fossils	89	2.25	2.54	54.33	
	siltstone	Gray,massive bedding,Microwave bedding in the bottom	92	2.30	2.50	56.83	
	Fine grained sandstone	Off-white,lumpy,Microwave bedding	90	1.90	2.10	58.93	
	medium grained sandstone	Gray,The main component is quartz,dark-colored mineral,massive bedding,Locally containing coal line	88	7.00	7.98	76.91	
	siltstone	Gray,Microwave bedding at upper part, massive bedding at middle part, oblique bedding at lower part	100	3.40	3.40	80.31	
	Fine grained sandstone	Gray,Massive bedding	100	0.80	0.80	81.11	
	muddy siltstone	Gray,Microwave bedding	99	2.55	2.57	83.68	
	mudstone	Dark gray, broken core	100	1.00	1.00	84.68	
	3-1 coal seam	Macroscopic type:1.60(Semi-bright type),0.4(Semi-darktype),0.46(Semi-bright type)	94	2.30	2.46	87.14	
	siltstone	Gray,Massive bedding clip microwave bedding,Top contains a large amount of carbonized plant debris	83	1.25	1.50	88.64	
	Fine grained sandstone	Gray,massive bedding	100	1.10	1.10	89.74	
	siltstone	Gray,massive bedding,The upper and lower parts contain carbonized plant fragments	96	1.00	1.04	90.78	

Fig. 10.3 Borehole histogram

Table 10.1 Mining grade and permissible mining degree of water body in mining area

Coal seam position	Water mining dynamic grade	Water body type	Permissible mining degree	Types of safe coal and rock pillars
Water body Under	I	1. All kinds of surface water bodies directly above bedrock or under the bottom interface without stable cohesive soil aquifer 2. Loose pore water body with strong and medium aquifer directly above bedrock or below the bottom interface without stable cohesive soil aquifer 3. Strong and medium aquifers of bedrock without stable mudstone aquifuge under bottom interface 4. Various surface water bodies and loose aquifer water bodies above steep coal seams 5. Water bodies required to be protected as important water sources and tourist destinations	The water-conducting fracture zone is not allowed to spread to the water body	Roof waterpro-ofing Safety coal pillar
Under	II	1. Under the bottom interface, there are loose layers with multi-layer structure, thick thickness and weak water content, or medium and small water bodies on the surface with strong aquifer in the upper part and weak aquifer in the lower part 2. Under the bottom interface, there are stable thick cohesive soil water-resisting layer or loose weak aquifer in the middle and upper pore of loose layer, and in the middle aquifer water body 3. Loose layer and bedrock weak aquifer water body with drainage conditions	The water-conducting fracture zone is allowed to affect the weak aquifer water body with loose pores, but the caving zone is not allowed to affect the water body	Roof sand control Safety coal pillar

(continued)

Table 10.1 (continued)

Coal seam position	Water mining dynamic grade	Water body type	Permissible mining degree	Types of safe coal and rock pillars
	III	1. Under the bottom interface is the loose layer of stable thick clay aquifuge, its middle and upper parts are water body of pore weak aquifer 2. Loose layer or bedrock water body that has been or is close to being drained	The water-conducting fracture zone is allowed to enter the weak aquifer with loose pores, and the caving zone is allowed to spread to the weak aquifer	Roof collapse prevention Safety coal pillar

the roof can pose a threat to mining depends on the development height of the fracture zone in the overlying strata. If the development height of the fracture zone in the overlying strata of the roof does not touch the water body after mining, the mining work is safe [2, 4, 5, 7].

When mining under a water body, the main consideration is whether the cracks in overlying strata caused by mining are interconnected and whether the interconnected cracks spread to water body. Once the overburden damage affects the water body, even if it only touches the edge of the water body, it will cause all the water in the water body to flow into the downhole. In this sense, the water body needs to be protected. Therefore, it is very important to study the movement and failure law of overburden strata, especially the development height and distribution form of water-conducting fracture zone. Relatively speaking, the study of earth's surface deformation is relegated to a secondary position at this time. Because in many cases, although the surface has great movement and deformation and even cracks, as long as these cracks close on their own at a certain depth and do not constitute a channel for water gushing, there will be no flooding accident [1, 8].

10.4.1 According to the "Regulations for Coal Pillar Setting and Coal Pressure Mining of Buildings, Water Bodies, Railways and Main Roadways"

According to the borehole data, the roof lithology of 3^{-1} coal seam mainly includes fine sandstone, medium sandstone, mudstone and siltstone, which belong to medium hard rock stratum. According to the Rules for Coal Pillar Setting and Coal Pressing Mining of Buildings, Water Bodies, Railways and Main Roadways promulgated in 2000 (hereinafter referred to as the Rules for Coal Pressing under Three Conditions), the height of water-conducting fracture zone after mining in 3^{-1} coal seam is predicted as follows according to medium hard rock stratum:

(1) The height of caving zone is calculated according to the following equation

$$H_m = \frac{100 \sum M}{4.7 \sum M + 19} \pm 2.2 \tag{10.1}$$

where:

H_m caving zone height, m;
$\sum M$ cumulative mining height, m.

The average thickness of 3^{-1} coal seam is approximately 3.2 m, and the height of spanning falling zone is:

$$H_m = 100 \times 3.2/(3.2 \times 4.7 + 19) \pm 2.2 = 9.4 - 11.6\,\text{m} \tag{10.2}$$

(2) The height of water-conducting fracture zone is calculated according to the following equation

$$H_{li} = \frac{100 \sum M}{1.6 \sum M + 3.6} \pm 5.6 \qquad (10.3)$$

where:

H_{li} the height of fracture zone, m;
$\sum M$ cumulative mining height, m.

The development height of roof water-conducting fracture zone after mining in 3^{-1} coal seam is as follows:

$$H_{li} = \frac{100 \sum M}{1.6 \sum M + 3.6} \pm 5.6 = \frac{100 \times 3.2}{1.6 \times 3.2 + 3.6} \pm 5.6 = 31.1 - 42.3 \, m \quad (10.4)$$

10.4.2 Hydrogeological and Engineering Geological Exploration Code for Mining Area

According to the empirical equation of maximum height of water-conducting fracture zone in caving zone provided in Appendix F of Code for Hydrogeological Engineering Geological Exploration in Mining Area (GB 12719-91), the height of water-conducting fracture zone is calculated according to the following equation:

$$H_{Li} = \frac{100M}{3.3n + 3.8} + 5.1 \qquad (10.5)$$

where:

H_{li} height of water-conducting fracture zone (m);
M cumulative mining thickness (m);
n the number of layers of coal, take 1 here.

Then, the development height of roof water-conducting fracture zone after mining in 3^{-1} coal seam is as follows:

$$H_{li} = \frac{100M}{3.3n + 3.8} + 5.1 = \frac{100 \times 3.2}{3.3 + 3.8} + 5.1 = 50.2 \, m \qquad (10.6)$$

10.5   Basis for Setting up Safe Coal and Rock Pillars

10.5.1   Basis for Setting up Safe Coal and Rock Pillars

According to Article 50 of "Three Coal Mining Regulations", when mining near a water body, the mining influence on water body must be strictly controlled. According to the type of water body, flow pattern, scale, storage condition and allowable mining influence degree, the water body affected by mining is divided into different mining grades, as shown in Table 10.1 for details.

10.5.2   Design Calculation of Safe Coal and Rock Pillar Retention

According to the hydrological engineering geological conditions of overlying strata, Quaternary loose strata and surface water body in Jinjie Coal Mine, the water mining grade of this working face is Grade I, and the safe coal and rock pillars are reserved according to the waterproof coal and rock pillars.

According to the Regulations of Coal Pillar Setting and Coal Pressing Mining for Buildings, Water Bodies, Railways and Main Roadways and the Regulations of Mine Hydrogeology, the calculation equation of safe coal and rock pillar is determined as follows:

$$H_{sh} \geq H_{li} + H_b \qquad (10.7)$$

where:

H_{sh} Waterproof safety coal and rock pillar, m;
H_{li} Maximum height of water-conducting fracture zone, m;
H_b Thickness of protective layer, m.

(1) Determination of the thickness of protective layer

According to the relevant provisions of the thickness of the protective layer of the waterproof safety coal and rock pillar in the 'three regulations' (Table 10.2), combined with the hydrological, geological and mining conditions of the Jinjie Coal Mine, the lower part of the loose layer is the burned metamorphic rock in the burning area, and the fracture is developed. It has good water permeability and cannot be used as a component of waterproof coal and rock pillar. Therefore, the thickness of the protective layer of the waterproof safety coal rock pillar is calculated as follows.

$$H_b = 7A \quad A = \sum M/n \qquad (10.8)$$

Table 10.2 Thickness of protective layer of waterproof and safety coal and rock pillar (unit: m)

Thickness of protective layer In Overburden lithology	Thickness of cohesive soil layer or weak aquifer at the bottom of loose layer greater than the cumulative mining thickness	Thickness of cohesive soil layer or weak aquifer at the bottom of loose layer less than the cumulative mining thickness	Total thickness of loose layer greater than the cumulative mining thickness	Absence of clay layer at the bottom of loose layer
Stiff	4A	5A	6A	7A
Medium hard	3A	4A	5A	6A
Weakness	2A	3A	4A	5A
Extremely weak	2A	2A	3A	4A

Note A = M/n; ΣM—Cumulative mining thickness; ΣN—layer number

where:

$\sum M$ mining height (m);

n number of layers.

Substituting corresponding parameters, the thickness of protective layer is as follows:

$$3^{-1} \text{ coal: } H_b = 3.2 \times 7/1 = 22.4 \, \text{m} \tag{10.9}$$

(2) Thickness of safe waterproof coal pillar

The thickness of safe waterproof coal pillar is generally:

$$H_{sh} \geq H_{li} + H_b \tag{10.10}$$

H_{li} According to the above calculation, considering the medium hard rock stratum, the maximum height of water-conducting fracture zone is 50.2 m when the mining height of 3^{-1} coal seam is 3.2 m.

H_b According to the hydrogeological characteristics of Jinjie Coal Mine, if the value is 7 times the mining height, the protective layer thickness of 22.4 m should be reserved.

According to the "three-under" mining regulations, the height of safe waterproof coal pillar to prevent Quaternary loose layer water and surface water is calculated as follows:

$$3^{-1} \text{ coal seam} : H_{sh} = 50.2 + 22.4 = 72. \, 6 \, \text{m} \tag{10.11}$$

10.6   Safe Mining Scheme Under Hezegou and Qingcaojiegou Water Bodies

Water–sand inrush is a potential safety hazard of coal mining under river in shallow depth, thin bedrock, and water-rich area. Therefore, the main problems of coal mining under river are river embankment protection and safe mining under river. Only on the basis of full demonstration, adopting feasible coal mining technology and mining methods can we achieve the goal of not only mining more coal, but also protecting the surface riverbank and safety better [3, 6, 9].

10.6.1   Main Coal Mining Methods Under River Pressure

Many mining areas in China have carried out production practice and scientific experiments of coal mining under water bodies, successfully mined a large number of coal seams under various types of water bodies, gained some regular understanding, and accumulated experience for the development of coal industry [4]. There are mainly the following coal mining methods under the river:

(1)   Entire roof subsidence

Considering that the bedrock in this area is thin and belongs to thin bedrock, the development of water-conducting fracture zone in caving mining is highly affected to bedrock aquifer, burnt rock aquifer and loose layer aquifer, so it is obviously inappropriate to adopt all caving mining method for coal mining under rivers.

(2)   Separated layer grouting method

Separation layer grouting mining technology is a new mining technology under construction, which was put forward from the study of internal movement law of rock strata in China since the end of 1980s. According to the theory of "arch-beam balance", it seeks the separation position of overlying strata, and fills some media in the separation layer, such as fly ash, sand, and gravel, to reduce the movement of overlying strata and control the surface subsidence. From the results of engineering practice, under some overburden conditions, the separated layer grouting method can obviously reduce the surface subsidence and effectively delay the surface subsidence speed. Through the analysis of several examples in grouting mining areas, the surface subsidence is reduced by approximately 60–70%. Because of the shallow mining depth of the first coal seam in this area, the separation position of overlying strata will not be obvious and even don't produce separation, so the separation grouting method should not be implemented in this mine.

(3)   Set up safe coal and rock pillars

Appropriate coal and rock pillars are reserved between the upper stoping limit and the water body to isolate the water source, so that the water body will not be affected after coal mining. The advantages are as follows: ① Generally, the extra displacement of the mine is not increased; ② There is no need to change the mining method. The disadvantage is that the loss of coal resources is large, so it is necessary to change the layout of mining projects to mine the coal resources separated by rivers on both sides.

(4) River diversion

River diversion leads the river passing through the mining area to the outside of the mining area, which is the most effective way to change the headwater supply, and the effect is remarkable under certain conditions. However, the problem of river diversion involves a wide range, with large engineering quantity and large investment. Therefore, it is necessary to make comprehensive planning and arrangement according to the topography of the mining area and the development needs of industrial and agricultural production and construction.

(5) Gob filling method

The mined-out area filling mining is to fill the mined-out area behind the working face to control the overlying strata movement and surface subsidence. Traditional filling mining methods are hydraulic filling, wind filling, mechanical filling, gangue self-sliding filling, etc. With the development of mine filling technology, filling materials and technology have been developed in the process of continuous transformation and innovation. The methods of water–sand filling, low-concentration cemented filling, high-concentration cemented filling, paste pumping filling and high-water quick-setting material consolidation filling have been experienced. The paste filling material is a material aggregate without water separation after filling and is generally formed when the concentration is 76%-85%. Coal mine gangue, fly ash, industrial slag, river sand, etc. are processed into non-dehydrated toothpaste paste. Under the action of filling pump, it is transported to the underground through pipeline, and the goaf is filled in time to form a cemented filling body mainly composed of paste to support the overlying strata, effectively control the surface subsidence and improve the recovery rate of resources. Paste filling is one of the key technologies of green mining technology system in coal mines.

(6) Room and pillar mining

Room-pillar mining is to excavate coal seams from mine room or roadway and keep some residual ore body as coal pillar between mine room or roadway to control the local working performance of direct roof rock and the overall reaction of surrounding rock. It not only has the advantages of small amount of mine development preparation, fast coal production, less equipment investment and flexible relocation of working face, but also has small roadway pressure, low damage degree of overlying rock mass and small surface subsidence. Especially when using room-pillar continuous mining technology, shearer, loader, shuttle car and bolt machine work in harmony, which has high production efficiency. Room and pillar mining is widely used in the United States, Australia, Canada, India and South Africa. For example, in the

United States, more than half of the coal mined by underground workers is mined by room and pillar. In China, room-pillar mining technology is seldom used. To recover the pressed coal under the tower of high voltage transmission line, the continuous mining equipment (LN800 shearer, IOSC-40B shuttle car, TD1-4S bolt machine) has been successfully used to mine 2.2 m thick near horizontal coal seam in Xiqu Coal Mine of Xishan Mining Bureau. Xiaohengshan Mine of Jixi Mining Bureau and Jiangjiawan Mine of Datong Mining Bureau have also adopted room-pillar mining. Since the 1990s, Daliuta Mine of Shenhua Group Company has achieved a monthly output of 90,000 tons. Nantun Mine of Yankuang Group Company adopts room-pillar mining of some corner coal. According to the research in the United States, under the condition that the coal pillar can ensure stability, when the recovery rate is lower than 50%, the surface subsidence is very small. When the recovery rate is greater than 50%, the surface subsidence law of room-pillar mining is similar to that of strip mining. Under the condition of mining houses and columns immediately, the law of surface subsidence is similar to that of all mining. The biggest potential danger of room-pillar mining is that the coal pillar does not have long-term stability.

10.6.2   Filling Mining Scheme of Longwall Working Face

10.6.2.1   Filling Mining Technology

According to the current mining practice of Jinjie Coal Mine, if the longwall mining method is adopted in the mining under the river of 3^{-1} coal mine to fill the goaf, the paste with relatively low price can be selected for filling, but the filling mining process is complicated, and the equipment investment is more. Coal mine paste filling process is divided into four parts: material preparation, batching pulping, pipeline pumping and working face filling (Fig. 10.4).

(1)   Material preparation

Coal mine paste filling materials are mainly wind-blown sand, coal gangue, fly ash, industrial slag, municipal solid waste, inferior sand, and other solid wastes. The materials with smaller granularity can enter the silo for standby after simple treatment (sifting out blocks and linear sundries larger than 20 mm, etc.), and when large materials such as coal gangue are used, they need to be properly crushed. To reduce the amount of binder, mechanical grinding and alkaline excitation can be used to stimulate the chemical activity of fly ash.

The paste filling cementing material is transported to the cementing bin for standby by wind, and the filling water is pumped from the pool or well to the mixer by water pump.

(2)   Batching pulping

The amount of cementing material in coal mine paste filling material is very small. According to the concept of general concrete, it is a kind of "extremely poor" concrete.

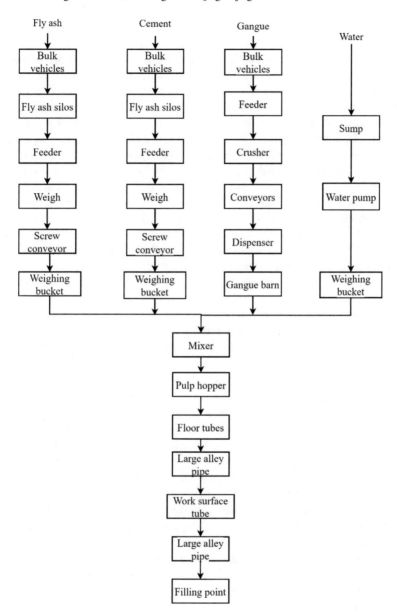

Fig. 10.4 Process flow of paste filling system

The filling paste must be accurately prepared according to the designed concentration and the ratio of various materials, and fully mixed evenly, to ensure the flowability and solidification performance of the filling material, and the filling of underground working face can achieve the expected control goal of surface subsidence.

According to the material proportioning experiment, to make the material flow performance stable, the change range of filling slurry weight concentration must be controlled within 0.5%, which requires the weighing error of materials to be less than 2%. In order to ensure the quality of filling slurry, the following technical measures have been taken: ① batch batching, weighing and measuring each material under static conditions; ② Adopt advanced and high-precision load cell with good stability; ③ The batching system is controlled by industrial microcomputer and PLC programmable controller, and the batching process is programmed and automated. In addition, due to the limitation of filling cost, it is impossible to remove mud and dehydrate filling materials. The mud content and water content of filling materials taken at different places and at different times will inevitably change and fluctuate greatly, which will affect the weight concentration of filling slurry.

In order to ensure the stability of fluidity and strength of paste filling materials, it is necessary to measure the quality changes of river sand and fly ash quickly and accurately and adjust the ratio in time. According to the characteristics of coal mine paste filling materials, the periodic concrete mixer is selected for forced mixing, and the feeding, mixing and discharging are carried out cyclically, so it is easy to control the ratio and mixing quality.

(3) Pipeline pumping

Paste filling slurry is transported by concrete pump pressurized pipeline. The slurry stirred by the mixer first enters the slurry buffer hopper, and then feeds to the feed hopper of the filling pump by the self-weight of the slurry. The paste slurry pressurized by the filling pump passes through the filling pipe, goes down the well through the filling borehole near the filling station, and then is transported to the filling working face along the filling pipe arranged in the roadway. The selection of filling pump and filling pipe should be determined according to filling capacity, filling pipeline length, paste slurry characteristics and flow rate. One of the important characteristics of paste filling slurry in pipeline transportation is that there is no critical flow rate, and it can be transported for a long distance under the condition of very low flow rate. If the flow rate is too high, the hydraulic gradient that the slurry needs to overcome is large, the wear speed of the pipeline is also large, and the requirement for pumping pressure is also high. If the flow rate is too small, the filling capacity cannot meet the production needs. Generally, the design flow rate of paste filling system is 0.7–1.0 m/ s.

10.6.2.2   Feasibility Analysis of Backfill Mining

(1) Equivalent mining height theory

As the mining area belongs to shallow coal seam, The water-conducting fissure formed after coal seam mining can easily communicate with the overlying aquifer, resulting in sand and water inrush in mine. Therefore, the full filling method will be adopted for mining. With the filling of backfill, the subsidence of roof and the compaction process of backfill, it is equivalent to reducing the mining thickness of coal seam and the falling space of goaf roof, thus reducing the height of water-conducting cracks. To study the change characteristics of overlying strata movement in solid filling mining, relevant scholars have established the equivalent mining height theory to analyze the mine pressure and overlying strata movement in solid filling mining, and its expression is as follows:

$$Hz = hz + hq + k(H - hz) \qquad (10.12)$$

Hz: the equivalent mining height; hz: Under-knot top quantity; hq: the advance subsidence of the roof; H: For mining height; Is the compressibility of filling body; k: The equivalent mining height model is shown in Fig. 10.5.

The average thickness of pressed coal seam under the surface river in Jinjie Coal Mine is 3 m. Practice has proved that the compressibility of paste after filling is generally 1–2%. For the sake of conservation, we take the compressibility $\varepsilon = 2\%$, the advance subsidence of roof hq is 240 mm according to experience. The amount of underfilling roof hz is generally 50–100 mm. For the sake of conservation, it takes 100 mm. Using the equation of equivalent mining height of paste filling, the equivalent mining height of paste filling mining Hz is 0.4 m. After compaction, the permeability of paste is low, which also provides a good anti-seepage foundation for

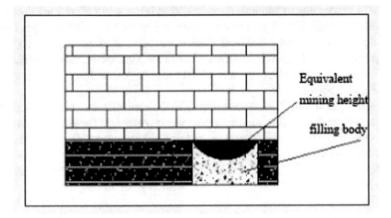

Fig. 10.5 Equivalent mining height model

the later mining of 4^{-2} coal. According to the theory of equivalent mining height, the development height of water-conducting fracture zone is 3.8–15 m.

(2) Numerical simulation software.

Based on the regional geological environment of mining 3^{-1} coal seam under the surface river in Jinjie Coal Mine, this book uses FLAC3D software and Mohr–Coulomb criterion. Simulate the development height of water-conducting fractures under equivalent mining height. In order to save the production cost and reduce the mining difficulty, on the basis of the original working face, filling mining is carried out in the section where the overlying strata are weak, and the mining difficulty is great.

FLAC3D (Fast Lagrangian Analysis of Continua) is a simulation calculation software with a three-dimensional finite differential program developed by Itasca Company of USA. It can be used to simulate the three-dimensional structural stress characteristics and plastic flow analysis of soil, rock, and other materials. By adjusting the polyhedral elements in the three-dimensional mesh, the actual structure is fitted. The element material can adopt linear or nonlinear constitutive model. Under the action of external force, when the material yields and flows, the grid can deform and move accordingly. FLAC3D adopts explicit Lagrange algorithm and hybrid-discrete partition technology, which can simulate plastic failure and flow of materials very accurately. Because there is no need to form a stiffness matrix, a large range of three-dimensional problems can be solved based on a small memory space.

(3) Filling mining of 31,310–31,314 working face

Filling mining is carried out in the area under the river where 31,310–31,314 working faces are located. Based on the borehole data of J1210, combined with regional analysis, the development height of water-conducting fractures in overlying bedrock after filling mining in 31,310–31,314 working faces is simulated and calculated. The strike length of the calculation model is 280 m, i.e., x = 280, and the width is 200 m, i.e., y = 200 m. According to borehole data, 143.32 m is taken in Z direction. According to the comprehensive histogram of coal seam, the calculation model adopts Mohr–Coulomb model. The computational mechanics model is shown in Fig. 10.6, and the physical and mechanical parameters of rocks used in the calculation are obtained according to the rock mechanics test results table of Jinjie Coal Mine, and the specific parameters are shown in Table 10.3.

The calculation model is divided into 120,000 elements and 128,371 nodes, in which the excavation part is simulated by zero-empty element. Horizontal constraints are imposed on the perimeter boundary and vertical constraints are imposed on the bottom boundary.

To better compare the movement and zoning characteristics of overlying strata of mining in paste filling mining under the same mining conditions, the caving mining simulation test was carried out. Excavation is carried out along the floor of coal seam, and the average mining height is 2.8 m. In the mining model of caving method, the circular mining footage is 10 m, and the thickness of complete bedrock roof is 40 m.

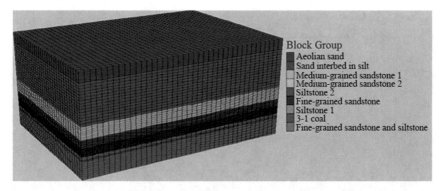

Fig. 10.6 Numerical calculation model

Table 10.3 Physical and mechanical parameters of J1210 coal strata

Name of coal strata	Bulk modulus (GPa)	Shear modulus (GPa)	Cohesion (MPa)	Angle of internal friction/°	Tensile strength (MPa)	Density (Kg/m³)
Aeolian sand	0.01	0.008	0.004	17	0	2050
Sandstone	0.02	0.18	0.003	38	0	1950
Medium-grained sandstone	0.33	0.38	0.49	43.2	0.18	2290
Siltstone	3.267	2.548	4.68	37.85	1.82	2470
Fine-grained sandstone	2.514	2.28	4.31	36.6	1.62	2440
Siltstone	3.267	2.548	3.46	37	1.36	2460
3^{-1} coal	2.8	1.701	2.05	36.5	0.737	1300
Siltstone	3.267	2.548	4.68	37.85	1.82	2470

According to experience, when the advancing distance of mining working face is the same as the dip length of working face, the failure of overlying strata on working face reaches the maximum, as shown in Figs. 10.7 and 10.8, and the model simulates caving mining. When the working face is pushed to 200 m, the destruction area of bedrock develops to the maximum, the destruction area of the working face basically penetrates the surface. The water-conducting fracture zone communicates with the overlying aquifer, which easily causes sand and water inrush in the working face and causes serious potential safety hazards to the mining of the working face. The surface settlement above the working face reaches 0.8–1 m.

Filling with full filling mining method: According to the principle of equivalent mining height, the average equivalent mining height is 0.4 m. In the filling mining model, the circulating footage is 2 m and the filling step is 2 m. According to experience, when the mining working face advances to the same distance as the inclination of the working face, the failure of the overlying strata on the working face reaches the

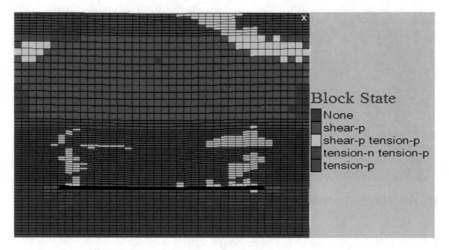

Fig. 10.7 Distribution of failure area after caving mining

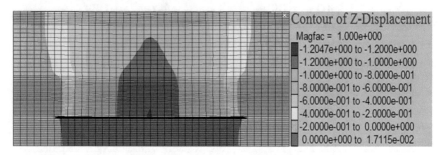

Fig. 10.8 Distribution of vertical deformation in caving mining

maximum. As shown in Fig. 10.9 to Fig. 10.10, the shape of the vertical deformation area is basically saddle-shaped. Under the influence of mining, plastic failure occurs in the loose layer of the surface, and the maximum settlement of the surface after filling is 160–170 mm. According to J1210 borehole drilling, the thickness of overlying complete bedrock is 40 m, and the development height of water-conducting fracture is 12 m, which does not penetrate the complete bedrock of roof, and will not cause the risk of water–sand inrush.

(4)   Filling mining of 31,304–31,309 working face

Filling mining is carried out in the area under the river where 31,304–31,309 working faces are located. According to J1212 borehole data and regional analysis, the development height of water-conducting fractures in overlying bedrock after filling mining in 31,304–31,309 working faces is simulated and calculated. The strike length of the model is 280 m, i.e., x = 280, and the width is 200 m, i.e., y = 200. According to the exposure of J1212 borehole, the z direction is 121.92 m, and the Mohr–Coulomb

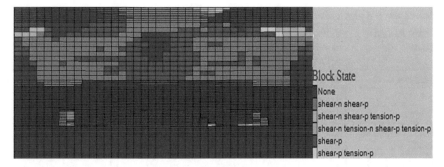

Fig. 10.9 Distribution of failure area after filling mining

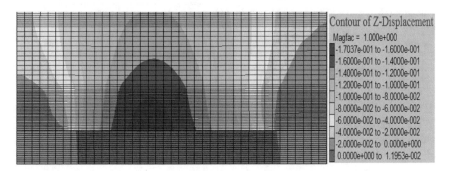

Fig. 10.10 Distribution of vertical deformation in filling mining

model is adopted. The model is shown in Fig. 10.11, and the physical and mechanical parameters of rocks are shown in Table 10.4.

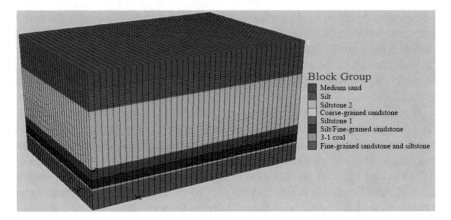

Fig. 10.11 Numerical simulation model mechanics diagram

Table 10.4 Physical and mechanical parameters of J1212 coal strata

Name of coal strata	Bulk modulus (GPa)	Shear modulus (GPa)	Cohesion (MPa)	Angle of internal friction/°	Tensile strength (MPa)	Density (Kg/m³)
Medium sand	0.028	0.025	0.002	17	0	2050
Silt	0.019	0.017	0.004	20	0	1950
Siltstone	0.29	0.21	0.49	37	0.18	2390
Medium-grained sandstone	2.683	2.18	4.37	34.9	1.83	2290
Coarse-grained sandstone	0.886	0.662	4.92	30.95	1.45	2390
Siltstone	3.267	2.548	4.31	36.6	1.62	2440
Fine-grained sandstone	2.514	2.28	3.46	37	1.36	2460
Siltstone	3.267	2.548	4.31	36.6	1.62	2440
3^{-1} coal	2.8	1.701	2.05	36.5	0.737	1300
Siltstone	3.267	2.548	4.31	36.6	1.62	2440

The calculation model is divided into 154,800 elements and 165,230 nodes, in which the excavation part is simulated by zero-empty element. Horizontal constraints are imposed on the perimeter boundary and vertical constraints are imposed on the bottom boundary.

Excavation is carried out along the floor of coal seam, with an average mining height of 2.9 m. In the caving mining model, the circular mining footage is 10 m, and the thickness of complete bedrock roof is 57 m. According to experience, when the advancing distance of mining working face is the same as the dip length of working face, the failure of overlying strata on working face reaches the maximum, as shown in Fig. 10.12 to Fig. 10.13, and the model simulates caving mining. When the working face is pushed to 200 m, the bedrock failure area develops to the maximum, the failure area of the working face basically penetrates the surface, and the water-conducting fracture zone communicates with the overlying aquifer, which easily causes sand and water inrush in the working face, causing serious potential safety hazards to the mining of the working face, and the surface settlement above the working face reaches 1–1.2 m.

The filling method is fully infill mined. According to the principle of equivalent mining height, the average equivalent mining height is 0.4 m. In the filling mining model, the circulating footage is 2 m and the filling step is 2 m. According to experience, when the mining working face advances to the same distance as the inclination of the working face, the failure of the overlying strata on the working face reaches the maximum. As shown in Fig. 10.14 to Fig. 10.15, the shape of the vertical deformation area is basically saddle-shaped. Under the influence of mining, plastic failure occurs in the loose layer of the ground surface, and the maximum settlement of the ground surface after filling is 150 mm. According to J1212 drilling, the thickness of

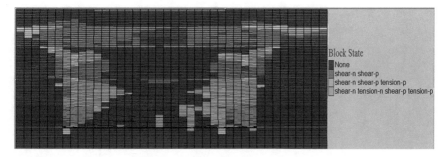

Fig. 10.12 Distribution of failure area after caving mining

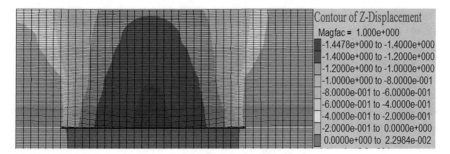

Fig. 10.13 Distribution of vertical deformation in caving mining

overlying complete bedrock is 57 m, and the development height of water-conducting fracture is 17.4 m, which does not penetrate the complete bedrock of roof, and will not cause the risk of water–sand inrush.

(5) Filling mining of 31,117–31,121 working face

Filling mining is carried out in the area under the river where 31,117–31,121 working faces are located. Figure 10.16 is the stratum profile of J606–J605 boreholes in this mining area, and the thickness of overlying complete bedrock is 28.18–30.83 m.

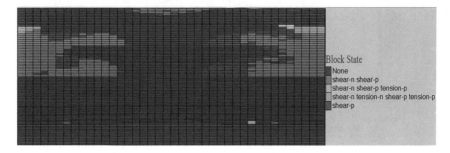

Fig. 10.14 Distribution of failure area after filling mining

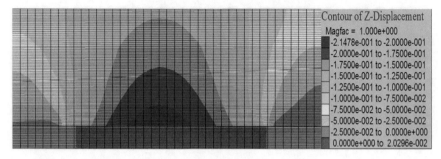

Fig. 10.15 Vertical deformation distribution diagram of filling mining

Combined with regional analysis, the movement law of overlying bedrock after mining and filling in 31,121 working face is simulated. Because the hydrogeological condition of 31,117—31,120 working face is similar to that of 31,121 working face, the calculation is based on the model of 31,121 working face. According to J605 borehole data, combined with regional analysis, the 31,117–31,121 working face is simulated and filled. The strike length of the calculation model is 280 m, i.e., x = 280, and the width is 200 m, i.e., y = 200. According to the borehole exposure, the z direction is 84.66 m. The calculation model adopts Mohr–Coulomb model, which is shown in Fig. 10.17. The physical and mechanical parameters of the rocks used in the calculation are obtained according to the rock mechanics test results table of Jinjie Coal Mine. See Table 10.5 for specific parameters.

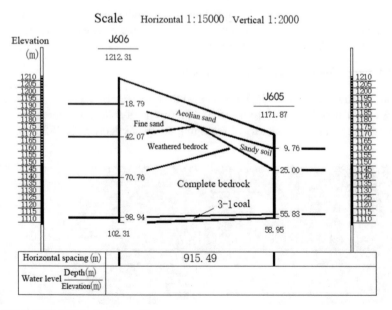

Fig. 10.16 J606-J605 formation profile

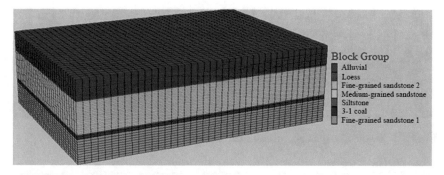

Fig. 10.17   Numerical calculation model

Table 10.5   Physical and mechanical parameters of J605 coal strata

Name of coal strata	Bulk modulus (GPa)	Shear modulus (GPa)	Cohesion (MPa)	Angle of internal friction/°	Tensile strength (MPa)	Density (Kg/m³)
Alluvial deposit	0.017	0.015	0.05	17	0	1600
Loess	0.028	0.03	0.2	26	0	1800
Fine-grained sandstone	0.29	0.21	0.49	37	0.18	2390
Medium-grained sandstone	2.351	1.755	1.2	43	0.891	2280
Siltstone	2.333	1.971	2.6	41.5	1.6	2450
3^{-1} coal	2.8	1.701	2.05	36.5	0.737	1300
Fine-grained sandstone	2.409	2.186	2.8	40	1.64	2280

The calculation model is divided into 102,000 elements and 109,306 nodes, in which the excavation part is simulated by zero-empty element. Horizontal constraints are imposed on the perimeter boundary and vertical constraints are imposed on the bottom boundary.

Excavation is carried out along the floor of coal seam, with an average mining height of 3.2 m. In the caving mining model, the circular mining footage is 10 m, and the thickness of competent bedrock roof is 30.83 m. According to experience, when the advancing distance of mining working face is the same as the dip length of working face, the failure of overlying strata on working face reaches the maximum, as shown in Fig. 10.18 to Fig. 10.19, and the model simulates caving mining. When the working face is pushed to 200 m, the bedrock failure area develops to the maximum, the failure area of the working face basically penetrates the surface, and the water-conducting fracture zone communicates with the overlying aquifer, which easily causes sand and water inrush in the working face and causes serious potential safety

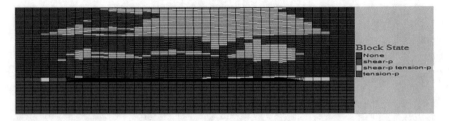

Fig. 10.18 Distribution of failure area after caving mining

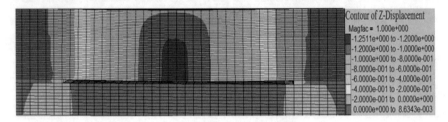

Fig. 10.19 Distribution of vertical deformation in caving mining

hazards to the mining of the working face. The surface settlement above the working face reaches 0.8–1 m.

The filling method is fully infill mined. According to the principle of equivalent mining height, the average equivalent mining height is 0.4 m. In the filling mining model, the circulating footage is 2 m and the filling step is 2 m. According to experience, when the mining working face advances to the same distance as the inclination of the working face, the failure of the overlying strata on the working face reaches the maximum. As shown in Fig. 10.20 to Fig. 10.21, the shape of the vertical deformation area is basically saddle-shaped. Under the influence of mining, plastic failure occurs in the loose layer of the surface, and the maximum settlement of the surface after filling is 200 mm. According to J605 drilling, the thickness of overlying complete bedrock is 28 m, and the height of water-conducting fracture is 10.31 m, which does not penetrate the complete bedrock of roof, and will not cause the risk of water–sand inrush.

(6)   Filling mining of 31,218–31,220 working faces

Filling mining is carried out in the area under the river where 31,218–31,220 working faces are located. According to the stratum profile line of J1307–30 borehole (Fig. 10.22), the thickness of overlying complete bedrock is 30.38–31.01 m. According to the drilling data of J1307, combined with regional analysis, the 31,218 working face is simulated and filled. The strike length of the calculation model is 280 m, i.e., x = 280, the dip length of the working face is 200 m, i.e., y = 200, and the Z direction is 124 m according to the borehole exposure. The calculation model adopts Mohr–Coulomb model. The calculation model is shown in Fig. 10.23. The

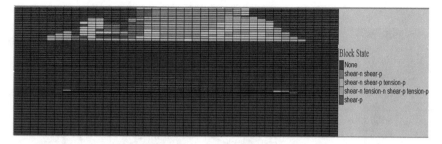

Fig. 10.20   Distribution of failure area after filling mining

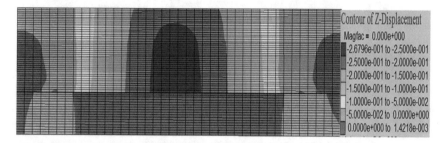

Fig. 10.21   Distribution of vertical deformation in filling mining

calculation model is divided into 116,400 elements and 124,558 nodes, in which the excavation part is simulated by zero-space element. Horizontal constraints are imposed on the perimeter boundary and vertical constraints are imposed on the bottom boundary. The physical and mechanical parameters of rocks used in calculation are obtained according to the table of rock mechanics test results of Jinjie Coal Mine, and the specific parameters are shown in Table 10.6.

Excavation is carried out along the floor of coal seam, and the mining height is 2.8 m. In the mining model of the carving method, the circular mining footage is 10 m, and the thickness of complete bedrock roof is 30 m. According to experience, when the advancing distance of mining working face is the same as the dip length of working face, the failure of overlying strata on working face reaches the maximum, as shown in Fig. 10.24 to Fig. 10.25, and the model simulates caving mining. When the working face is pushed to 200 m, the destruction area of bedrock develops to the maximum, the destruction area of the working face basically penetrates the surface, and the water-conducting fracture zone communicates with the overlying aquifer, which easily causes sand and water inrush in the working face and causes serious potential safety hazards to the mining of the working face. The surface settlement above the working face reaches 1.2–1.4 m.

The filling method is fully infill mined. According to the principle of equivalent mining height, the average equivalent mining height is 0.4 m. In the filling mining model, the circulating footage is 2 m and the filling step is 2 m. According to experience, when the mining working face advances to the same distance as the inclination

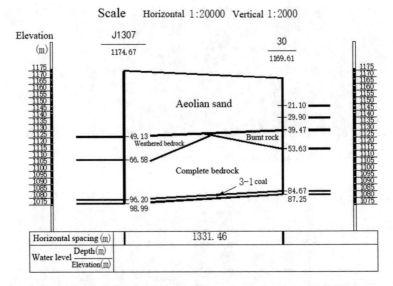

Fig. 10.22 Stratigraphic section of J1307–30

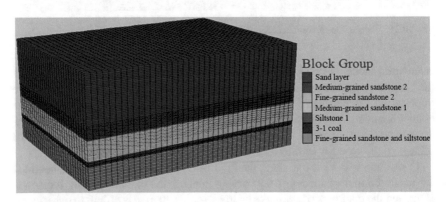

Fig. 10.23 Numerical computational mechanics model

of the working face, the failure of the overlying strata on the working face reaches the maximum. As shown in Figs. 10.26 and 10.27, the shape of the vertical deformation area is basically saddle-shaped. Under the influence of mining, plastic failure occurs in the loose layer of the surface, and the maximum settlement of the surface after filling is 270 mm. According to J1307 borehole drilling, the thickness of overlying complete bedrock is 30.38 m, and the development height of water-conducting fracture is 10.58 m, which does not penetrate the complete bedrock of roof, and will not cause the risk of water–sand inrush.

Table 10.6 J1307 physical and mechanical parameters of coal strata

Name of coal strata	Bulk modulus (GPa)	Shear modulus (GPa)	Cohesion (MPa)	Angle of internal friction/°	Tensile strength (MPa)	Density (kg/m³)
Sand layer	0.028	0.025	0.002	17	0	2050
Sandy mudstone	0.29	0.21	0.49	37	0.18	2600
Siltstone	1.739	1.524	2.6	41.5	1.6	2450
Medium-grained sandstone	3.048	2.477	3.4	41	2.12	2530
Siltstone	3.5	3.175	3.5	40.5	1.45	2400
3^{-1} coal	2.8	1.701	2.05	36.5	0.737	1300
Fine-grained sandstone	2.514	2.28	4.68	37.85	1.82	2470

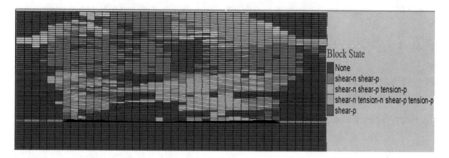

Fig. 10.24 Distribution of failure area after caving mining

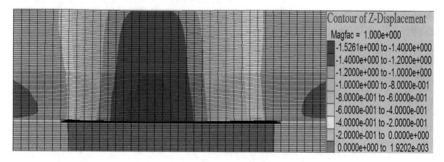

Fig. 10.25 Distribution of vertical deformation in caving mining

(7) There are problems

Paste filling technology not only solves the problem of coal mining under rivers, but also opens a new treatment way for coal gangue, washing gangue, fly ash, industrial

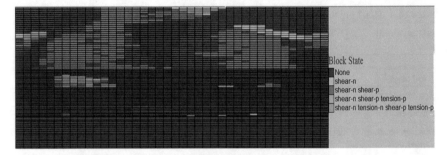

Fig. 10.26 Distribution of failure area after filling mining

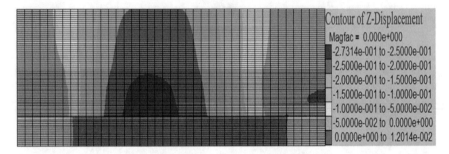

Fig. 10.27 Vertical deformation distribution diagram of filling mining

slag, urban construction waste and other wastes, which creates conditions for effectively promoting green mining in coal mines and has far-reaching economic and social significance. But it is for the effective bedrock only 20–40 m under the geological conditions of the river mining. To ensure the safe use of the ground river embankment and the safe mining of the underground, whether full filling or partial filling will face many problems. Filling mining should strictly control the convergence of roof and floor before filling, the amount of unconnected roof and the compression height of filling body.

When fully mechanized mining method is adopted, no matter what materials are used and what filling methods are adopted, the technical difficulty of large-scale filling will be very great, and it is very complex in material transportation and organization, and the realization method of technology, so it is difficult to operate. According to the application of filling mining method in metal mines, many factors should be considered when using filling mining method, including the cost of filling materials, filling technology, material source, filling effect and engineering management.

10.6.2.3 Analysis of Room and Pillar Mining Scheme

1. Theoretical design of coal room width

(1) Calculation of load Q of multi-layer rock beam under combined action

Before design, the load Q on the roof rock beam should be determined. The roof is generally composed of more than one layer of rock strata, so the load selected when calculating the limit span of the first layer should be determined according to the mutual influence between the rock strata above the roof. The load $(q_n)_1$ caused by the combined influence of the nth layer on the first layer can be calculated by the following equation:

$$(q_n)_1 = \frac{E_1 h_1^3 (\gamma_1 h_1 + \gamma_2 h_2 + \cdots \cdots + \gamma_n h_n)}{E_1 h_1^3 + E_2 h_2^3 + \cdots \cdots + E_n h_n^3} \tag{10.13}$$

where:

$E_1, E_2, \ldots, E_n$ elastic modulus of each rock layer, n is the number of rock layers;
$h_1, h_2, \ldots, h_n$ Thickness of each stratum;
$\gamma_1, \gamma_2, \ldots, \gamma_n$ Bulk density of each rock stratum.

When $(q_{n+1})_1 < (q_n)_1$ is calculated, that the $(q_n)_1$ is used as the load applied to the first stratum, and the weight of the stratum above the $n + 1$ layer will not affect the first stratum. At this time, the results of the above equation can be used as the load on rock beams to calculate the limit span of coal houses.

The thickness, bulk density and elastic modulus of each rock stratum are shown in Table 10.7.

The load value of roof stratum is calculated from the first roof stratification:

$$(q_1)_1 = \gamma_1 h_1 = 1.11 \text{ kg/cm}^2 \tag{10.14}$$

$$(q_2)_1 = \frac{E_1 h_1^3 (\gamma_1 h_1 + \gamma_2 h_2)}{E_1 h_1^3 + E_2 h_2^3} = 2.30 \text{ kg/cm}^2 \tag{10.14a}$$

Table 10.7 3^{-1} Physical and mechanical properties of coal roof and floor strata

Nature Rock stratum	Thickness/m	Bulk densityKN/ m³	Elastic modulus/ GPA	Tensile strength/ MPa
Siltstone	16.42	24.5	6.5	
Medium sandstone	7.47	25.3	10.0	
Siltstone	4.62	24.0	6.0	2.8
3^{-1} coal	2.8	13.6	2.0	0.7

$$(q_3)_1 = \frac{E_1 h_1^3 (\gamma_1 h_1 + \gamma_2 h_2 + \gamma_3 h_3)}{E_1 h_1^3 + E_2 h_2^3 + E_3 h_3^3} = 2.20 \text{ kg/cm}^2 \qquad (10.14\text{b})$$

Because $(q_3)_1 < (q_2)_1$, the second layer roof will be separated from its lower part, and the distributed load value on the rock beam can be taken as $q = 2.30$ kg/cm².

(2) Determine the width of coal room according to the theory of "beam"

According to the constraint conditions of coal pillars on both sides of roadway to roof rock beam, roof rock beam can be analyzed according to the situation of "simply supported beam" or "fixed beam". Generally, when the coal seam exists in a shallow depth, the abutment pressure generated in the coal pillars on both sides is not too large after the roadway or coal room is excavated, or both sides of the coal pillars are mined out in a large area, the "clamping" effect of the coal pillars on the roof is small, and the rock beam can be treated as a "simply supported beam". On the contrary, if the coal seam is buried deeply, both sides of the coal pillar are not mined out, and the "clamping" effect of the coal pillar on the roof rock beam is greater, it is more reasonable to treat it as "fixed beam".

① The roof rock beam is simplified as a simply supported beam

Let the permissible normal stress and shear stress of rock beam be σ_e and τ_e separately, and the tensile strength and shear strength be σ_c and τ_c, respectively and replace σ_{max} in the equation with σ_e. Take $q = 2$, 30 kg/cm², $\sigma_c = 2.0$ MPa, $\tau_c = 5.2$ MPa, and take the maximum safety factor, i.e., $F = 4$. It can be obtained that the limit span to ensure that the rock beam will not be damaged because the maximum tensile stress in the center of the span exceeds its tensile strength is as follows:

$$L = \sqrt{\frac{4h^2 \sigma_e}{3q}} = \sqrt{\frac{4h^2 \sigma_c}{3q F}} = 7.8 \text{ m} \qquad (10.15)$$

Let replace τ_{max} in the equation with τ_e. the limit span to ensure that the rock beam will not be damaged due to the maximum shear stress in the beam exceeding its shear strength is:

$$L = \frac{4h \tau_e}{3q} = \frac{4h \tau_c}{3q F} = 34.8 \text{ m} \qquad (10.16)$$

In the actual design of coal room span, the smaller value in the calculation results should be taken. According to the geological conditions in this area, the limit span of coal house is 7.8 m when the roof rock beam is simplified as a simply supported beam.

② The roof rock beam is simplified as a fixed beam

The limit span distance to ensure that the rock beam is not destroyed by the maximum tensile stress exceeding its strength limit is:

$$L = \sqrt{\frac{2h^2 \sigma_e}{q}} = \sqrt{\frac{2h^2 \sigma_c}{qF}} = 9.6\ \text{m} \qquad (10.17)$$

The ultimate span distance to ensure that rock beams will not be damaged because the maximum shear stress exceeds its shear strength is:

$$L = \frac{4h\tau_e}{3q} = \frac{4h\tau_c}{3qF} = 34.8\ \text{m} \qquad (10.18)$$

In the actual design of coal room span, the smaller value in the calculation results should be taken. According to the geological conditions in this area, the limit span of coal house is 9.6 m when the roof rock beam is simplified as a fixed beam.

Comparing the calculation results of simply supported beam and fixed supported beam, the safe limit span of roof slab of coal house is 7.8 m only from the angle of coal house. For simplicity, the size of coal house is designed as integer 5, 6 and 7 m.

The roof span determined by the above equation is the maximum allowable roof span without any reinforcement measures or roof support. When reinforcement or support measures are taken, the parameters such as load Q and beam thickness H in the above equation should be modified according to the mechanism of support and reinforcement.

2. Theoretical calculation of coal pillar width

Coal pillars are used to support the weight of overlying strata and protect the integrity of adjacent roadways and traverses, thus allowing miners to mine and pass safely between coal pillars. According to the action mechanism, coal pillars can be divided into two types: rigid coal pillars and yield coal pillars. The designed rigid coal pillar represents the coal pillar to support the expected load during the whole service life; The designed yielding coal pillar indicates that the coal pillar yields at the appropriate time and deformation rate, transferring an appropriate amount of load to the adjacent coal body. In other words, the yield pillar is not designed to withstand loads at any stage of its service life. There are three main steps in coal pillar design: ① determining the expected load of coal pillar; ② Determine the strength of coal pillar; ③ Determine the safety factor.

(1) Stress of coal pillar

Under the condition of large area mining and horizontal burial of coal seam, when the shape of coal pillar is the same, each coal pillar will uniformly bear the weight of overlying strata above the roadway or coal room within 1/2 span range above and around the coal pillar, as shown in Fig. 10.28.

Considering rectangular coal pillars, at this time, each coal pillar will share the load, and the total load it bears is:

$$P_t = (W_o + W_p)(W_o + L_p)\gamma H \qquad (10.19)$$

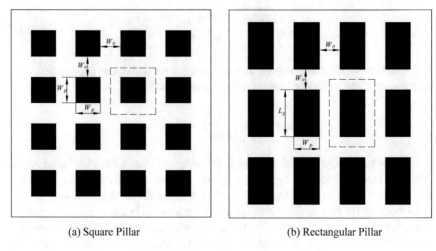

(a) Square Pillar   (b) Rectangular Pillar

Fig. 10.28 Schematic diagram of coal pillar size and coal pillar load

The average load on the coal pillar is:

$$\sigma_a = \frac{(W_o + W_p)(W_o + L_p)\gamma H}{W_p \cdot L_p} = \left(1 + \frac{W_o}{W_p}\right)\left(1 + \frac{W_o}{L_p}\right)\sigma_v \qquad (10.20)$$

where:

σ_a the average stress acting on the coal pillar, MPa;

W_p, L_P the width and length of coal pillar, m;

W_o the width of roadway or coal room, m;

γ average bulk density of overlying strata, 25KN/m^3;

H   mining depth, m;

σ_v the vertical stress of the original rock before excavation at the stratum with depth H, MPa.

Because of the rupture and loosening of the edge of coal pillar, the effective bearing area of coal pillar decreases, and the stress coefficient of coal pillar should increase by 1.1. Stress coefficient is the ratio between the stress of strip coal pillar under the specified minimum surrounding rock pressure and the stress under the actual working pressure of surrounding rock, which is a dimensionless parameter. The coal pillar load calculation equation is:

$$\sigma_a = \frac{1.1\sigma_v}{1 - R_e} = \frac{1.1\gamma H}{1 - R_e} \qquad (10.21)$$

The above equation gives the upper limit of coal pillar load, i.e., the safety limit of coal pillar load. Because this method is simple and the estimated safety factor of coal pillar load is large, it can meet the engineering requirements and is widely used to calculate the average load on coal pillar. Wang Mingli, Zhang Yuzhuo, Zhang Jincai and others think that the average stress calculated by the auxiliary area method is the upper limit of the vertical stress on coal pillars, and its value is generally approximately 50% larger than the actual stress, which is safe. In this chapter, the coal pillar load is reduced by 50% based on the auxiliary area method.

(2) Strength of coal pillar

With the development of room and pillar mining methods, many countries have done extensive laboratory and field tests to estimate coal pillar strength. Up to now, more than ten equations for estimating coal pillar strength have been put forward, which are mainly the following linear and nonlinear calculation equations:

$$\sigma_p = K_1 \left[A + B \frac{W_p}{h} \right] \tag{10.22}$$

$$\sigma_p = K_2 \frac{W_p^\alpha}{h^\beta} \tag{10.23}$$

where:

σ_p coal pillar strength with width Wp and height h, MPa;
K_1, K_2 Strength of ultimate cubic coal pillar in site, MPa;
A, B   are all constants;
α, β are all constants.

The linear and exponential equations are based on the ultimate cube strength. Firstly, the uniaxial compressive strength of coal rock is obtained by laboratory tests, and then converted into the ultimate cube strength by using the equation. Then the coal pillar strength is obtained by the equation of the relationship between the ultimate cube strength and the coal pillar strength.

Because the mechanical properties of coal and rock with different sizes and shapes are different, the strength and mechanical parameters of coal samples measured in laboratory cannot be directly used as the overall strength of coal and rock in site, and the indoor test results must be properly converted or tested in site. Hustrulid suggested the following methods to convert the strength of laboratory coal samples into the strength of field limit cubic coal samples, to eliminate the size effect of coal sample strength.

① When the side length d of the limit cube sample is less than 0.9 m:

$$\sigma_m = \sigma_c \sqrt{\frac{D}{d}} \tag{10.24}$$

② When the side length d of the limit cube sample is greater than 0.9 m:

$$\sigma_m = \sigma_c \sqrt{\frac{D}{0.9}} \tag{10.25}$$

where:

σ_c uniaxial compressive strength of coal samples measured in laboratory, MPa;
σ_m site limit cubic coal strength, MPa;
D   laboratory test size, m.

After obtaining the strength of the limit cube, the strength of coal pillar can be calculated by empirical equation. Commonly used empirical equations of coal pillar strength are shown in Table 10.8.

(3)   Safety factor

Safety factor is defined as the ratio of coal pillar strength to coal pillar stress, namely:

$$F = \frac{\sigma_p}{\sigma_a} \tag{10.26}$$

where:

F   safety factor, the recommended safety factor is 1.3–5.0, generally 1.5–2.0;
σ_p Strength of coal pillar;
σ_a Stress acting on coal pillar.

In engineering problems, safety factors are used to explain uncertain factors. The safety factor is most suitable for the design of underground coal pillar, because there are great uncertainties in the main methods to determine the load and strength of coal pillar, and the greater the safety factor, the lower the failure probability.

(4)   Coal pillar design

Considering the influence of test block size on strength, the ultimate cubic coal strength is determined. The uniaxial compressive strength of coal sample is 23.3 MPa, and the ultimate cubic coal strength $\sigma_m = 0.2357\sigma_c$, $\sigma_c = 5.5$ MPa.

Table 10.8 Empirical Equation of Common Coal Pillar Strength

Equation Type	Source	K_1 or K_2/ MPa	A or *Alpha*	B or *Beta*	Applicable conditions	Study country	Age
Linear	Obert-Duvall	σ_m	0.778	0.222	$w_p/h = $ 1–8a	United States	1967
	Bieniawski-Hairton	σ_m	0.64	0.36		South Africa	1969
	UNSW	5.36	0.64	0.36	$w_p/h = $ 1–10	Australia	1996
Nonlinearity	Holland-Gaddy	$\sigma_c\sqrt{D}$	0.5	1	$w_p/h = $ 2–8	United States	1984
	Salamon-Munro	7.2	0.46	0.66		South Africa	1976

Considering the influence of geometric shape on coal pillar strength, the coal pillar strength is determined, and the mining height is 2.8 m. The calculation results are shown in Table 10.9.

By substituting the obtained 1 and the calculated 2 into the equation, we can get:

$$F = \frac{\sigma_p}{\sigma_a} = (1 - R_e)\frac{\sigma_p}{1.1\gamma H} \tag{10.27}$$

According to the width of coal room and coal pillar, the safety factor is shown in Tables 10.10, 10.11 and 10.12.

Table 10.9 Calculation results of coal pillar strength

Source	σ_p/MPA under different coal pillar widths Wp/m				
	5.0	6.0	7.0	8.0	9.0
Obert-Duvall	6.46	6.90	7.33	7.77	8.20
Bieniawski-Hairton	7.06	7.76	8.47	9.18	9.88

Table 10.10 3^{-1} Safety factor of different coal room pillar size scheme when coal depth is 90 m

Program	Coal pillar strength σ_p/MPa	Coal pillar stress σ_a/ MPa	Safety factor/F	Remarks
Scheme 1 adopts 4 and leaves 5	6.46	4.00	1.6	
Scheme 2: take 5 and leave 5	6.46	4.95	1.3	
Scheme 3: adopt 5 and leave 6	6.90	4.16	1.7	
Scheme 4: adopt 6 and leave 5	6.46	5.99	1.1	
Scheme 5: take 6 and leave 6	6.90	4.95	1.4	
Scheme 6: adopt 6 and leave 7	7.33	4.27	1.7	
Scheme 7: take 7 and leave 6	6.90	5.80	1.2	
Scheme 8: take 7 and leave 7	7.33	4.95	1.5	
Scheme 9: take 7 and leave 8	7.77	4.35	1.8	

Remarks According to the existing engineering practice experience, the adopted safety factor F is 1.3–5.0. Considering the recovery rate and safety, it is recommended to take 1.5–2.0

Table 10.11 3^{-1} safety factor of different coal room pillar size schemes when coal depth is 120 m

Program	Coal pillar strength σ_p/MPa	Coal pillar stress σ_a/MPa	Safety factor/F	Remarks
Scheme 1 adopts 4 and leaves 5	6.46	5.32	1.2	
Scheme 2 adopts 4 and leaves 6	6.90	4.58	1.5	
Scheme 3: take 5 and leave 5	6.46	6.58	1.0	
Scheme 4: take 5 and leave 6	6.90	5.53	1.3	
Scheme 5: take 5 and leave 7	7.33	4.85	1.5	
Scheme 6: take 6 and leave 5	6.46	7.97	0.8	
Scheme 7: take 6 and leave 6	6.90	6.58	1.0	
Scheme 8: take 6 and leave 7	7.33	5.68	1.3	
Scheme 9: take 7 and leave 6	6.90	7.71	0.9	
Scheme 10: take 7 and leave 7	7.33	6.58	1.1	
Scheme 11: take 7 and leave 8	7.77	5.79	1.4	

10.7   Summary

(1) Regardless ofof the selected mining scheme, the problem of dike restoration in Qingcaojiegou and Hezegou should be involved. Therefore, if the mine wants to mine, it must apply to the water conservancy, river affairs and other relevant departments for the record in time and formulate the re-dike plan according to the requirements of the competent department.

(2) Production practice shows that exploratory mining is an important technical principle of coal mining under water. Trial mining is to first mine the coal seam away from the water body and then mine the coal seam below the adjacent water body; first mining aquifuge thick, after mining aquifuge thin coal seam; first mining coal seam with simple geological conditions, then mining coal seam with complex geological conditions; first mining deeper, later mining shallower coal seam. Through the first easy to difficult exploratory mining, gradually getting close to the water body. In this way, not only can we accurately understand the damage degree of mining to waterproof coal and rock pillars in trial mining, but also can constantly find out the best mining methods and measures suitable for the mining area.

Table 10.12 3^{-1} Safety factor of different coal room pillar size schemes when coal depth is 150 m

Program	Coal pillar strength σ_p/MPa	Coal pillar stress σ_a/MPa	Safety factor F	Remarks
Scheme 1 adopts 4 and leaves 5	6.46	6.64	1.0	
Scheme 2 adopts 4 and leaves 6	6.90	5.50	1.3	
Scheme 3: take 5 and leave 5	6.46	8.22	0.8	
Scheme 4: take 5 and leave 6	6.90	6.91	1.0	
Scheme 5: take 5 and leave 7	7.33	6.06	1.2	
Scheme 6: take 5 and leave 8	7.77	5.45	1.4	
Scheme 7: take 6 and leave 5	6.46	9.94	0.7	
Scheme 8: take 6 and leave 6	6.90	8.22	0.8	
Scheme 9: take 6 and leave 7	7.33	7.09	1.0	
Scheme 10: adopts 6 and leaves 8	7.77	6.32	1.2	
Scheme 11: take 7 and leave 7	7.33	8.22	0.9	
Scheme 12: take 7 and leave 8	7.77	7.22	1.1	
Scheme 3: take 5 and leave 5	6.46	8.22	0.8	

(3) When the river diversion scheme is adopted, it needs to be approved by the water department. The advantage is that the safety is good, and the river diversion problem involves a wide range, a large amount of engineering and a large investment. Because this area is the birthplace of the river, it is a perennial river formed by springs above the working face, and it is not easy to take river diversion. When the caving method is used to manage the roof, it is necessary to set up a safe waterproof coal and rock pillar. The advantage is that the process is simple and does not increase the drainage cost. The disadvantage is that the waterproof coal pillar is lost, and the number of moving times is increased, which affects the reasonable layout of the overall mining project.

When the filling method is used to manage the roof, the caving zone is generally not formed, and the development height of the fracture zone is obviously reduced. Its advantage is that it can liberate the coal under the river, reduce the development of the water-conducting fracture of the overlying strata, prevent the sand and water inrush from the working face, increase the recovery rate of coal, reduce the waste of coal resources, prevent the destruction of the original aquifer, and protect the original fragile ecological environment in the local area. It is conducive to the formation of a new waterproof roof and lays a good foundation for the next 4–2 coal mining. The disadvantage is that the paste filling technology process is complex, once the equipment is put into use, it will increase the cost of tons of coal mining.

When the use of room and pillar mining method, taking into account the recovery rate and safety generally take the safety factor F is 1.5–2.0. In the process of mining, it should be in accordance with the principle of trial mining. If it is found that the coal pillar and roof overburden are seriously damaged or the water inflow in the goaf increases, the block section should be closed and a mining scheme with greater safety factor should be replaced in the next block section. If according to the room and pillar mining, it is suggested that the coal pillar stress should be monitored in the mining process to verify whether the coal pillar stress conforms to the theoretical value determined by this subject. The disadvantage is that the recovery rate is low, the safety factor and the recovery rate cannot be obtained at the same time, and because the residual coal pillar in the room mining goaf does not have long-term stability, it may cause the roof of the room mining goaf to collapse suddenly at any time, the large area of the surface to sink suddenly, causing surface damage, or causing rock burst and roof accidents in the nearby mining face, resulting in major safety accidents or casualties. Due to the long-term existence of rivers, this method has certain potential dangers. For shallow coal seams, the biggest potential threat of this mining method is the large-scale roof caving disaster. The air in the goaf is compressed, and the high-pressure air flow is quickly discharged from the goaf, forming a huge storm, which should be paid attention to.

(4) Whether the coal under the river is mined should be investigated in detail to prevent decision-making errors. In addition, according to the current data, solving the problem of mining under thin bedrock river is not much at home and abroad, risking huge investment risks and potential safety hazards. Once the expected purpose is not achieved, it will cause huge losses to the enterprise. Considering that there is also a risk of water–sand inrush in the mining of non-river coal seams in the thin bedrock area, and considering the safety, according to the theoretical calculation and evaluation of the research group, it is suggested that the paste filling mining method should be used in the river and thin bedrock area according to the principle of trial mining.

References

1. Fan H, Wang LG, Lu YL, Li ZL, Li WS, Wang K (2020) Height of water-conducting fractured zone in a coal seam overlain by thin bedrock and thick clay layer: a case study from the Sanyuan coal mine in North China. Environ Earth Sci 79(6)
2. Ju Y, Zhu Y, Xie HP, Nie XD, Zhang Y, Lu C, Gao F (2019) Fluidized mining and in-situ transformation of deep underground coal resources: a novel approach to ensuring safe, environmentally friendly, low-carbon, and clean utilisation. Int J Coal Sci Technol 6(2)

3. Li L, Li FM, Zhang Y, Yang DM, Liu X (2020) Formation mechanism and height calculation of the caved zone and water-conducting fracture zone in solid backfill mining. Int J Coal Sci Technol 7(5)
4. Liu Y, Yuan SC, Yang BB, Liu JW, Ye ZY (2019) Predicting the height of the water-conducting fractured zone using multiple regression analysis and GIS. Environ Earth Sci 78(14)
5. Liu Z, Wang WY, Yang H, Yu SJ, Xin L (2020) Experimental study on the fractal pattern of a coal body pore structure around a water injection bore. J Energy Resour Technol 142(1)
6. Liu JW, Liu CY, Li XH (2020) Determination of fracture location of double-sided directional fracturing pressure relief for hard roof of large upper goaf-side coal pillars. Energy Explor Expl 38(1)
7. Niu GG, Zhang K, Yu BS, Chen YL, Wu Y, Liu JF (2019) Experimental study on comprehensive real-time methods to determine geological condition of rock mass along the boreholes while drilling in underground coal mines. Shock Vibr
8. Xie XS, Hou EK, Wang SM, Sun XY, Hou PF, Wang SB, Xie YL, Huang YA (2021) formation mechanism and the height of the water-conducting fractured zone induced by middle deep coal seam mining in a sandy region: a case study from the Xiaobaodang coal mine. Adv Civ Eng
9. Zhang Y, Cao SG, Guo S, Wan T, Wang JJ (2018) Mechanisms of the development of water-conducting fracture zone in overlying strata during shortwall block backfill mining: a case study in Northwestern China. Environ Earth Sci 77(14)

Chapter 11
"Coal-Water" Dual Resource Coordinated Mining Technology System

According to the results obtained in the research content of the above chapters, if the mine in the Yushenfu mining area adopts unreasonable mining intensity when mining the coal seam, it will easily lead to further deterioration of the ecologically fragile environment very much, and eventually lead to the occurrence of mine water disaster accidents. Therefore, under the ecologically fragile environment in the Yushenfu mining area, how to ensure the coordinated development of "coal resources-ecological environment", and finally realize mine safety water control mining, is still a major difficulty faced by the western mines and long-term, and must be Do in-depth research on it.

11.1 Analysis of Contradiction Between Coal Mining and Stratigraphic Ecological Environment

11.1.1 Coexistence Characteristics of Coal and Water in Yushenfu Mining Area

According to the strata analysis of various coal-accumulating basins in my country, all coal seams in my country have a common feature, i.e., the coexistence of coal seams and aquifers. According to the positional relationship between the coal seam and the aquifer and the characteristics of the influence on the safe mining of the coal seam, it can be divided into coal measure roof loose layer phreatic aquifer and coal measure roof and floor confined aquifer. At the same time, the overall pattern of the distribution of coal resources in China shows the characteristics of "rich in the north and poor in the south, more in the west and less in the east" [7, 10]; however, the distribution pattern of water resources in China is opposite to the distribution of coal resources, showing that "the west is less, and the east is more" characteristics. The surface

Fig. 11.1 Surface vegetation map

ecological landforms in the mining area are mainly covered by fixed sand dunes, semi-fixed sand dunes, mobile sand dunes and loess gully landforms. The study area is affected by the stratigraphic sedimentary environment, and each area exhibits different hydrological characteristics, and the groundwater level required by different vegetation is also different, resulting in different vegetation types in the water-rich area unit and the weak water-rich area unit. The Yushenfu mining area belongs to the dry and semi-arid season all the year round, and the annual rainfall in the area is low. Surface vegetation is mainly composed of Artemisia desertorum, Hedysarum scoparium, Caragana korshinskii, Amorpha fruticosa, Astragalus adsurgens, Salix psammophila and other drought-tolerant plants (Fig. 11.1). Therefore, the protection of water resources is of great significance to the development of the region [3, 4, 7, 10].

Salawusu Formation aquifer is the main water supply in the region and an important water system for maintaining the sustainable development of the surface ecology, and its regional water resources are relatively abundant. As the direct water-filled aquifer in the mining area, the weathered bedrock aquifer has a large thickness and strong water permeability, which are the main control objects for water-controlled coal mining in the mining area. The Quaternary Lishi Formation loess and Neogene Baode Formation red soil in the area are important aquifers to ensure whether the Salawusu Formation can directly recharge with the lower aquifer. The thickness of both of them gradually increases from southwest to northeast as a whole. The Salawusu Formation aquifer and weathered bedrock aquifer discussed above are the main control objects for water-controlled coal mining in Yushenfu mining area [3, 4, 9]. According to the existing exploration results of coal measures in Yushenfu mining area and mine mining situation, the mining area generally presents a monoclinic layer that dips gently inward to northwest, with a dip angle of approximately 1°, and no major faults are found in the mining area. The main coal-bearing strata in the mining area is the Middle Jurassic Yan'an Formation. The middle and upper parts of this section have good coal content and large thickness, while the middle and lower coal seams are thinner. The main mineable coal seams are 5 layers: 1^{-2}, 2^{-2}, 3^{-1}, 4^{-2}, 5^{-2}; now the mining area mainly mines the 2^{-2} coal seam of Yan'an

Formation. Therefore, this chapter mainly discusses the coexistence characteristics between the 2^{-2} # coal seam and the above two layers of aquifers.

2^{-2} coal seam is owned by the entire Yushenfu mining area, but its thickness varies widely. The thickness of the coal seam in the Dabaodang—Jinjitan area is relatively large, and the average coal seam is approximately 11 m. And the position of the coal seam roof is approximately 110 m away from the above two layers of aquifers; the development height of the water-conducting fracture zone after mining affects the above-mentioned aquifers, and its mining damage intensity is relatively large. However, the area near the Hongliulin mine field in Shenmu is affected by burnt rocks. The average distance from the coal measure to the top of the weathered bedrock is approximately 30 m. The mining in this area is greatly affected by the two aquifers, which can easily lead to flood accidents [1, 2, 5, 8].

To sum up, as an important coal energy producing area in the west, the Yushenfu mining area has the characteristics of abundant resource reserves, large variation of hydrological units, fragile surface ecology, and single type of surface vegetation. Therefore, how to ensure the sustainable and coordinated development of "water resources-ecological environment-coal resources" in the mining area is the main problem faced by the mines in the mining area, and it is also one of the scientific problems to be solved urgently in the ecologically fragile mining area.

11.1.2 Ecological and Environmental Problems Caused by High-Intensity Mining in Yushenfu Mining Area

Since the 1980s, the Yushenfu mining area, as one of the main coalfield bases in my country, has made great contributions to the local economic development. At present, the "high efficiency" mining methods such as fully mechanized mining and top coal caving are mainly used in the Yushenfu mining area. However, the cracks in the roof of the coal seam have already affected the bedrock water barrier horizontally and vertically to varying degrees after mining with this method; at the same time, the overlying rock has also caused incalculable effects on its aquifer during the process of strong mining. In particular, the overburden fissures directly cut through the surface during the mining of shallow coal seams, resulting in the destruction of the most important aquifer of the Salawusu Formation in the area. Therefore, the use of this high-intensity recovery process without considering the ultimate bearing capacity of the formation is a mining method that destroys the ecological environment for short-term interests and can be summarized as a "high-efficiency but low-efficiency" recovery method. However, in contrast to the backfill and short-wall mining systems, in the application process of the system, the working face length is short, the mine pressure is weak, the disturbance range of the overlying rock is small, the height of the water-conducting fracture zone is low, and the influence range of the surface damage is small, but the recovery efficiency is low and the process is complicated. It can be summarized as a kind of "High-efficiency and low-efficiency" mining method. However, when mining coal resources, the mining area does not consider the coordination relationship between "water resources-ecological environment-coal resources", and simply pursues "high-efficiency" mining, thus causing great negative impacts the ecological environment and water resources of the mining area. They are mainly manifested in: ① The vegetation in the mining area withered due to the decline of the groundwater level; ② The scope of land desertification expanded; ③ The "three wastes" produced by the mine production affected the ecological environment; ④ There are large-scale subsidences on the ground and the groundwater level drops. The occurrence of the above problems greatly restricts the green and safe mining of mines; at the same time, due to the continuous development of coal seams, a large number of rivers have dried up, springs have dried up and the ecological water level has dropped in a wide range, as shown in Fig. 11.2. In short, the mine side only considers "high-efficiency" predatory mining and "mining thick and discarding thin" mining, but in the mining process, the mine side did not consider the economic losses caused by the secondary disasters such as environmental governance, water environment damage, and land subsidence caused by high-intensity mining. In particular, a large number of falling funnels formed by high-intensity mining of thick coal seams have caused many environmental problems such as groundwater level decline and water runoff changes, which have caused environmental problems such as surface vegetation.

According to the coexistence characteristics of coal and water in the Yushenfu mining area and the regional ecological environment problems caused by high-intensity mining, the main contradiction between coal seam mining and stratigraphic ecological environment in the Yushenfu mining area is the contradiction between coal seam mining intensity and ecological geological environment carrying capacity. Therefore, each mine in the Yushenfu mining area must formulate a recovery method suitable for the mine during the mining process to ensure the realization of "high-efficiency" recovery of coal resources, thereby avoiding the formation of irreversible "secondary disasters" in the mine, and limiting the sustainable development of the mine.

11.2   Discussion on the "Coal-Water" Dual Resource Coordinated Mining Mode

11.2.1   The Theory of "Coal-Water" Dual Resource Coordinated Exploitation

"Coal-water" dual resource coordinated mining technology and method [6, 11], which is based on the "three-in-one" optimization and combination mode of mine water drainage, supply, and ecological environmental protection, and the "three-in-one" optimization and combination mode of mine groundwater control, utilization, and ecological environmental protection. While it is also based on mine water control, treatment, utilization, recharge, and ecological environmental protection are based

(a) Surface subsidence   (b) Land desertification

(c) rivers dry up   (d) Water level drop

Fig. 11.2 Disasters caused by high-intensity mining in Yushenfu mining area

on the "five-in-one" optimization and combination mode theory, combined with the constraints of development and construction environmental objectives, so that coal resources, water resources and ecological environmental protection become a comprehensive system of development technology.

Connotation of "coal-water" dual-funded mine mining technology: the main purpose of this technology is to solve the contradictions and conflicts that appear in the current coal mining process, such as those in Sect. 11.1, so as to realize safe and efficient mines, green and coordinated mining win goal. The technology can be generally summarized into two different types, one is a mining method that does not change the hydrogeological characteristics of the overlying rock, and the other is a mining method that changes the hydrogeological characteristics. The mining method that does not change the hydrogeological characteristics of the overlying rock is mainly to select an appropriate mining method according to the character- istics of the geological conditions to ensure that the hydrogeological conditions of the overlying rock do not change during the mining process. The mining methods that change the hydrogeological characteristics of the overlying rock are mainly by transforming the key water-repellent layer, pumping, and draining the water body of the aquifer to change its water richness, etc., so as to make full use of the positive effect of its resources to avoid the negative effects of its resources, so as to ensure the realization of "coal-water" dual resource coordinated exploitation.

As an important coal production base in western my country, the Yushenfu mining area has good coal seam occurrence conditions and large coal seam mining thickness, which creates favorable conditions for its realization of high-intensity mining. However, the Yushenfu mining area is located in an ecologically fragile area in the arid and semi-arid regions of China. The Salawusu Formation which is the most important aquifer in the mining area is the main source of water for local residents and industries. In recent years, due to the high-intensity mining in the mining area, been severely affected. Therefore, mining in mining areas should practice the connotation of "coal-water" dual-resource mines to achieve maximum utilization of resources.

11.2.2 Determine the Ecological Environment Bearing Characteristics of the Mining Area

According to the contradiction analysis between coal seam mining and ecological environment in the existing Yushenfu mining area described in Sect. 11.1 above and the connotation of building a "coal-water" dual-resource mine described in Sect. 11.2.1, it can be seen that the ecological environment bearing characteristics of the mining area are to ensure that the main basis for realizing the coordinated development of "coal resources-ecological environment" in mining areas. The ecological environment carrying capacity of the mining area has a great correlation with the ecological vegetation characteristics of the mining area and the self-repairing ability of the mining area. However, under the premise that the vegetation characteristics and self-repairing ability of the mining area are determined, the damage intensity of the water-resistant stratum and the water loss intensity of the aquifer directly determine the ecological damage degree of the mining area. Therefore, to determine the ecological environment bearing characteristics of the mining area is to determine the maximum damage intensity that the water-resistant stratum can bear and the maximum water loss that the aquifer can bear during the mining process.

According to the geological model of water–sand inrush under strong mining conditions established in Sect. 9.2 above (Fig. 9.1), and the simplified model of the fixed beam for the water-resistant soil layer (Fig. 9.2), the analysis shows that as long as the water-conducting fault zone is Without penetrating the key aquifuge, it can still carry the upper aquifer, maintain the ecological water level, and ensure the survival of surface vegetation. Therefore, determining the characteristics of ecological carrying capacity requires determining the ultimate carrying capacity of the critical aquifuge and the maximum possible water loss of the aquifer.

According to the theoretical calculation of the force of beam theory, the maximum bending moment M_{max} occurs at both ends of the beam, $M \frac{qL^2}{12}{}_{max}$. Therefore, the maximum tensile stress there σ_{max} is:

$$\sigma \frac{qL^2}{2h^2}_{\max} \tag{11.1}$$

where: q is the bearing load of the water-proof key layer, MPa; L is the limit length of the rock beam cantilever span, m; h is the thickness of the water-proof key layer, m.

According to Eq. (11.1), it can be known that if the maximum tensile stress of the water-proof key layer exceeds its ultimate tensile strength, the water-proof key layer will lose its water-proof effect, resulting in a drop in the water level of the aquifer, and ultimately destroying the groundwater environment. At that time $\sigma T_{\max}$, i.e., the normal stress on the water-proof key layer reaches the ultimate tensile strength of the rock beam R_T, and the water-proof key layer breaks. The load $q = \gamma h_1$ on the water-proof key layer, and the ultimate span of the L rock beam when the break occurs is:

$$L = \sqrt{\frac{2R_T}{\gamma h_1}} h \tag{11.2}$$

where: γ is the average bulk density of the rock formation, which is taken as 25 KN/m^3 here; h_1 is the sum of the thicknesses of the overlying rock layers in the water-proof key layer.

According to Fan Limin's research on the response of groundwater to high-intensity coal mining in ecologically fragile mining areas in the west, it can be seen that after the conduction of the water-conducting fracture zone, the water source is mainly composed of static reserves and dynamic reserves, i.e., the elastic water release of the confined aquifer, the gravity water release of the submerged aquifer, and the amount of water supplied by the descending funnel formed by the overlying rock after mining. The equations for calculating the reserves of each water source are:

$$Q_j = MF(\mu_s \Delta H + \mu) \tag{11.3}$$

$$Q_d = \int_0^t \left[K \frac{M(2H_0 - M) - H_W^2}{2L} \right] (n_1 b + n_2 vt) dt \tag{11.4}$$

where: Q_j is the static reserve; μ_s is the water storage coefficient; F is the goaf area, $F = vt * L * M$; ΔH is the water level drawdown; μ is the specific yield; Q_d is the dynamic reserve; K is the hydraulic conductivity; H_0 is the initial water head; H_W is the water head in the well; n_1, n_2 is the coefficient, taking 2 for the water well and 1 for one side of the well; v is the advancing speed of the working face; M is the mining thickness of the working face.

From Eqs. (11.1), (11.2), (11.3) and (11.4), it can be seen that the premise of water loss in the aquifer is that after the water-proof key layer reaches its limit span during the advancing process of the working face, the aquifer begins to lose water in the form of static reserves and dynamic reserves along the water-conducting fracture channel, and the total water loss calculation equation is:

$$Q_{\text{all}} = MF\left(\mu_s \Delta H + \mu\right) + \left[K \frac{M\left(2H_0 - M\right) - H_W^2}{2\sqrt{\frac{2R_T}{\gamma h_1}} h}\right] \int_0^t \left(n_1 b + n_2 vt\right) dt \qquad (11.5)$$

Under the premise of the determination of vegetation characteristics and landforms in the mining area, the failure strength of the water-proof formation and the water loss strength of the aquifer are mainly controlled by the advancing speed of the working face, the thickness and ultimate tensile strength of the water-proof key layer, and the hydrological coefficient of each aquifer.

11.2.3 Connotation of Water Control and Coal Mining in Ecologically Fragile Mining Areas

To study the mining mode and mining mode of water -controlled coal mining in ecologically fragile mining areas with high-intensity mining, to provide theoretical support for coordinated mining between "coal resources and ecological environment" in ecologically fragile mining areas under the condition of high-intensity mining. Based on the eco-environmental problems caused by high intensity mining in the Yushenfu mining area, this book presents a generalized geological model for full fracture of overburden strata under high intensity mining (Fig. 11.3).

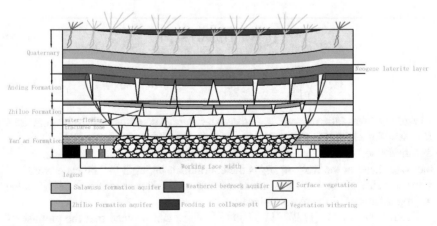

Fig. 11.3 Geological generalization model

From the geological generalized model shown in Fig. 11.3, the mining of the coal seam destroys the equilibrium state of the original rock layer, and the overlying rock layer on the roof of the coal seam breaks with the mining of the coal seam, and finally forms a water conduction channel. When the top interface of the water conduction channel communicates with the confined aquifer, the water body will infiltrate from the water conduction channel to the goaf. If the instantaneous water infiltration into the goaf is greater than the drainage of the mine, the mine will have a water inrush disaster. If the top interface of the water conduction channel communicates with the aquifer of the Salawusu Formation, the water level will drop sharply; if the water level cannot be restored to the threshold value of the ecological water level required by the vegetation after a certain period of time, the vegetation will die due to lack of water. When the water in the vadose zone accumulates in the low-lying part of the collapsed land, the vegetation roots die due to prolonged immersion in the water, which will eventually further deteriorate the ecologically fragile mine landform. Therefore, when high-intensity mining is carried out in ecologically fragile areas, the contradiction between the fragile ecological environment and high-intensity coal mining must be resolved, to achieve the purpose of water control and coal mining in ecologically fragile mining areas.

11.2.4 Water-Controlled Coal Mining Mode in Ecologically Fragile Mining Areas

From the above analysis, it can be seen that if the coal seam is mined in the ecologically fragile area, if the coal production is purely pursued, and the mine safety and ecological environment problems caused by coal mining are not considered, the return of coal resources will definitely increase the degree of damage to the ecologically fragile mining area. Therefore, the research on the mining mode of water -controlled coal mining in ecologically fragile areas has important application value for the multi-win goals of maximizing coal recovery rate, minimizing aquifer water loss rate, and maximizing ecological landform protection rate in ecologically vulnerable mining areas.

Based on the connotation of water-controlled coal mining in ecologically fragile areas, and combined with the mining status of mining areas with fragile surface ecological environment, this section designs a collaborative mining model for water -controlled coal mining in ecologically fragile areas as shown in Fig. 11.4. The collaborative mining model is divided into four parts. The main content of the first part consists of the following three aspects, namely, checking the ecological landforms of the mining area, checking the hydrogeological conditions of the mining area, and checking the height of the water-conducting fracture zone. Therefore, the content of the first part is briefly described as "three investigations". The main content of the second part is to determine the coal mining methods and process parameters used in the coal seam mining under the premise of the first part. The main reason is that the carrying capacity of different aquifers and surface ecology is different, and different coal mining methods and process parameters have different degrees of damage to the aquifer and surface ecology. At the same time, the damage degree of coal mining methods and process parameters to strata must be maintained within the carrying capacity range of water-resisting strata and surface ecology. Therefore, the content of the second part is briefly described as 'must'. The main content of the third part is to use laboratory research methods to demonstrate the rationality of the coal mining methods and mining processes determined in the second part. If the verification results are difficult to meet the requirements of mine water control and coal mining, the second part will be redone. The same is true. The research methods used in the third part are mainly physical simulation and numerical simulation. Therefore, the content of the third part is briefly described as "two verifications". The main content of the fourth part is to track and analyze whether the surface damage degree of the finally determined coal mining method and process parameters during mining and whether the water level change of the aquifer meets the requirements of water-controlled coal mining during the mining process. The content of its tracking analysis mainly includes the degree of surface ecological damage after mining and the change of water level in the observation well. Therefore, the fourth part is briefly described as 'two tracking'.

In summary, the content of the collaborative mining mode of water-controlled coal mining in ecologically fragile areas shown in Fig. 11.4 can be briefly described as the ten words policy of three checks, certain, two verifications, 'two tracking'. The four parts of the collaborative mining model complement each other and restrict each other, which abandons the inappropriate practice of studying "mine geology, mining methods and ecological environment" separately. Therefore, using this model can not only maximize the utilization rate of coal resources in the mining area, but also avoid irreversible "secondary disasters" in mines, thereby achieving the goal of green and safe mining.

11.3 Summary

Based on systematically summarizing the contradiction between coal mining and stratigraphic ecological environment in the Yushenfu mining area, and the connotation of "coal-water" dual-resource mine construction, this chapter proposes a water-controlled coal mining mode in ecologically fragile mining areas. The main conclusions are as follows:

(1) The Yushenfu mining area, as an important coal energy producing area in the west, has the characteristics of abundant resource reserves, large variation range of hydrological units, fragile surface ecology, and single type of surface vegetation. A large number of falling funnels formed by high-intensity mining of thick coal seams have caused many water-related environmental problems such

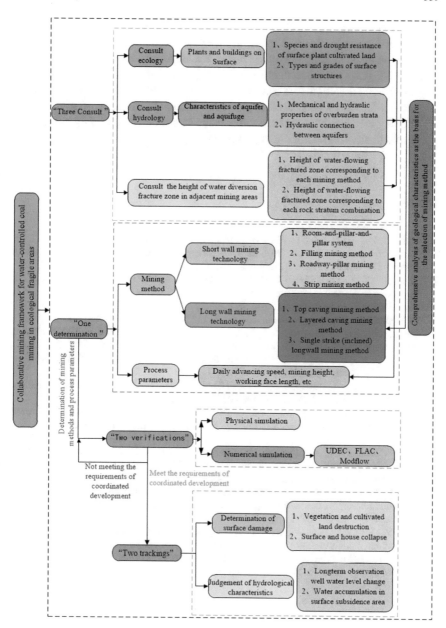

Fig. 11.4 Collaborative mining model of water-controlled coal mining in ecologically fragile areas

as groundwater level decline and water runoff changes, which have caused other environmental problems such as surface vegetation.

(2) This chapter systematically discusses the important connotation of building a "coal-water" dual-resource mine in the Yushenfu mining area, and based on the contradiction between the current coal mining and the stratum ecological environment, and the ecological environment bearing characteristics of the mining area are determined, and the water-controlled coal mining mode in the ecologically fragile mining area is proposed.

References

1. Chi MB, Zhang DS, Liu HL, Wang HZ, Zhou YZ, Zhang S, Yu W, Liang SS, Zhao Q (2019) Simulation analysis of water resource damage feature and development degree of mining-induced fracture at ecologically fragile mining area. Environ Earth Sci 78(3)
2. Dong SN, Xu B, Yin SX, Han Y, Zhang XD, Dai ZX (2019) Water Resources utilization and protection in the coal mining area of Northern China. Sci Rep 9(1)
3. Fan LM, Ma XD, Jiang H et al (2016) Risk zoning of water inrush and sand burst in mines in ecologically fragile mining areas in the west. Coal Sci Newspaper 41(3):531–537
4. Fan LM, Ma XD, Li YH et al (2017) The status quo and prevention and control technology of mine geological disasters in the western high-intensity mining area. J Coal 41(2):276–285
5. Gao YB, Gao HN, Zhang XY (2020) Synergetic system for water body detection in coal mine: a case study. Geotech Geol Eng An Int J 38(S1)
6. Shi HY, Shi JJ, Tian D (2016) Distribution of floor stress field and fracture echanism of aquifuge above confined water. Coal Sci Technol 44(11):047–050
7. Wu LX (2003) Analysis of the contradiction between northwest mining development and water resources and its countermeasures. South-to-North Water Transf Water Conserv Sci Technol 1(1):3
8. Xiao W, Zhang WK, Lv XJ et al (2020) Temporal and spatial changes of ecosystem services under different mining intensities in mines in ecologically fragile areas in the west—taking Shenfu mining area as an example. J Nat Resour 35(1):68–81
9. Ye GJ, Zhang L, Li WP et al (2000) Main hydro-environmental problems and prevention measures for coal resources development in Yushenfu mining area, northern Shaanxi Zhi. J Eng Geol 08(04):0446–0456
10. Zeng YF, Pang ZZ, Qiang WQ et al (2022) Study on water-controlled and environment-friendly coal mining models in ecologically fragile area of northwest China. Mine Water Environ 1–15
11. Zhang YJ (2020) Principle and key technology of water-controlled mining technology and practice of fully mechanized caving mining under sandstone aquifer. J Coal 45(10):3380–3388

Printed in the United States
by Baker & Taylor Publisher Services